W9-CCX-639

Library of Congress Catalog Card Number 66-14 573

Printed in Germany

Titel No. 5114

TOPOLOGICAL METHODS IN ALGEBRAIC GEOMETRY

F. HIRZEBRUCH
UNIVERSITY OF BONN

THIRD ENLARGED EDITION

NEW APPENDIX AND TRANSLATION
FROM THE SECOND GERMAN EDITION

BY R. L. E. SCHWARZENBERGER
UNIVERSITY OF WARWICK

WITH AN ADDITIONAL SECTION

BY A. BOREL
INSTITUTE FOR ADVANCED STUDY, PRINCETON

SPRINGER-VERLAG
BERLIN · HEIDELBERG · NEW YORK
1966

DIE GRUNDLEHREN DER

MATHEMATISCHEN WISSENSCHAFTEN

IN EINZELDARSTELLUNGEN MIT BESONDERER BERÜCKSICHTIGUNG DER ANWENDUNGSGEBIETE

BAND 131

SPRINGER-VERLAG
BERLIN · HEIDELBERG · NEW YORK
1966

To my teachers

Heinrich Behnke and Heinz Hopf

Preface to the first edition

In recent years new topological methods, especially the theory of sheaves founded by J. LERAY, have been applied successfully to algebraic geometry and to the theory of functions of several complex variables. H. CARTAN and J.-P. SERRE have shown how fundamental theorems on holomorphically complete manifolds (STEIN manifolds) can be formulated in terms of sheaf theory. These theorems imply many facts of function theory because the domains of holomorphy are holomorphically complete. They can also be applied to algebraic geometry because the complement of a hyperplane section of an algebraic manifold is holomorphically complete. J.-P. SERRE has obtained important results on algebraic manifolds by these and other methods. Recently many of his results have been proved for algebraic varieties defined over a field of arbitrary characteristic. K. KODAIRA and D. C. SPENCER have also applied sheaf theory to algebraic geometry with great success. Their methods differ from those of SERRE in that they use techniques from differential geometry (harmonic integrals etc.) but do not make any use of the theory of STEIN manifolds. M. F. ATIYAH and W. V. D. HODGE have dealt successfully with problems on integrals of the second kind on algebraic manifolds with the help of sheaf theory.

I was able to work together with K. KODAIRA and D. C. SPENCER during a stay at the Institute for Advanced Study at Princeton from 1952 to 1954. My aim was to apply, alongside the theory of sheaves, the theory of characteristic classes and the new results of R. THOM on differentiable manifolds. In connection with the applications to algebraic geometry I studied the earlier research of J. A. TODD. During this time at the Institute I collaborated with A. BOREL, conducted a long correspondence with THOM and was able to see the correspondence of KODAIRA and SPENCER with SERRE. I thus received much stimulating help at Princeton and I wish to express my sincere thanks to A. BOREL, K. KODAIRA, J.-P. SERRE, D. C. SPENCER and R. THOM.

This book grew out of a manuscript which was intended for publication in a journal and which contained an exposition of the results obtained during my stay in Princeton. Professor F. K. SCHMIDT invited me to use it by writing a report for the "Ergebnisse der Mathematik". Large parts of the original manuscript have been taken over unchanged, while other parts of a more expository nature have been expanded. In this way the book has become a mixture between a report, a textbook and an original article. I wish to thank Professor F. K. SCHMIDT for his great interest in my work.

I must thank especially the Institute for Advanced Study at Princeton for the award of a scholarship which allowed me two years of undisturbed work in a particularly stimulating mathematical atmosphere. I wish to thank the University of Erlangen which gave me leave of absence during this period and which has supported me in every way; the Science Faculty of the University of Münster, especially Professor H. BEHNKE, for accepting this book as a Habilitationsschrift; and the Society for the Advancement of the University of Münster for financial help during the final preparation of the manuscript. I am indebted to R. REMMERT and G. SCHEJA for their help with the proofs, and to H.-J. NASTOLD for preparing the index. Last, but not least, I wish to thank the publishers who have generously complied with all my wishes.

Fine Hall, Princeton F. HIRZEBRUCH
23 January 1956

Preface to the third edition

In the ten years since the publication of the first edition, the main results have been extended in several directions. On the one hand the RIEMANN-ROCH theorem for algebraic manifolds has been generalised by GROTHENDIECK to a theorem on maps of projective algebraic varieties over a ground field of arbitrary characteristic. On the other hand ATIYAH and SINGER have proved an index theorem for elliptic differential operators on differentiable manifolds which includes, as a special case, the RIEMANN-ROCH theorem for arbitrary compact complex manifolds.

There has been a parallel development of the integrality theorems for characteristic classes. At first these were proved for differentiable manifolds by complicated deductions from the almost complex and algebraic cases. Now they can be deduced directly from theorems on maps of compact differentiable manifolds which are analogous to the RIEMANN-ROCH theorem of GROTHENDIECK. A basic tool is the ring $K(X)$ formed from the semi-ring of all isomorphism classes of complex vector bundles over a topological space X, together with the BOTT periodicity theorem which describes $K(X)$ when X is a sphere. The integrality theorems also follow from the ATIYAH-SINGER index theorem in the same way that the integrality of the TODD genus for algebraic manifolds follows from the RIEMANN-ROCH theorem.

Very recently ATIYAH and BOTT obtained fixed point theorems of the type first proved by LEFSCHETZ. A holomorphic map of a compact complex manifold V operates, under certain conditions, on the cohomology groups of V with coefficients in the sheaf of local holomorphic sections of a complex analytic vector bundle W over V. For a special class of holomorphic maps, ATIYAH and BOTT express the alternating sum of the traces of these operations in terms of the fixed point set of the map. For the identity map this reduces to the RIEMANN-ROCH theorem. Another application yields the formulae of LANGLANDS (see 22.3) for the dimensions of spaces of automorphic forms. ATIYAH and BOTT carry out these investigations for arbitrary elliptic operators and differentiable maps, obtaining a trace formula which generalises the index theorem. Their results have a topological counterpart which generalises the integrality theorems.

The aim of the translation has been to take account of these developments – especially those which directly involve the TODD genus – within the framework of the original text. The translator has done this chiefly by the addition of bibliographical notes to each chapter and by a new appendix containing a survey, mostly without proofs, of some of the applications and generalisations of the RIEMANN-ROCH theorem made since 1956. The fixed point theorems of ATIYAH and BOTT could be mentioned only very briefly, since they became known after the manuscript for the appendix had been finished. A second appendix consists of a paper by A. BOREL which was quoted in the first edition but which has not previously been published. Certain amendments to the text have been made in order to increase the usefulness of the book as a work of reference. Except for Theorems 2.8.4, 2.9.2, 2.11.2, 4.11.1–4.11.4, 10.1.1, 16.2.1 and 16.2.2 in the new text, all theorems are numbered as in the first edition.

The author thanks R. L. E. SCHWARZENBERGER for his efficient work in translating and editing this new edition, and for writing the new appendix, and A. BOREL for allowing his paper to be added to the book.

We are also grateful to Professor F. K. SCHMIDT for suggesting that this edition should appear in the "Grundlehren der mathematischen Wissenschaften", to D. ARLT, E. BRIESKORN and K. H. MAYER for checking the manuscript, and to ANN GARFIELD for preparing the typescript. Finally we wish to thank the publishers for their continued cooperation.

Bonn and Coventry F. HIRZEBRUCH
23 January 1966 R. L. E. SCHWARZENBERGER

Contents

Introduction

The theory of sheaves, developed and applied to various topological problems by LERAY [1], [2][1]), has recently been applied to algebraic geometry and to the theory of functions of several complex variables. These applications, due chiefly to CARTAN, SERRE, KODAIRA, SPENCER, ATIYAH and HODGE have made possible a common systematic approach to both subjects. This book makes a further contribution to this development for algebraic geometry. In addition it contains applications of the results of THOM on cobordism of differentiable manifolds which are of independent interest. Sheaf theory and cobordism theory together provide the foundations for the present results on algebraic manifolds. This introduction gives an outline (0.1—0.8) of the results in the book. It does not contain precise definitions; these can be found by reference to the index. Remarks on terminology and notations used throughout the book are at the end of the introduction (0.9).

0.1. A compact complex manifold V (not necessarily connected) is called an *algebraic manifold* if it admits a complex analytic embedding as a submanifold of a complex projective space of some dimension. By a theorem of CHOW [1] this definition is equivalent to the classical definition of a non-singular algebraic variety. Algebraic manifolds in this sense are often also called non-singular projective varieties. In 0.1—0.6 we consider only algebraic manifolds.

Let V_n be an algebraic manifold of complex dimension n. The arithmetic genus of V_n has been defined in four distinct ways. The postulation formula (HILBERT characteristic function) can be used to define integers $p_a(V_n)$ and $P_a(V_n)$. These are the first two definitions. SEVERI conjectured that

$$p_a(V_n) = P_a(V_n) = g_n - g_{n-1} + \cdots + (-1)^{n-1} g_1 , \tag{1}$$

where g_i is the number of complex-linearly independent holomorphic differential forms on V_n of degree i (i-fold differentials of the first kind). The alternating sum of the g_i can be regarded as a third definition of the arithmetic genus. Further details can be found, for instance, in SEVERI [1]. Equation (1) can be proved easily by means of sheaf theory (KODAIRA-SPENCER [1]) and therefore the first three definitions of the arithmetic genus agree.

1) Numbers in square brackets refer to the bibliography at the end of the book.

The form of the alternating sum of g_i in (1) is inconvenient and we modify the classical definition slightly. We call

$$\chi(V_n) = \sum_{i=0}^{n} (-1)^i g_i \tag{2}$$

the *arithmetic genus* of the algebraic manifold V_n. The integer g_0 in (2) is the number of linearly independent holomorphic functions on V_n and is therefore equal to the number of connected components of V_n. It is usual to call g_n the *geometric genus* of V_n and g_1 the *irregularity* of V_n. In the case $n = 1$ a connected algebraic curve V_1 is a compact RIEMANN surface homeomorphic to a sphere with p handles. Then $g_n = g_1 = p$ and the arithmetic genus of V_1 is $1 - p$. The arithmetic genus and the geometric genus behave multiplicatively:

The genus of the cartesian product $V \times W$ of two algebraic manifolds is the product of the genus of V and the genus of W.

Under the old terminology the arithmetic genus clearly does not have this property. The arithmetic genus $\chi(V_n)$ is a birational invariant because each g_i is a birational invariant (KÄHLER [1] and VAN DER WAERDEN [1], [2]). Under the old terminology the arithmetic genus of a rational variety is 0. According to the present definition it is 1.

0.2. The fourth definition of the arithmetic genus is due to TODD [1]. He showed in 1937 that the arithmetic genus could be represented in terms of the canonical classes of EGER-TODD (TODD [3]). The proof is however incomplete: it relies on a lemma of SEVERI for which no complete proof exists in the literature.

The EGER-TODD class K_i of V_n is by definition an equivalence class of algebraic cycles of real dimension $2n - 2i$. The equivalence relation implies, but does not in general coincide with, the relation of homology equivalence. For example K_1 $(= K)$ is the class of canonical divisors of V_n. (A divisor is canonical if it is the divisor of a meromorphic n-form.) The equivalence relation for $i = 1$ is linear equivalence of divisors. The class K_i defines a $(2n - 2i)$-dimensional homology class. This determines a $2i$-dimensional cohomology class which agrees (up to sign) with the CHERN class c_i of V_n. This "agreement" between the EGER-TODD classes and the CHERN classes was proved by NAKANO [2] (see also CHERN [2], HODGE [3] and ATIYAH [3]).

Remark: The sign of the $2i$-dimensional cohomology class determined by K_i depends on the orientation of V_n. We shall always use the natural orientation of V_n. If $z_1, z_2, \ldots, z_n$ are local coordinates with $z_k = x_k + i\, y_k$ then this orientation is given by the ordering $x_1, y_1, x_2, y_2, \ldots, x_n, y_n$ or in other words by the positive volume element $d x_1 \wedge d y_1 \wedge d x_2 \wedge d y_2 \wedge \cdots \wedge d x_n \wedge d y_n$. In this case K_i determines the cohomology class $(-1)^i c_i$.

In this book we use only the CHERN *classes* and so the fact that the EGER-TODD classes agree with the CHERN classes is not needed. The definition of the TODD genus $T(V_n)$ is given in terms of the CHERN classes. One of the chief purposes of this book is then to prove that $\chi(V_n) = T(V_n)$.

0.3. The natural orientation of V_n defines an element of the $2n$-dimensional integral homology group $H_{2n}(V_n, \mathbf{Z})$ called the fundamental cycle of V_n. The value of a $2n$-dimensional cohomology class b on the fundamental cycle is denoted by $b[V_n]$.

The definition of $T(V_n)$ is in terms of a certain polynomial T_n of weight n in the CHERN classes c_i of V_n, the products being taken in the cohomology ring of V_n. This polynomial is defined algebraically in § 1; it is a rational $2n$-dimensional cohomology class whose value on the fundamental cycle is by definition $T(V_n)$. For small n (see 1.7)

$$T(V_1) = \tfrac{1}{2} c_1[V_1], \quad T(V_2) = \tfrac{1}{12}(c_1^2 + c_2)[V_2], \quad T(V_3) = \tfrac{1}{24} c_1 c_2 [V_3]. \quad (3)$$

The definition implies that $T(V_n)$ is a rational number. The equation $\chi(V_n) = T(V_n)$ implies the non-trivial fact that $T(V_n)$ is an integer and that $T(V_n)$ is a birational invariant. The sequence of polynomials $\{T_n\}$ must be chosen so that, like the arithmetic genus, $T(V_n)$ behaves multiplicatively on cartesian products. There are many sequences with this property: it is sufficient for $\{T_n\}$ to be a multiplicative sequence (§ 1). The sequence $\{T_n\}$ must be further chosen so that $T(V_n)$ agrees with $\chi(V_n)$ whenever possible. In particular if $\mathbf{P}_n(\mathbf{C})$ denotes the n-dimensional complex projective space then $T(\mathbf{P}_n(\mathbf{C})) = 1$ for all n. This condition is used in § 1 to determine the multiplicative sequence $\{T_n\}$ uniquely (Lemma 1.7.1).

For fixed n the polynomial T_n is determined uniquely by the following property: $T_n[V_n] = 1$ *if* $V = \mathbf{P}_{j_1}(\mathbf{C}) \times \cdots \times \mathbf{P}_{j_r}(\mathbf{C})$ *is a cartesian product of complex projective spaces with* $j_1 + \cdots + j_r = n$. *Therefore* T_n *is the unique polynomial which takes the value* 1 *on all rational manifolds of dimension* n.

0.4. The divisors of the algebraic manifold V_n can be formed into equivalence classes with respect to linear equivalence. A divisor is linearly equivalent to zero if it is the divisor (f) of a meromorphic function f on V_n. This equivalence is compatible with addition of divisors and therefore the divisor classes form an additive group. We can also consider complex analytic line bundles (with fibre $\mathbf{C}$ and group $\mathbf{C}^*$; see 0.9) over V_n. In this introduction we identify isomorphic line bundles (see 0.9). Then the line bundles form an abelian group with respect to the tensor product $\otimes$. The identity element, denoted by $\mathbf{1}$, is the trivial complex line bundle $X \times \mathbf{C}$. The inverse of a complex line bundle F is denoted by F^{-1}. *The group of line bundles is isomorphic to the group of divisor classes*:

Every divisor determines a line bundle. The sum of two divisors determines the tensor product of the corresponding line bundles. Two divisors determine the same line bundle if and only if they are linearly equivalent. Finally, every line bundle is determined by some divisor (KODAIRA-SPENCER [2]). Denote by $H^0(V_n, D)$ the complex vector space of all meromorphic functions f on V_n such that $D + (f)$ is a divisor with no poles. $H^0(V_n, D)$ is the "RIEMANN-ROCH space" of D and is *finite dimensional.* The dimension $\dim H^0(V_n, D)$ depends only on the divisor class of D. The determination of $\dim H^0(V_n, D)$ for a given divisor D is the RIEMANN-ROCH problem. If F is the line bundle corresponding to the divisor D then $H^0(V_n, D)$ is isomorphic to $H^0(V_n, F)$, the complex vector space of holomorphic sections of F.

0.5. It has already been said that one aim of this work is to prove the equation

$$\chi(V_n) = T(V_n) . \tag{4}$$

The CHERN number $c_n[V_n]$ is equal to the EULER-POINCARÉ characteristic of V_n. Therefore equation (4) gives, for a connected algebraic curve V homeomorphic to a sphere with p handles:

$$\chi(V_1) = T(V_1) = \tfrac{1}{2} c_1[V_1] = \tfrac{1}{2}(2-2p) . \tag{4_1}$$

The RIEMANN-ROCH theorem for algebraic curves states (see for instance WEYL [1]):

$$\dim H^0(V_1, D) - \dim H^0(V_1, K - D) = d + 1 - p \tag{4_1^*}$$

where d is the degree of the divisor D and K is a canonical divisor of V_1. Since $\dim H^0(V_1, K) = g_1$ the substitution $D = 0$ in (4_1^*) gives (4_1). It will be shown that for algebraic manifolds of arbitrary dimension equation (4) admits a generalisation which corresponds precisely to the generalisation (4_1^*) of (4_1). This generalisation will be given in terms of line bundles rather than divisors.

Let F be a complex analytic line bundle and let $H^i(V_n, F)$ be the i-th cohomology group of V_n with coefficients in the sheaf of germs of local holomorphic sections of F. In the case $F = 1$ this is the sheaf of germs of local holomorphic functions. The cohomology "group" $H^i(V_n, F)$ is a complex vector space which, by results of CARTAN-SERRE [1] (see also CARTAN [4]) and KODAIRA [3], is of finite dimension. The vector space $H^0(V_n, F)$ is the "RIEMANN-ROCH space" of F defined in 0.4. A theorem of DOLBEAULT [1] implies that $\dim H^i(V_n, 1) = g_i$. The integer $\dim H^i(V_n, F)$ depends only on the isomorphism class of F and is zero for $i > n$. It is therefore possible to define

$$\chi(V_n, F) = \sum_{i=0}^{n} (-1)^i \dim H^i(V_n, F) . \tag{5}$$

This is the required generalisation of the left hand side of (4). It will be shown that $\chi(V_n, F)$ can be expressed as a certain polynomial in the CHERN classes of V_n and a 2-dimensional cohomology class f determined by the line bundle F. Here f is the first CHERN class of F (the cohomology obstruction to the existence of a continuous never zero section of F). If F is represented by a divisor D then f is also determined by the $(2n-2)$-dimensional homology class corresponding to D. For small n,

$$\chi(V_1, F) = \left(f + \tfrac{1}{2} c_1\right)[V_1], \quad \chi(V_2, F) = \left(\tfrac{1}{2}(f^2 + f c_1) + \tfrac{1}{12}(c_1^2 + c_2)\right)[V_2],$$

$$\chi(V_3, F) = \left(\tfrac{1}{6} f^3 + \tfrac{1}{4} f^2 c_1 + \tfrac{1}{12} f(c_1^2 + c_2) + \tfrac{1}{24} c_1 c_2\right)[V_3].$$

This is the generalisation of the RIEMANN-ROCH theorem to algebraic manifolds of arbitrary dimension (Theorem 20.3.2). By the SERRE duality theorem (see 15.4.2) $\dim H^1(V_1, F) = \dim H^0(V_1, K \otimes F^{-1})$ and $\dim H^2(V_2, F) = \dim H^0(V_2, K \otimes F^{-1})$ where K denotes the line bundle determined by canonical divisors. It follows that the equations for $\chi(V_1, F)$ and $\chi(V_2, F)$ imply the classical RIEMANN-ROCH theorem for an algebraic curve and for an algebraic surface. Full details are given in 19.2 and 20.7.

KODAIRA [4] and SERRE have given conditions under which $\dim H^i(V_n, F) = 0$ for $i > 0$ (see Theorem 18.2.2 and CARTAN [4], Exposé XVIII). The definition of $\chi(V_n, F)$ in (5) then shows that our formula for $\chi(V_n, F)$ yields a formula for $H^0(V_n, F)$. In such cases the "RIEMANN-ROCH problem" stated in 0.4 is completely solved. This corresponds for algebraic curves to the well known fact that the term $\dim H^0(V_1, K - D)$ in (4_1^*) is zero if $d > 2p - 2$.

0.6. There is a further generalisation of equation (4). Let W be a complex analytic vector bundle over V_n [with fibre $\mathbf{C}_q$ and group $\mathbf{GL}(q, \mathbf{C})$; see 0.9]. Let $H^i(V_n, W)$ be the i-th cohomology group of V_n with coefficients in the sheaf of germs of local holomorphic sections of W. Then $H^i(V_n, W)$ is again a complex vector space of finite dimension and $\dim H^i(V_n, W)$ is zero for $i > n$. It is therefore possible to define

$$\chi(V_n, W) = \sum_{i=0}^{n} (-1)^i \dim H^i(V_n, W). \tag{6}$$

It was conjectured by SERRE, in a letter to KODAIRA and SPENCER (29 September 1953), that $\chi(V_n, W)$ could be expressed as a polynomial in the CHERN classes of V_n and the CHERN classes of W. We shall obtain an explicit formula for the polynomial of $\chi(V_n, W)$. This is the RIEMANN-ROCH theorem for vector bundles (Theorem 21.1.1). A corollary in the case $n = 1$ (algebraic curves) is the generalisation of the RIEMANN-ROCH theorem due to WEIL [1]. Full details are given in 21.1.

The result on $\chi(V_n, W)$ can be applied to particular vector bundles over V_n. We define (see KODAIRA-SPENCER [3])

$$\chi^p(V_n) = \chi(V_n, \lambda^p T) \tag{7}$$

where $\lambda^p T$ is the vector bundle of covariant p-vectors of V_n. The CHERN classes of $\lambda^p T$ can be expressed in terms of the CHERN classes of V_n (Theorem 4.4.3). Therefore $\chi^p(V_n)$ is a polynomial of weight n in the CHERN classes of V_n. By a theorem of DOLBEAULT [1], $\dim H^q(V_n, \lambda^p T)$ is the number $h^{p,q}$ of complex-linearly independent harmonic forms on V_n of type (p, q). Therefore $\chi^p(V_n) = \sum_{q=0}^{n} (-1)^q h^{p,q}$. For example, in the case $n = 4$, there is an equation

$$\chi^1(V_4) = h^{1,0} - h^{1,1} + h^{1,2} - h^{1,3} + h^{1,4} = 4\chi(V_4) - \tfrac{1}{12}(2c_4 + c_3 c_1)[V_4]. \tag{8}$$

The sum $\sum_{p=0}^{n} \chi^p(V_n)$ is clearly zero for n odd. The alternating sum $\sum_{p=0}^{n} (-1)^p \chi^p(V_n)$ is by theorems of DE RHAM and HODGE equal to the EULER-POINCARÉ characteristic $c_n[V_n]$ of V_n. The polynomials for $\chi^p(V_n)$ have the same properties. HODGE [4] proved that for n even the sum $\sum_{p=0}^{n} \chi^p(V_n)$ is equal to the index of V_n. By definition the index of V_n is the signature (number of positive eigenvalues minus number of negative eigenvalues) of the bilinear symmetric form $xy[V_n]$ $(x, y \in H^n(V_n, \mathbf{R}))$ on the n-dimensional real cohomology group of V_n. Therefore the index of V_n is a polynomial in the CHERN classes of V_n. This polynomial can actually be expressed as a polynomial in the PONTRJAGIN classes of V_n and is therefore defined for any oriented differentiable manifold.

0.7. We have just remarked that the main result of this book [the expression of $\chi(V_n, W)$ as a certain polynomial in the CHERN classes of V_n and W] implies that the index of an algebraic manifold V_{2k} can be expressed as a polynomial in the PONTRJAGIN classes of V_{2k}. In fact this theorem is the starting point of our investigation. Let M^{4k} be an oriented differentiable manifold of real dimension $4k$. In this work "differentiable" always means "C^∞-differentiable" so that all partial derivatives exist and are continuous. The orientation of M^{4k} defines a $4k$-dimensional fundamental cycle. The value of a $4k$-dimensional cohomology class b on the fundamental cycle is denoted by $b[M^{4k}]$. In Chapter Two the cobordism theory of THOM is used to express the index $\tau(M^{4k})$ of M^{4k} as a polynomial of weight k in the PONTRJAGIN classes of M^{4k}. For example,

$$\tau(M^4) = \tfrac{1}{3} p_1[M^4], \quad \tau(M^8) = \tfrac{1}{45}(7p_2 - p_1^2)[M^8]. \tag{9}$$

The formula for $\tau(M^4)$ was conjectured by WU. The formulae for $\tau(M^4)$ and $\tau(M^8)$ were both proved by THOM [2]. A brief summary of the deduction of the formula for $\chi(V_n, W)$ from that for $\tau(M^{4k})$ can be found in HIRZEBRUCH [2].

0.8. The definitions in 0.1–0.6 were formulated only for algebraic manifolds. In the proof of the RIEMANN-ROCH theorem we make this restriction only when necessary. The index theorem described in 0.7 is proved in Chapter Two for arbitrary oriented differentiable manifolds. The main results of THOM on cobordism are quoted: the proofs, which make use of differentiable approximation theorems and algebraic homotopy theory, are outside the scope of this work.

In Chapter Three the formal theory of the TODD genus and of the associated polynomials is developed for arbitrary compact almost complex manifolds (T-theory). In particular we obtain an integrality theorem (14.3.2). This theorem has actually little to do with almost complex manifolds; its relation to subsequent integrality theorems for differentiable manifolds is discussed in the bibliographical note to Chapter Three and in the Appendix.

In Chapter Four the theory of the integers $\chi(V_n, W)$ is developed as far as possible for arbitrary compact complex manifolds (χ-theory). The necessary results on sheaf cohomology due to CARTAN, DOLBEAULT, KODAIRA, SERRE and SPENCER are described briefly. In the course of the proof it is necessary to assume first that V_n is a KÄHLER manifold. Finally, if V_n is an algebraic manifold, we are able to identify the χ-theory with the T-theory (RIEMANN-ROCH theorem for vector bundles; Theorem 21.1.1).

The Appendix contains a review of applications and generalisations of the RIEMANN-ROCH theorem. In particular it is now known that the identification of the χ-theory with the T-theory holds for any compact complex manifold V_n (see § 25).

The author has tried to make the book as independent of other works as is possible within a limited length. The necessary preparatory material on multiplicative sequences, sheaves, fibre bundles and characteristic classes can be found in Chapter One.

0.9. Remarks on notation and terminology

The following notations are used throughout the book.

$\mathbf{Z}$: integers, $\mathbf{Q}$: rational numbers, $\mathbf{R}$: real numbers, $\mathbf{C}$: complex numbers, $\mathbf{R}^q$: vector space over $\mathbf{R}$ of q-ples $(x_1, \ldots, x_q)$ of real numbers, $\mathbf{C}_q$: vector space over $\mathbf{C}$ of q-ples of complex numbers. $\mathbf{GL}(q, \mathbf{R})$ denotes the group of invertible $q \times q$ matrices (a_{ik}) with real coefficients a_{ik}, *i. e.* the group of automorphisms of $\mathbf{R}^q$

$$x_i' = \sum_{k=1}^{q} a_{ik} x_k .$$

$\mathbf{GL}^+(q, \mathbf{R})$ denotes the subgroup of $\mathbf{GL}(q, \mathbf{R})$ consisting of matrices with positive determinant (the group of orientation preserving automorphisms). $\mathbf{O}(q)$ denotes the subgroup of orthogonal matrices of $\mathbf{GL}(q, \mathbf{R})$ and $\mathbf{SO}(q) = \mathbf{O}(q) \cap \mathbf{GL}^+(q, \mathbf{R})$. Similarly $\mathbf{GL}(q, \mathbf{C})$ denotes the group of invertible $q \times q$ matrices with complex coefficients, and $\mathbf{U}(q)$ the subgroup of unitary matrices of $\mathbf{GL}(q, \mathbf{C})$. We write $\mathbf{C}^* = \mathbf{GL}(1, \mathbf{C})$, the multiplicative group of non-zero complex numbers. $\mathbf{P}_{q-1}(\mathbf{C})$ denotes the complex projective space of complex dimension $q-1$ (the space of complex lines through the origin of $\mathbf{C}_q$). We shall often denote real dimension by an upper suffix (for example M^{4k}, $\mathbf{R}^q$) and complex dimension by a lower suffix (for example V_n, $\mathbf{C}_q$).

We have adopted one slight modification of the usual terminology. An isomorphism class of principal fibre bundles with structure group G is called a G-bundle. Thus a G-bundle is an element of a certain cohomology set. On the other hand, we use the words fibre bundle, line bundle and vector bundle to mean a particular fibre space and not an isomorphism class of such spaces (see 3.2). In Chapter Four all constructions depend only on the isomorphism class of the vector bundles involved and it is possible to drop this distinction (see 15.1).

The book is divided into chapters and then into paragraphs, which are numbered consecutively throughout the book. Formulae are numbered consecutively within each paragraph. The paragraphs are divided into sections. Thus 4.1 means section 1 of § 4; 4.1 (5) means formula (5) of § 4, which occurs in section 4.1; 4.1.1 refers to Theorem 1 of section 4.1. The index includes references to the first occurrence of any symbol.

Chapter One

Preparatory material

The elementary and formal algebraic theory of multiplicative sequences is contained in § 1. In particular the TODD polynomials T_j, and also the polynomials L_j used in the index theorem, are defined. Results on sheaves needed in the sequel are collected in § 2. The basic properties of fibre bundles are given in § 3. In § 4 these are applied to obtain characteristic classes. In particular, the CHERN classes and PONTRJAGIN classes are defined. The results of § 1 are not used until § 8. The reader is therefore advised to begin with § 2 and to refer to § 1 only when necessary.

§ 1. Multiplicative sequences

1.1. Let B be a commutative ring with identity element 1. Let $p_0 = 1$ and let $p_1, p_2, \ldots$ be indeterminates. Consider the ring $\mathfrak{B} = B[p_1, p_2, \ldots]$ obtained by adjoining the indeterminates p_i to B. Then $\mathfrak{B}$ is the ring of polynomials in the p_i with coefficients in B, and is graded in the following way:

The product $p_{j_1} p_{j_2} \ldots p_{j_r}$ has weight $j_1 + j_2 + \cdots + j_r$ and

$$\mathfrak{B} = \sum_{k=0}^{\infty} \mathfrak{B}_k, \tag{1}$$

where $\mathfrak{B}_k$ is the additive group of those polynomials which contain only terms of weight k and $\mathfrak{B}_0 = B$. The group $\mathfrak{B}_k$ is a module over B whose rank is equal to the number $\pi(k)$ of partitions of k. Clearly

$$\mathfrak{B}_r \mathfrak{B}_s \subset \mathfrak{B}_{r+s}. \tag{2}$$

1.2. Let $\{K_j\}$ be a sequence of polynomials in the indeterminates p_i with $K_0 = 1$ and $K_j \in \mathfrak{B}_j$ $(j = 0, 1, 2, \ldots)$. The sequence $\{K_j\}$ is called a *multiplicative sequence* (or m-sequence) if every identity of the form

$$\begin{aligned} &1 + p_1 z + p_2 z^2 + \cdots \\ &\quad = (1 + p_1' z + p_2' z^2 + \cdots)(1 + p_1'' z + p_2'' z^2 + \cdots) \end{aligned} \tag{3}$$

with z, p_i', p_i'' indeterminate implies an identity

$$\begin{aligned} &\sum_{j=0}^{\infty} K_j(p_1, p_2, \ldots, p_j) z^j \\ &= \sum_{i=0}^{\infty} K_i(p_1', p_2', \ldots, p_i') z^i \sum_{j=0}^{\infty} K_j(p_1'', p_2'', \ldots, p_j'') z^j . \end{aligned} \tag{4}$$

In abbreviated notation we write

$$K\left(\sum_{j=0}^{\infty} p_i z^i\right) = \sum_{j=0}^{\infty} K_j(p_1, \ldots, p_j) z^j$$

both when the p_i are indeterminates and when they are replaced by particular values. The power series

$$K(1+z) = \sum_{i=0}^{\infty} b_i z^i \quad (b_0 = 1,\ b_i = K_i(1, 0, \ldots, 0) \in B)$$

is called the *characteristic power series* of the m-sequence $\{K_j\}$.

In the sequel we consider formal factorisations

$$1 + p_1 z + \cdots + p_m z^m = \prod_{i=1}^{m} (1 + \beta_i z). \tag{5_m}$$

That is, the elements p_i are regarded as the elementary symmetric functions in $\beta_1, \ldots, \beta_m$. The ring $\mathfrak{B}$ is then the ring of all symmetric polynomials in $\beta_1, \ldots, \beta_m$ with coefficients in B. The following two lemmas give a complete description of all possible m-sequences.

Lemma 1.2.1. *The m-sequence $\{K_j\}$ is completely determined by its characteristic power series $Q(z) = K(1+z)$.*

Proof: By (3), (4) and (5)

$$\sum_{j=0}^{m} K_j(p_1, \ldots, p_j)\, z^j + \sum_{j=m+1}^{\infty} K_j(p_1, \ldots, p_m, 0, \ldots, 0)\, z^j = \prod_{i=1}^{m} Q(\beta_i z)\,. \quad (6_m)$$

Therefore any polynomial K_j with $j \leqq m$ is determined as a symmetric polynomial in the β_i and hence as a polynomial in the p_i. This holds for arbitrary m and so completes the proof.

Lemma 1.2.2. *To every formal power series $Q(z) = \sum_{i=0}^{\infty} b_i z^i$ ($b_0 = 1$, $b_i \in B$) there is associated an m-sequence $\{K_j\}$ with $K(1+z) = Q(z)$.*

Proof: We apply (5_m) and consider the product $\prod_{i=1}^{m} Q(\beta_i z)$. The coefficient of z^j in this product is symmetric in the β_i and homogeneous of weight j. It can therefore be expressed as a polynomial $K_j^{(m)}(p_1, \ldots, p_j)$ of weight j in a unique way. It follows easily that $K_j^{(m)}$ does not depend on m for $m \geqq j$. Define $K_j = K_j^{(m)}$ for $m \geqq j$. The sequence $\{K_j\}$ is the required multiplicative sequence. For the proof note that (6_m) holds. This implies that the multiplicative property "(3) implies (4)" is true if the indeterminates p_i', p_i'' are replaced by 0 for large values of i. Hence $\{K_j\}$ is an m-sequence. Finally (6_m) for $m = 1$ shows that $K(1+z) = Q(z)$.

Lemmas 1.2.1 and 1.2.2 together show that there is a one-one correspondence between m-sequences and formal power series with constant term 1. For instance the m-sequence $\{p_j\}$ has $1+z$ as its characteristic power series.

1.3. It is convenient to reformulate 1.1 and 1.2 with the indeterminates p_i replaced by c_i, the indeterminate z by x, and the roots β_i in (5_m) by γ_i. The two formulations will be linked by putting $c_0 = p_0 = 1$, $z = x^2$ and $\beta_i = \gamma_i^2$. In other words we introduce the relations

$$z = x^2 \quad \text{and} \quad \sum_{i=0}^{\infty} p_i(-z)^i = \left(\sum_{j=0}^{\infty} c_j(-x)^j\right)\left(\sum_{i=0}^{\infty} c_i x^i\right). \quad (7)$$

We have the following trivial

Lemma 1.3.1. *Let $\{K_j(p_1, \ldots, p_j)\}$ be the m-sequence with $Q(z)$ as characteristic power series and $\{\tilde{K}_j(c_1, \ldots, c_j)\}$ the m-sequence with $\tilde{Q}(x) = Q(x^2)$ as characteristic power series. Then the relations (7) imply*

$$K_j(p_1, \ldots, p_j) = \tilde{K}_{2j}(c_1, \ldots, c_{2j})$$
$$0 = \tilde{K}_{2j+1}(c_1, \ldots, c_{2j+1})\,.$$

In particular the m-sequence in the c_i with $1 + x^2$ as characteristic power series is $1, 0, p_1, 0, p_2, \ldots$.

Note that

$$p_1 = -2c_2 + c_1^2, \quad p_2 = 2c_4 - 2c_3 c_1 + c_2^2, \quad p_3 = -2c_6 + 2c_5 c_1 - 2c_4 c_2 + c_3^2.$$

1.4. Given a power series $Q(z) = \sum_{i=0}^{\infty} b_i z^i \; (b_i \in B, b_0 = 1)$ we can consider the formal factorisation

$$1 + b_1 z + b_2 z^2 + \cdots + b_m z^m = (1 + \beta_1' z)(1 + \beta_2' z) \cdots (1 + \beta_m' z). \tag{8}$$

The sum

$$\Sigma (\beta_1')^{j_1} (\beta_2')^{j_2} \cdots (\beta_r')^{j_r}, \quad (j_1 \geqq j_2 \geqq \cdots \geqq j_r \geqq 1; \; \sum_{s=1}^{r} j_s = k \leqq m), \tag{9}$$

denotes as usual the symmetric function in the β_i' which is the sum of all pairwise distinct monomials obtained by applying a permutation of $(\beta_1', \beta_2', \ldots, \beta_m')$ to the monomial $(\beta_1')^{j_1} (\beta_2')^{j_2} \cdots (\beta_r')^{j_r}$. The number of monomials in the sum is $m!/h$ where h is the number of permutations of $(\beta_1', \beta_2', \ldots \beta_m')$ which leave $(\beta_1')^{j_1} (\beta_2')^{j_2} \ldots (\beta_r')^{j_r}$ fixed. The conditions on $j_1, j_2, \ldots, j_r$ in (9) imply that the symmetric function $\Sigma (\beta_1')^{j_1} (\beta_2')^{j_2} \ldots (\beta_r')^{j_r}$ is a polynomial of weight k in the b_i with integer coefficients. It does not depend on m and can be denoted simply by $\Sigma(j_1, j_2, \ldots, j_r)$. We can now formulate a lemma which facilitates the explicit calculation of the polynomials of an m-sequence.

Lemma 1.4.1. *Let $\{K_j(p_1, \ldots, p_j)\}$ be the m-sequence corresponding to the power series $Q(z) = \sum_{i=0}^{\infty} b_i z^i$. Then the coefficient of*

$$p_{j_1} p_{j_2} \cdots p_{j_r} \quad \text{in} \quad K_k \; (j_1 \geqq j_2 \geqq \cdots \geqq j_r \geqq 1, \; \sum_{s=1}^{r} j_s = k)$$

is equal to $\Sigma(j_1, j_2, \ldots, j_r)$.

The proof uses (6) and (8). The details are left to the reader. As an example, the coefficient of p_k in K_k is equal to $s_k = \Sigma(k)$.

$$s_0 = 1, \; s_1 = b_1, \; s_2 = -2b_2 + b_1^2, \; s_3 = 3b_3 - 3b_2 b_1 + b_1^3, \text{ etc.}$$

Similarly the coefficient of p_k^2 in K_{2k} is $\Sigma(k, k) = \frac{1}{2}(s_k^2 - s_{2k})$. The s_k can be calculated by a formula of CAUCHY:

$$1 - z \frac{d}{dz} \log Q(z) = Q(z) \frac{d}{dz}\left(\frac{z}{Q(z)}\right) = \sum_{j=0}^{\infty} (-1)^j s_j z^j. \tag{10}$$

1.5. In this section, and in the following sections 1.6—1.8, we define the particular m-sequences which arise in the present work. We consider first the power series

$$Q(z) = \frac{\sqrt{z}}{\tanh \sqrt{z}} = 1 + \sum_{k=1}^{\infty} (-1)^{k-1} \frac{2^{2k}}{(2k)!} B_k z^k.$$

Here the coefficient ring B is the field $\mathbf{Q}$ of rational numbers. The coefficients B_k are the BERNOULLI numbers (in the notation for which $B_k > 0$ and $\neq \frac{1}{2}$ for all k):

$$B_1 = \frac{1}{6}, \quad B_2 = \frac{1}{30}, \quad B_3 = \frac{1}{42}, \quad B_4 = \frac{1}{30},$$

$$B_5 = \frac{5}{66}, \quad B_6 = \frac{691}{2730}, \quad B_7 = \frac{7}{6}, \quad B_8 = \frac{3617}{510}.$$

The m-sequence with $Q(z)$ as characteristic power series is denoted by $\{L_j(p_1, \ldots, p_j)\}$. The methods of 1.4 can be used to calculate the first few polynomials L_j:

$$L_1 = \frac{1}{3} p_1,$$

$$L_2 = \frac{1}{45}(7p_2 - p_1^2),$$

$$L_3 = \frac{1}{3^3 \cdot 5 \cdot 7}(62p_3 - 13p_2 p_1 + 2p_1^3),$$

$$L_4 = \frac{1}{3^4 \cdot 5^2 \cdot 7}(381p_4 - 71p_3 p_1 - 19p_2^2 + 22p_2 p_1^2 - 3p_1^4),$$

$$L_5 = \frac{1}{3^5 \cdot 5^2 \cdot 7 \cdot 11} \times$$

$$\times (5110p_5 - 919p_4 p_1 - 336p_3 p_2 + 237p_3 p_1^2 + 127p_2^2 p_1 - 83p_2 p_1^3 + 10p_1^5).$$

The coefficient s_k of p_k in L_k can be calculated by 1.4 (10)

$$\sum_{j=0}^{\infty} (-1)^j \, s_j \, z^j = \frac{1}{2} + \frac{1}{2} \frac{2\sqrt{z}}{\sinh 2\sqrt{z}}$$

and therefore

$$s_0 = 1 \text{ and } s_k = \frac{2^{2k}(2^{2k-1} - 1)}{(2k)!} B_k \text{ for } k \geqq 1. \tag{11}$$

The following lemma shows that the substitution $p_i = \binom{2k+1}{i}$ defined by $1 + p_1 z + p_2 z^2 + \cdots + p_k z^k = (1+z)^{2k+1} \pmod{z^{k+1}}$ gives the value $L_k(p_1, \ldots, p_k) = 1$.

Lemma 1.5.1. *Let* $Q(z) = \frac{\sqrt{z}}{\tanh \sqrt{z}}$. *Then for every* k *the coefficient* J_k *of* z^k *in* $(Q(z))^{2k+1}$ *is equal to* 1, *and* $Q(z)$ *is the only power series with rational coefficients which has this property.*

Proof: By the CAUCHY integral formula

$$J_k = \frac{1}{2\pi i} \int \frac{1}{z^{k+1}} \left(\frac{\sqrt{z}}{\tanh \sqrt{z}} \right)^{2k+1} dz.$$

The substitution $t = \tanh \sqrt{z}$ gives

$$J_k = \frac{1}{2\pi i} \int \frac{dt}{(1-t^2)\, t^{2k+1}} = 1.$$

In both cases the integral is over a small circle with centre the origin in the z-plane and t-plane. Note that under the substitution a circuit of the circle in the t-plane corresponds to two circuits in the z-plane. The equations $J_k = 1$ can then be used to calculate the coefficients of $Q(z)$ inductively.

The following lemma is not used in the present work but is nevertheless important for applications of the polynomials L_k to cohomology operations. A proof is given in ATIYAH-HIRZEBRUCH [4].

Lemma 1.5.2. *The polynomial L_k can be written in a unique way as a polynomial with coprime integer coefficients, divided by a positive integer $\mu(L_k)$, where*

$$\mu(L_k) = \Pi q^{\left[\frac{2k}{q-1}\right]}$$

is a product over all primes q with $3 \leqq q \leqq 2k+1$.

1.6. The m-sequence with $Q(z) = \frac{2\sqrt{z}}{\sinh 2\sqrt{z}}$ as characteristic power series is denoted by $\{A_k(p_1, \ldots, p_k)\}$. The methods of 1.4 give

$$A_1 = -\tfrac{2}{3} p_1, A_2 = \tfrac{2}{45}(-4p_2 + 7p_1^2), A_3 = \tfrac{-4}{3^3 \cdot 5 \cdot 7}(16p_3 - 44p_2p_1 + 31p_1^3).$$

Remark: The polynomial A_k can be written in a unique way as a polynomial with coprime integer coefficients multiplied by $2^{\alpha(k)}/\mu(L_k)$. Here $\alpha(k)$ is the number of non-zero terms in the dyadic expansion of k (see ATIYAH-HIRZEBRUCH [2]).

1.7. The last two particular m-sequences which are needed in the sequel will be given in the (c_i, x, γ_i) formulation (see 1.3). Let the coefficient ring B be the field $\mathbf{Q}$ of rational numbers. Consider the m-sequence $\{T_k(c_1, \ldots, c_k)\}$ with characteristic power series

$$Q(x) = \frac{x}{1 - e^{-x}} = 1 + \frac{1}{2}x + \sum_{k=1}^{\infty} (-1)^{k-1} \frac{B_k}{(2k)!} x^{2k}.$$

The polynomials T_k are called TODD polynomials. The identity

$$\frac{x}{1 - e^{-x}} = \exp\left(\frac{1}{2}x\right) \frac{\frac{1}{2}x}{\sinh \frac{1}{2}x} \quad (\text{we write } \exp(a) = e^a)$$

is useful for the calculation of the first few TODD polynomials. It implies, using Lemma 1.3.1, formula (6_m) in the (c_i, x, γ_i) formulation and the relations (7), that

$$T_k(c_1, \ldots, c_k) = \Sigma \frac{1}{2^{4s} r!} \left(\frac{1}{2} c_1\right)^r A_s(p_1, \ldots, p_s) \tag{12}$$

where the sum is over all non-negative integers r, s with $r + 2s = k$.

The result (compare TODD [1]) is

$$T_1 = \frac{1}{2} c_1 ,$$
$$T_2 = \frac{1}{12} (c_2 + c_1^2) ,$$
$$T_3 = \frac{1}{24} c_2 c_1 ,$$
$$T_4 = \frac{1}{720} (-c_4 + c_3 c_1 + 3c_2^2 + 4c_2 c_1^2 - c_1^4) ,$$
$$T_5 = \frac{1}{1440} (-c_4 c_1 + c_3 c_1^2 + 3c_2^2 c_1 - c_2 c_1^3) ,$$
$$T_6 = \frac{1}{60480} (2c_6 - 2c_5 c_1 - 9c_4 c_2 - 5c_4 c_1^2 - c_3^2 + 11 c_3 c_2 c_1 + \\ + 5c_3 c_1^3 + 10 c_2^3 + 11 c_2^2 c_1^2 - 12 c_2 c_1^4 + 2c_1^6) .$$

Remarks: 1). Formula (12) implies that T_k is divisible by c_1 for k odd.

2). It follows from formula 1.4 (10) applied to the m-sequence $\{T_k\}$ that the coefficients of c_k and c_1^k in the TODD polynomial T_k are equal. It is easy to see that $\{T_k\}$ is the only m-sequence which has this property and for which $T_1 = \frac{1}{2} c_1$.

The following lemma shows that the substitution $c_i = \binom{n+1}{i}$ defined by

$$1 + c_1 x + \cdots + c_n x^n = (1+x)^{n+1} \pmod{x^{n+1}}$$

gives the value $T_n(c_1, \ldots, c_n) = 1$.

Lemma 1.7.1. *Let* $Q(x) = \frac{x}{1 - e^{-x}}$. *Then for every* k *the coefficient of* x^k *in* $(Q(x))^{k+1}$ *is equal to* 1, *and* $Q(x)$ *is the only power series with rational coefficients which has this property.*

Proof: By the CAUCHY integral formula as in Lemma 1.5.1. A similar proof gives

Lemma 1.7.2. *Substitute in* $T_k(c_1, \ldots, c_k)$ *the values* c_i *given by*

$$1 + c_1 x + \cdots + c_k x^k = (1+x)^k (1-x) \pmod{x^{k+1}} .$$

Then $T_k = 0$ *for* $k \geqq 1$.

There is a result analogous to Lemma 1.5.2 which is proved in ATIYAH-HIRZEBRUCH [4]:

Lemma 1.7.3. *The polynomial* T_k *can be written in a unique way as a polynomial with coprime integer coefficients, divided by a positive integer* $\mu(T_k)$, *where*

$$\mu(T_k) = \prod q^{\left[\frac{k}{q-1}\right]}$$

is a product over all primes q *with* $2 \leqq q \leqq k+1$. *Moreover* (see Lemma 1.5.2) $\mu(T_{2k+1}) = 2\mu(T_{2k}) = 2^{2k+1} \mu(L_k)$.

1.8. Now let the coefficient ring B be the ring $\mathbf{Q}[y]$ of polynomials in an indeterminate y with rational coefficients. Consider the m-sequence

$T_j(y; c_1, \ldots, c_j)$ with characteristic power series

$$Q(y; x) = \frac{x(y+1)}{1 - e^{-x(y+1)}} - yx = \frac{x(y+1)}{e^{x(y+1)} - 1} + x .$$

The following generalisation of Lemma 1.7.1 shows that the substitution $c_i = \binom{n+1}{i}$ gives

$$T_n(y; c_1, \ldots, c_n) = 1 - y + y^2 - \cdots + (-1)^n y^n .$$

Lemma 1.8.1. *For every n the coefficient of x^n in $(Q(y; x))^{n+1}$ is equal to $\sum_{i=0}^{n} (-1)^i y^i$, and $Q(y, x)$ is the only power series with coefficients in $\mathbf{Q}[y]$ which has this property.*

The polynomial $T_n(y; c_1, \ldots, c_n)$ can be written in a unique way in the form

$$T_n(y; c_1, \ldots, c_n) = \sum_{p=0}^{n} T_n^p(c_1, \ldots, c_n)\, y^p .$$

The polynomials $T_n^p(c_1, \ldots, c_n)$ satisfy

$$T_n^p(c_1, \ldots, c_n) = (-1)^n\, T_n^{n-p}(c_1, \ldots, c_n) . \tag{13}$$

Proof of (13): $Q\left(\frac{1}{y}; y\,x\right) = Q(y; -x)$ and therefore

$$y^n\, T_n\left(\frac{1}{y}; c_1, \ldots, c_n\right) = (-1)^n\, T_n(y; c_1, \ldots, c_n). \text{ Q. E. D.}$$

Consider the formal factorisation

$$1 + c_1 x + \cdots + c_n x^n = \prod_{i=1}^{n} (1 + \gamma_i x) \tag{14}$$

where x is an indeterminate. Then (again writing $\exp(a) = e^a$)

$$T_n^p(c_1, \ldots, c_n) = \varkappa_n \left[\Sigma \exp(-\gamma_{j_1} - \cdots - \gamma_{j_p}) \prod_{i=1}^{n} \frac{\gamma_i}{1 - \exp(-\gamma_i)}\right] . \tag{15}$$

The sum is over all $\binom{n}{p}$ combinations of p pairwise distinct γ_i. $\varkappa_n[\;]$ denotes the sum of all homogeneous terms of degree n in the γ_i which occur in $[\;]$. By (14) this sum can be written as a polynomial in the c_i of weight n.

Proof of (15): Denote temporarily the expression on the right hand side of (15) by $\tilde{T}_n^p$. Then

$$\sum_{p=0}^{n} \tilde{T}_n^p y^p = \varkappa_n \left[\prod_{i=1}^{n} \left((1 + y\exp(-\gamma_i))\, \frac{\gamma_i}{1 - \exp(-\gamma_i)}\right)\right]$$

$$= \varkappa_n \left[\prod_{i=1}^{n} \frac{(1 + y\exp(-(1+y)\gamma_i))}{1+y} \cdot \frac{(1+y)\gamma_i}{1 - \exp(-(1+y)\gamma_i)}\right]$$

$$= \varkappa_n \left[\prod_{i=1}^{n} Q(y; \gamma_i)\right] = \sum_{p=0}^{n} T_n^p(c_1, \ldots, c_n)\, y^p . \qquad \text{Q. E. D.}$$

Finally note that $Q(0; x) = \frac{x}{1-e^{-x}}$, $Q(-1; x) = 1 + x$ and $Q(1; x) = \frac{x}{\tanh x}$. Therefore (see 1.5 and Lemma 1.3.1)

$$T_n^0(c_1, \ldots, c_n) = T_n(c_1, \ldots, c_n) \quad \text{(TODD polynomial)},$$

$$\sum_{p=0}^{n} (-1)^p \, T_n^p(c_1, \ldots, c_n) = c_n,$$

$$\sum_{p=0}^{n} T_n^p(c_1, \ldots, c_n) = \tilde{L}_n(c_1, \ldots, c_n), \text{ i. e.} \tag{16}$$

$$\sum_{p=0}^{n} T_n^p(c_1, \ldots, c_n) \begin{cases} = 0 \text{ for } n \text{ odd} \\ = L_k(p_1, \ldots, p_k) \text{ for } n = 2k. \end{cases}$$

1.9. The TODD polynomials are essentially the BERNOULLI polynomials of higher order defined by NÖRLUND (see N. E. NÖRLUND, Differenzenrechnung, Berlin, Springer-Verlag, 1924, especially p. 143). The BERNOULLI polynomial $B_j^{(n)}(\gamma_1, \ldots, \gamma_n)$ is defined by

$$\prod_{i=1}^{n} \frac{\gamma_i x}{\exp(\gamma_i x) - 1} = \sum_{j=0}^{\infty} \frac{x^j}{j!} B_j^{(n)}(\gamma_1, \ldots, \gamma_n).$$

If the c_i are regarded as the elementary symmetric functions in $\gamma_1, \ldots, \gamma_n$ [see 1.8 (14)] then

$$T_k(c_1, \ldots, c_k) = \frac{(-1)^k}{k!} B_k^{(n)}(\gamma_1, \ldots, \gamma_n) \quad \text{for } k \leqq n.$$

A corresponding remark holds for the polynomials A_k defined in 1.6. They are essentially the polynomials D_k considered by NÖRLUND. In the notation of 1.3 and 1.6

$$A_k(p_1, \ldots, p_k) = \tilde{A}_{2k}(c_1, \ldots, c_{2k}) = \frac{2^{2k}}{(2k)!} D_{2k}^{(n)}(\gamma_1, \ldots, \gamma_n) \quad \text{for } 2k \leqq n.$$

§ 2. Sheaves

This paragraph contains the basic results of sheaf theory needed in the present work (see also CARTAN [2], SERRE [2] and GRAUERT-REMMERT [1]). The book by GODEMENT [1] is strongly recommended as a self-contained introduction to algebraic topology and sheaf theory.

We use the following terminology. A *topological space* X is a set in which certain subsets are distinguished and called open sets of X. It is required that the empty set, and X itself, be open sets and that arbitrary unions and finite intersections of open sets be open. An *open neighbourhood* of a point $x \in X$ is an open set U such that $x \in U$. A system of open sets of X is called a *basis* (for the topology of X) if every open set of X is a union of sets in the system. X is a HAUSDORFF space if, given any two distinct points x_1, x_2 in X, there is an open neighbourhood U_1 of x_1 and an open neighbourhood U_2 of x_2 such that $U_1 \cap U_2$ is empty.

An *open covering* $\mathfrak{U}$ of X is an indexed system $\mathfrak{U} = \{U_i\}_{i \in I}$ of open sets of X whose union is equal to X. The index i runs through the given index set I and so it is possible for the same open set to occur several times in the covering. Since the index set is arbitrary there are logical difficulties in discussing the set of all open coverings of X. These difficulties can be avoided by considering the set of all proper coverings of X. An open covering $\mathfrak{U} = \{U_i\}_{i \in I}$ is *proper* if distinct indices $i, j \in I$ determine distinct open sets U_i, U_j and if the index set is chosen, in the natural way, as the set of all open sets of the covering. Each proper covering is then a subset of the set of all subsets of X.

An open covering $\mathfrak{V} = \{V_j\}_{j \in J}$ of X is a *refinement* of $\mathfrak{U} = \{U_i\}_{i \in I}$ if each V_j is contained in at least one U_i. Two open coverings are *cofine* if each is a refinement of the other. It is clear that, given any open covering $\mathfrak{U}$, there is a proper covering $\mathfrak{V}$ such that $\mathfrak{U}$ and $\mathfrak{V}$ are cofine. A HAUSDORFF space is *compact* if for every open covering $\mathfrak{U} = \{U_i\}_{i \in I}$ of X there is a finite subcollection $\{U_{i_1}, \ldots, U_{i_n}\}$ which is an open covering of X.

2.1. Definition of sheaves and homomorphisms

Definition: A *sheaf* $\mathfrak{S}$ (of abelian groups) over X is a triple $\mathfrak{S} = (S, \pi, X)$ which satisfies the following three properties:

I) *S and X are topological spaces and $\pi : S \to X$ is an onto continuous map.*

II) *Every point $\alpha \in S$ has an open neighbourhood N in S such that $\pi|N$ is a homeomorphism between N and an open neighbourhood of $\pi(\alpha)$ in X.*

The counterimage $\pi^{-1}(x)$ of a point $x \in X$ is called the *stalk* over x and denoted by S_x. Every point of S belongs to a unique stalk. Property II) states that π is a local homeomorphism and implies that the topology of S induces the discrete topology on every stalk.

III) *Every stalk has the structure of an abelian group. The group operations associate to points $\alpha, \beta \in S_x$ the sum $\alpha + \beta \in S_x$ and the difference $\alpha - \beta \in S_x$. The difference depends continuously on α and β.*

In III), "continuously" means that, if $S \oplus S$ is the subset $\{(\alpha, \beta) \in S \times S;\ \pi(\alpha) = \pi(\beta)\}$ of $S \times S$ with the induced topology, the map $S \oplus S \to S$ defined by $(\alpha, \beta) \to (\alpha - \beta)$ is continuous. Properties I), II), III) imply that the zero element 0_x of the abelian group S_x depends continuously on x, *i. e.*, the map $X \to S$ defined by $x \to 0_x$ is continuous. Similarly the sum $\alpha + \beta$ depends continuously on α, β.

Remark: Property III) can be modified to give a definition of a sheaf with any other algebraic structure on each stalk. It is sufficient to require that the algebraic operations be continuous. It will often happen that each stalk of S is a K-module (for some fixed ring K). In this case III) must be modified to include the condition: the module multiplication associates to $\alpha \in S_x$, $k \in K$ a point $k\alpha \in S_x$ and the map $S \to S$ defined by $\alpha \to k\alpha$ is continuous for each $k \in K$. In the sequel we shall tacitly assume that all sheaves are sheaves of abelian groups or sheaves of K-modules (fixed ring K). All definitions and theorems are formulated for sheaves of abelian groups only but they remain true for sheaves of K-modules with the appropriate modifications (*e. g.* with "homomorphism" replaced by "K-homomorphism"). In many cases $K = \mathbf{C}$ (field of complex numbers). Sections 2.1—2.4 can be carried over for sheaves with arbitrary algebraic structures. However, the definition of cohomology groups of a topological space X with coefficients in a sheaf $\mathfrak{S}$ given in 2.6 depends essentially on the fact that each stalk of $\mathfrak{S}$ is an abelian group or a K-module. The cohomology groups themselves are then abelian groups or K-modules. Part of the cohomology theory in dimension 1 holds also in the non-abelian case (see 3.1).

Definition: Let $\mathfrak{S} = (S, \pi, X)$ and $\tilde{\mathfrak{S}} = (\tilde{S}, \tilde{\pi}, X)$ be sheaves over the same space X. A *homomorphism* $h : \mathfrak{S} \to \tilde{\mathfrak{S}}$ is defined if

a) *h is a continuous map from S to $\tilde{S}$.*

b) $\pi = \tilde{\pi} h$, i. e. *h maps the stalk S_x to the stalk $\tilde{S}_x$ for each $x \in X$.*

c) *For each $x \in X$ the restriction*

$$h_x : S_x \to \tilde{S}_x \tag{1}$$

is a homomorphism of abelian groups.

By a) and b), h is a local homeomorphism from S to $\tilde{S}$.

If h_x is one-one for each point $x \in X$ we call h *a monomorphism.* Similarly h is an *epimorphism* if h_x is onto for each $x \in X$, and an *isomorphism* if h_x is an isomorphism for each $x \in X$. Further elementary properties of sheaves are discussed in 2.4.

2.2. Presheaves

In many concrete cases a sheaf over a topological space X is constructed by means of a presheaf.

Definition: A *presheaf* over X consists of an abelian group S_U for each open set U of X and a homomorphism $r_V^U : S_U \to S_V$ for each pair of open sets U, V of X with $V \subset U$. These groups and homomorphisms satisfy the following properties:

I) *If U is empty then $S_U = 0$ is the zero group.*

II) *The homomorphism $r_U^U : S_U \to S_U$ is the identity. If $W \subset V \subset U$ then $r_W^U = r_W^V r_V^U$.*

Remark: By I) it suffices to define S_U and r_V^U only for non-empty open sets U, V.

Every presheaf over X determines a sheaf over X by the following construction:

a) For each point $x \in X$ let S_x be the direct limit of the abelian groups S_U, $x \in U$, with respect to the homomorphisms r_V^U (see for instance EILENBERG-STEENROD [1], Chapter VIII). In other words: U runs through all open neighbourhoods of x. Each element $f \in S_U$ determines an element $f_x \in S_x$ called the *germ* of f at x. Every point of S_x is a germ. If U, V are open neighbourhoods of x and $f \in S_U$, $g \in S_V$ then $f_x = g_x$ if and only if there is an open neighbourhood W of x such that $W \subset U$, $W \subset V$ and $r_W^U f = r_W^V g$.

b) The direct limit S_x of the abelian groups S_U is itself an abelian group. Let S be the union of the groups S_x for distinct $x \in X$ and let $\pi: S \to X$ map points of S_x to x. Then S is a set in which the group operations of 2.1 III) are defined.

c) The topology of S is defined by means of a basis. An element $f \in S_U$ defines a germ $f_y \in S_y$ for each point $y \in U$. The points f_y, $y \in U$, form a subset f_U of S. The sets f_U (as U runs through all open sets of X, and f through all elements of S_U) form the required basis for the topology of S.

It is easy to check that by a), b) and c) the triple $\mathfrak{S} = (S, \pi, X)$ is a sheaf of abelian groups over X. This sheaf is called *the sheaf constructed from the presheaf* $\{S_U, r_V^U\}$.

Let $\mathfrak{G} = \{S_U, r_V^U\}$ and $\tilde{\mathfrak{G}} = \{\tilde{S}_U, \tilde{r}_V^U\}$ be presheaves over X. A *homomorphism* h from $\mathfrak{G}$ to $\tilde{\mathfrak{G}}$ is a system $\{h_U\}$ of homomorphisms $h_U: S_U \to \tilde{S}_U$ which commute with the homomorphisms r_V^U, $\tilde{r}_V^U$, *i. e.* $\tilde{r}_V^U h_U = h_V r_V^U$ for $V \subset U$.

The homomorphism h is called a *monomorphism (epimorphism, isomorphism)* if each homomorphism h_U is a monomorphism (epimorphism, isomorphism). $\mathfrak{G}$ is a *subpresheaf* of $\tilde{\mathfrak{G}}$ if, for each open set U, the group S_U is a subgroup of $\tilde{S}_U$ and r_V^U is the restriction of $\tilde{r}_V^U$ to S_U for $V \subset U$. If $\mathfrak{G}$ is a subpresheaf of $\tilde{\mathfrak{G}}$ then the *quotient presheaf* $\tilde{\mathfrak{G}} / \mathfrak{G}$ is defined. This assigns to each open set U the quotient group $\tilde{S}_U / S_U$. If h is a homomorphism from the presheaf $\mathfrak{G}$ to the presheaf $\tilde{\mathfrak{G}}$ then the kernel of h and the image of h are defined in the natural way. The kernel of h is a subpresheaf of $\mathfrak{G}$ and associates to each open set U the kernel of h_U. The image of h is a subpresheaf of $\tilde{\mathfrak{G}}$ and associates to each open set U the image of h_U.

Let $\mathfrak{S} = (S, \pi, X)$ and $\tilde{\mathfrak{S}} = (\tilde{S}, \tilde{\pi}, X)$ be the sheaves constructed from the presheaves $\mathfrak{G}$ and $\tilde{\mathfrak{G}}$. The homomorphism $h: \mathfrak{G} \to \tilde{\mathfrak{G}}$ induces a homomorphism from $\mathfrak{S}$ to $\tilde{\mathfrak{S}}$ which is also denoted by h. In order to define this homomorphism it is sufficient to define the homomorphisms $h_x: S_x \to \tilde{S}_x$ [see 2.1 (1)]: if $\alpha \in S_x$ is the germ at x of an element $f \in S_U$ then $h_x(\alpha)$ is the germ at x of the element $h_U(f) \in S_U$. This rule gives a well defined homomorphism $h_x: S_x \to \tilde{S}_x$ called the *direct limit* of the homomorphisms h_U.

2.3. The canonical presheaf of a sheaf

A *section* of a sheaf $\mathfrak{S} = (S, \pi, X)$ over an open set U is a continuous map $s: U \to S$ for which $\pi s: U \to U$ is the identity. By 2.1 III) the set of all sections of $\mathfrak{S}$ over U is an abelian group which we denote by $\Gamma(U, \mathfrak{S})$. The zero element of this group is the zero section $x \to 0_x$. If s is a section of S over U the image set $s(U) \subset S$ cuts each stalk S_x, $x \in U$, in exactly one point.

Now associate to each open set U of X the group $\Gamma(U, \mathfrak{S})$ of sections of $\mathfrak{S}$ over U, where if U is empty $\Gamma(U, \mathfrak{S})$ is understood to be the zero group. If $V \subset U$ let $r_V^U: \Gamma(U, \mathfrak{S}) \to \Gamma(V, \mathfrak{S})$ be the homomorphism which associates, to each section of $\mathfrak{S}$ over U, its restriction to V (if V is empty put $r_V^U = 0$). The presheaf $\{\Gamma(U, \mathfrak{S}), r_V^U\}$ is called the *canonical presheaf* of the sheaf $\mathfrak{S}$. By the construction of 2.2 a), b), c) the presheaf $\{\Gamma(U, \mathfrak{S}), r_V^U\}$ defines a sheaf; this is again the sheaf $\mathfrak{S}$. In fact by 2.1 I), II) every point $\alpha \in S$ belongs to at least one image set $s(U)$, where s is a section of $\mathfrak{S}$ over some open set U. If s, s' are sections over U, U' with $\alpha \in s(U) \cap s'(U')$ then s agrees with s' in an open neighbourhood of $x = \pi(\alpha)$. Therefore germs at x of sections of $\mathfrak{S}$ over open neighbourhoods of x [see 2.2 a)] are in one-one correspondence with points of the stalk S_x. Further the system of all image sets $s(U)$ is, by 2.1 I), II), a complete system of open sets for the topology of S, in agreement with 2.2 c).

Let $\mathfrak{S}$ be the sheaf constructed from a presheaf $\mathfrak{G} = \{S_U, r_V^U\}$. An element $f \in S_U$ has a germ f_x at x for each point $x \in U$ [2.2 a)]. Let $h_U(f)$ be the section $x \to f_x$ of $\mathfrak{S}$ over U. This defines a homomorphism $h_U: S_U \to \Gamma(U, \mathfrak{S})$ and hence a homomorphism h from $\mathfrak{G}$ to the canonical presheaf of $\mathfrak{S}$. In general h is neither a monomorphism nor an epimorphism (for details see SERRE [2], § 1, Propositions 1 and 2). The homomorphism $\{h_U\}$ from $\mathfrak{G}$ to the canonical presheaf of $\mathfrak{S}$ induces the identity isomorphism $h: \mathfrak{S} \to \mathfrak{S}$ (see the end of 2.2).

2.4. Subsheaves. Exact sequences. Quotient sheaves. Restriction and trivial extension of sheaves

We now come to further algebraic concepts of sheaf theory.

Definition: $\mathfrak{S}' = (S', \pi', X)$ is a *subsheaf* of $\mathfrak{S} = (S, \pi, X)$ if

I) *S' is an open set of S.*

II) *π' is the restriction of π to S' and maps S' onto X.*

III) *The stalk $\pi'^{-1}(x) = S' \cap \pi^{-1}(x)$ is a subgroup of the stalk $\pi^{-1}(x)$ for all $x \in X$.*

Condition I) is equivalent to

I*) *Let $s(U) \subset S$ be the image set of a section of $\mathfrak{S}$ over U and $\alpha \in s(U) \cap S'$. Then U contains an open neighbourhood V of $\pi(\alpha)$ such that $s(x) \in S'$ for all $x \in V$.*

Conditions I*), II) imply that π' is a local homeomorphism and III) implies that the group operations in $\mathfrak{S}'$ are continuous. Therefore the triple (S', π', X) is itself a sheaf. The inclusion of S' in S defines a monomorphism from $\mathfrak{S}'$ to $\mathfrak{S}$ (see **2.1**) called the *embedding* of $\mathfrak{S}'$ in $\mathfrak{S}$.

The *zero sheaf* 0 over X can be defined up to isomorphism as the triple (X, π, X) where π is the identity map and each stalk is the zero group. The zero sheaf is a subsheaf of every sheaf $\mathfrak{S}$ over X: let S' be the set $0(\mathfrak{S})$ of zero elements of stalks of $\mathfrak{S}$, i. e. $0(\mathfrak{S}) = s(X)$ where s is the zero element of $\Gamma(X, \mathfrak{S})$.

Let $\mathfrak{S} = (S, \pi, X)$ and $\tilde{\mathfrak{S}} = (\tilde{S}, \tilde{\pi}, X)$ be sheaves over X and $h: \mathfrak{S} \to \tilde{\mathfrak{S}}$ a homomorphism. If $S' = h^{-1}(0(\tilde{\mathfrak{S}}))$ and $\pi' = \pi|S'$ then (S', π', X) gives a subsheaf $h^{-1}(0)$ of $\mathfrak{S}$ called the *kernel* of h. The stalk of the sheaf $h^{-1}(0)$ over x is the kernel of the homomorphism $h_x: S_x \to \tilde{S}_x$ [see **2.1** (1)]. If $\tilde{S}' = h(S)$ and $\tilde{\pi}' = \tilde{\pi}|\tilde{S}'$ then $(\tilde{S}', \tilde{\pi}', X)$ gives a subsheaf $h(\mathfrak{S})$ of $\tilde{\mathfrak{S}}$ called the *image* of h. The stalk of the sheaf $h(\mathfrak{S})$ over x is the image of the homomorphism h_x.

Let $\{A_i\}$ be a sequence of groups (or presheaves or sheaves) and $\{h_i\}$ a sequence of homomorphisms $h_i: A_i \to A_{i+1}$. (The index i takes all integral values between two bounds n_0, n_1 which may also be $-\infty$ or $+\infty$. Thus A_i is defined for $n_0 < i < n_1$ and h_i for $n_0 < i < n_1 - 1$.) The sequence A_i, h_i is an *exact sequence* if the kernel of each homomorphism is equal to the image of the previous homomorphism, provided the latter is defined. If the A_i are presheaves $\{S_U^{(i)}\}$ over the topological space X, then the exactness means that for each open set U of X there is an exact sequence of groups

$$\cdots \to S_U^{(i)} \to S_U^{(i+1)} \to S_U^{(i+2)} \to \cdots. \tag{2}$$

If the A_i are sheaves over X then the exactness means that at each point $x \in X$ the stalks of the sheaves A_i form an exact sequence. Since the direct limit of exact sequences is again an exact sequence (EILENBERG-STEENROD [1], Chapter VIII, Theorem 5.4) we have

L e m m a 2.4.1. *Consider an exact sequence*

$$\cdots \to \mathfrak{G}_n \to \mathfrak{G}_{n+1} \to \mathfrak{G}_{n+2} \to \cdots \tag{3}$$

of presheaves over X. Then the induced sequence of sheaves $\mathfrak{S}_i$ constructed from $\mathfrak{G}_i$ is an exact sequence of sheaves over X.

For example let

$$0 \to \mathfrak{S}' \xrightarrow{h'} \mathfrak{S} \xrightarrow{h} \mathfrak{S}'' \to 0 \tag{4}$$

be an exact sequence of sheaves $\mathfrak{S}' = (S', \pi', X)$, $\mathfrak{S} = (S, \pi, X)$ and $\mathfrak{S}'' = (S'', \pi'', X)$ over X.

The first 0 denotes the zero subsheaf of $\mathfrak{S}'$, the first arrow the embedding of 0 in $\mathfrak{S}'$. Therefore exactness implies that h' is a monomorphism and can be regarded as the embedding of the subsheaf $\mathfrak{S}'$ in $\mathfrak{S}$. The final 0 denotes the zero subsheaf of $\mathfrak{S}''$, the final arrow the trivial homomorphism which maps each stalk of $\mathfrak{S}''$ to its zero element. Therefore exactness implies that h is an epimorphism. For each point $x \in X$ the exact sequence (4) gives a corresponding exact sequence of stalks over x:

$$0 \to S'_x \xrightarrow{h'_x} S_x \xrightarrow{h_x} S''_x \to 0 . \tag{5}$$

The group S''_x is isomorphic to the quotient group S_x/S'_x. It is easy to check that S'' has the quotient topology with respect to the map $h: S \to S''$ (a subset of S'' is open if and only if its counterimage under h is an open set in S). This shows that given the sheaf $\mathfrak{S}$ and the subsheaf $\mathfrak{S}'$ there is (up to isomorphism) at most one sheaf $\mathfrak{S}''$ for which the sequence (4) is exact. It is possible to prove the existence of such an $\mathfrak{S}''$ directly so that we may speak of *the* quotient sheaf $\mathfrak{S}'' = \mathfrak{S}/\mathfrak{S}'$. We shall obtain the existence of $\mathfrak{S}''$ in a slightly different manner by defining first a presheaf for $\mathfrak{S}''$.

Let $\mathfrak{S}'$ be a subsheaf of $\mathfrak{S}$ and U an open set of X. The group $\Gamma(U, \mathfrak{S}')$ of sections of $\mathfrak{S}'$ over U is then a subgroup of $\Gamma(U, \mathfrak{S})$, the group of sections of $\mathfrak{S}$ over U. We define $S''_U = \Gamma(U, \mathfrak{S})/\Gamma(U, \mathfrak{S}')$ so that there is an exact sequence

$$0 \to \Gamma(U, \mathfrak{S}') \to \Gamma(U, \mathfrak{S}) \to S''_U \to 0 . \tag{6}$$

If V is an open set contained in U the restriction homomorphism $\Gamma(U, \mathfrak{S}) \to \Gamma(V, \mathfrak{S})$ maps the subgroup $\Gamma(U, \mathfrak{S}')$ of $\Gamma(U, \mathfrak{S})$ to the subgroup $\Gamma(V, \mathfrak{S}')$ of $\Gamma(V, \mathfrak{S})$ and induces homomorphisms $r^U_V: S''_U \to S''_V$. The presheaf $\{S''_U, r^U_V\}$ is the quotient of the canonical presheaf of $\mathfrak{S}$ by the subpresheaf given by the canonical presheaf of $\mathfrak{S}'$. Let $\mathfrak{S}''$ be the sheaf constructed from the presheaf $\{S''_U, r^U_V\}$. The exact sequence (6) of presheaves induces, by Lemma 2.4.1, an exact sequence of sheaves as required. We collect our results in the following theorem:

Theorem 2.4.2. *Let $\mathfrak{S}$ be a sheaf over a topological space X and $\mathfrak{S}'$ a subsheaf of $\mathfrak{S}$ with embedding $h': \mathfrak{S}' \to \mathfrak{S}$. There exists a sheaf $\mathfrak{S}''$ over X, unique up to isomorphism, for which there is an exact sequence*

$$0 \to \mathfrak{S}' \xrightarrow{h'} \mathfrak{S} \xrightarrow{h} \mathfrak{S}'' \to 0 . \tag{7}$$

At each point $x \in X$ the homomorphism h_x gives an isomorphism between the quotient group S_x/S'_x and the stalk S''_x of $\mathfrak{S}''$ over x.

Remark: From (7) one obtains the exact sequence

$$0 \to \Gamma(U, \mathfrak{S}') \to \Gamma(U, \mathfrak{S}) \to \Gamma(U, \mathfrak{S}'') . \tag{8}$$

In general $\Gamma(U, \mathfrak{S}) \to \Gamma(U, \mathfrak{S}'')$ is not an epimorphism. By (6), S''_U is the subgroup of $\Gamma(U, \mathfrak{S}'')$ consisting of all sections of $\mathfrak{S}''$ over U which are images of sections of $\mathfrak{S}$ over U.

Let $\mathfrak{S} = (S, \pi, X)$ be a sheaf over X and let Y be a subset of X. If the subset $\pi^{-1}(Y)$ of S is given the induced topology the triple $(\pi^{-1}(Y), \pi|\pi^{-1}(Y), Y)$ defines in a natural manner a sheaf $\mathfrak{S}|Y$ over Y called the *restriction* of $\mathfrak{S}$ to Y.

Theorem 2.4.3. *Let Y be a closed subset of the topological space X and $\mathfrak{S} = (S, \pi, Y)$ a sheaf over Y. There exists a sheaf $\hat{\mathfrak{S}}$ over X, unique up to isomorphism, such that $\hat{\mathfrak{S}}|Y = \mathfrak{S}$ and $\hat{\mathfrak{S}}|(X - Y) = 0$. The groups $\Gamma(U, \hat{\mathfrak{S}})$ and $\Gamma(U \cap Y, \mathfrak{S})$ are isomorphic for any open set U of X.* ($\hat{\mathfrak{S}}$ is called the *(trivial) extension* of $\mathfrak{S}$ to X.)

Proof: Uniqueness follows immediately from the properties of $\hat{\mathfrak{S}}$: if $\hat{\mathfrak{S}} = (\hat{S}, \hat{\pi}, X)$ then $\hat{S} = S \cup ((X - Y) \times 0)$, $\hat{\pi}(\alpha) = \pi(\alpha)$ for $\alpha \in S$, $\hat{\pi}(a \times 0) = a$ for $a \in X - Y$ and therefore the stalk $\hat{S}_x = \hat{\pi}^{-1}(x)$ is equal to $\pi^{-1}(x)$ for $x \in Y$ and equal to the zero group for $x \in X - Y$. The sets $s(U \cap Y) \cup ((U \cap (X - Y)) \times 0)$, for arbitrary open sets U of X and arbitrary sections s of $\mathfrak{S}$ over U, define a basis for the topology of $\hat{S}$. This completes the construction of $\hat{\mathfrak{S}}$. It is also possible to define $\hat{\mathfrak{S}}$ by means of a presheaf: associate to each open set U of X the group $\hat{S}_U = \Gamma(U \cap Y, \mathfrak{S})$ and to each pair of open sets U, V with $V \subset U$ the restriction homomorphism $r_V^U : \Gamma(U \cap Y, \mathfrak{S}) \to \Gamma(V \cap Y, \mathfrak{S})$. Since Y is closed, each point $x \in X - Y$ has an open neighbourhood U for which $U \cap Y$ is empty and $\hat{S}_U = 0$. Therefore the sheaf $\hat{\mathfrak{S}}$ constructed from the presheaf $\{\hat{S}_U, r_V^U\}$ has $\hat{\mathfrak{S}}|Y = \mathfrak{S}$ and $\hat{\mathfrak{S}}|X - Y = 0$. In fact $\{\hat{S}_U, r_V^U\}$ is the canonical presheaf of $\hat{\mathfrak{S}}$.

Remark: Suppose that at some boundary point of Y the stalk of $\mathfrak{S}$ has a non-zero element. Then $\hat{S}$ is a non-Hausdorff space.

2.5. Examples

1) Let X be a topological space and A an abelian group. The *constant sheaf* over X with stalk A is defined by the triple $(X \times A, \pi, X)$ and is also denoted simply by A. Here $\pi : X \times A \to X$ is the projection from the cartesian product $X \times A$ where A is given the discrete topology. The sum and difference of points (x, a) and (x, a') in $X \times A$ are equal to $(x, a \pm a')$.

2) Let X be a topological space. Associate to each non-empty open set U of X the additive group S_U of all continuous complex valued functions defined on U. For $V \subset U$ the homomorphism $r_V^U : S_U \to S_V$ is defined by taking the restriction to V of each function defined on U. Let $\mathbf{C}_c$ be the sheaf constructed from the presheaf $\{S_U, r_V^U\}$ as in 2.2. Then $\mathbf{C}_c$ is called the *sheaf of germs of local complex valued continuous functions*.

The sheaf $\mathbf{C}_c^*$ of germs of local never zero complex valued continuous functions is defined similarly: associate to each non-empty open set U the abelian group S_U^* of never zero complex valued continuous functions defined on U. The group operation is ordinary multiplication. There is a homomorphism $S_U \to S_U^*$ which associates to each function $f \in S_U$ the function $e^{2\pi i f} \in S_U^*$. This defines a homomorphism $\{S_U, r_V^U\} \to \{S_U^*, r_V^U\}$ of presheaves and hence a homomorphism $\mathbf{C}_c \to \mathbf{C}_c^*$ of sheaves (see 2.2). The kernel of the homomorphism $\mathbf{C}_c \to \mathbf{C}_c^*$ is a subsheaf of $\mathbf{C}_c$ isomorphic to the constant sheaf over X with stalk the additive group $\mathbf{Z}$ of integers. Every point z_0 in the multiplicative group $\mathbf{C}^*$ of non-zero complex numbers has an open neighbourhood in which a single branch can be chosen for $\log z$. If k is a germ of $\mathbf{C}_c^*$ then $(2\pi i)^{-1} \log k$ is a germ of $\mathbf{C}_c$ which maps to k under $\mathbf{C}_c \to \mathbf{C}_c^*$. Therefore there is an exact sequence of sheaves over X

$$0 \to \mathbf{Z} \to \mathbf{C}_c \to \mathbf{C}_c^* \to 0 . \tag{9}$$

3) Now let X be a n-dimensional differentiable manifold. We adopt the following definition (see DE RHAM [1], § 1, and LANG [1]). X is a HAUSDORFF space with a countable basis. At each point $x \in X$ certain real valued functions are distinguished and called differentiable at x. Each function is defined on some open neighbourhood of x and the following axiom is satisfied:

There is an open neighbourhood U of x and a homeomorphism g from U on to an open subset of $\mathbf{R}^n$ such that, for all $y \in U$, if f is a real valued function defined on a neighbourhood V of y and $h = g|U \cap V$, then f is differentiable at y if and only if $f h^{-1}$ is C^∞-differentiable at $g(y)$.

Here $f h^{-1}$ is a real valued function defined on an open neighbourhood of $g(x)$ in $\mathbf{R}^n$. It is C^∞-differentiable at $g(x)$ if all the partial derivatives exist and are continuous in some neighbourhood of $g(x)$.

A homeomorphism g which satisfies this axiom is called an *admissible chart* of the differentiable manifold X.

If X is a differentiable manifold, and U is an open set of X, let S_U be the additive group of complex valued functions differentiable in U (a complex valued function is differentiable if and only if its real and imaginary parts are differentiable). Just as in 2), the presheaf $\{S_U, r_V^U\}$ defines a sheaf $\mathbf{C}_\mathfrak{b}$: the *sheaf of germs of local complex valued differentiable functions*. Similarly the sheaf $\mathbf{C}_\mathfrak{b}^*$ of germs of local never zero complex valued differentiable functions is defined. As in 2) there is an exact sequence of sheaves over X

$$0 \to \mathbf{Z} \to \mathbf{C}_\mathfrak{b} \to \mathbf{C}_\mathfrak{b}^* \to 0 . \tag{10}$$

4) Now let X be a n-dimensional complex manifold. The definition is analogous to that of a differentiable manifold (see WEIL [2]). X is a HAUSDORFF space with a countable basis. At each point $x \in X$ certain complex valued functions are distinguished and called *holomorphic* or *complex analytic* at x. Each function is defined on some open neighbourhood of x and the following axiom is satisfied:

There is an open neighbourhood U of x and a homeomorphism g from U on to an open subset of $\mathbf{C}_n$ such that, for all $y \in U$, if f is a complex valued function defined on an open neighbourhood V of y and $h = g|U \cap V$, then f is holomorphic at y if and only if $f\, h^{-1}$ is holomorphic at $g(y)$.

A homeomorphism g which satisfies this axiom is called an *admissible chart* of the complex manifold X. The admissible charts of a n-dimensional complex manifold X can be used, in a natural way, to define a $2n$-dimensional differentiable manifold with the same underlying space X.

If X is a complex manifold let S_U be the additive group of (complex valued) functions holomorphic in U. Just as in 2) and 3) these groups define a sheaf $\mathbf{C}_\omega$: the *sheaf of germs of local holomorphic functions.* Similarly the sheaf $\mathbf{C}_\omega^*$ of germs of local never zero holomorphic functions is defined and there is an exact sequence of sheaves over X

$$0 \rightarrow \mathbf{Z} \rightarrow \mathbf{C}_\omega \rightarrow \mathbf{C}_\omega^* \rightarrow 0\,. \tag{11}$$

Remarks: The sheaves $\mathbf{C}_c$, $\mathbf{C}_\mathfrak{d}$, $\mathbf{C}_\omega$ can also be regarded as sheaves of $\mathbf{C}$-modules. In the exact sequences (9), (10), (11) all sheaves are however to be regarded as sheaves of abelian groups. The presheaves used to construct $\mathbf{C}_c$, $\mathbf{C}_c^*$, $\mathbf{C}_\mathfrak{d}$, $\mathbf{C}_\mathfrak{d}^*$, $\mathbf{C}_\omega$, $\mathbf{C}_\omega^*$ are all canonical presheaves. For instance $\Gamma(U, \mathbf{C}_c)$ is the additive group of all complex valued continuous functions defined on U.

2.6. Cohomology groups with coefficients in a sheaf

The aim of this section is to define, for each integer $q \geqq 0$, the cohomology group $H^q(X, \mathfrak{S})$ of the topological space X with coefficients in a sheaf $\mathfrak{S}$ over X. As a first step we define the cohomology groups $H^q(\mathfrak{U}, \mathfrak{G})$ of an open covering $\mathfrak{U} = \{U_i\}_{i \in I}$ of X with coefficients in a presheaf $\mathfrak{G}$. The cohomology groups $H^q(\mathfrak{U}, \mathfrak{S})$ of $\mathfrak{U}$ with coefficients in a sheaf $\mathfrak{S}$ are defined to be the cohomology groups of $\mathfrak{U}$ with coefficients in the canonical presheaf of $\mathfrak{S}$. Finally the cohomology groups $H^q(X, \mathfrak{G})$, $H^q(X, \mathfrak{S})$ are defined as the direct limit of all groups $H^q(\mathfrak{U}, \mathfrak{G})$, $H^q(\mathfrak{U}, \mathfrak{S})$ as $\mathfrak{U}$ runs through "all" open coverings of X.

Cohomology groups $H^q(\mathfrak{U}, \mathfrak{G})$, $H^q(\mathfrak{U}, \mathfrak{S})$:

Let $\mathfrak{G} = \{S_U, r_V^U\}$ be a presheaf over X and $\mathfrak{U} = \{U_i\}_{i \in I}$ an open covering of X. A q-cochain is a function f which associates to each $(q+1)$-ple $(i_0, \ldots, i_q)$ of indices in I an element $f(i_0, \ldots, i_q)$ of $S_{(U_{i_0} \cap \cdots \cap U_{i_q})}$. The q-cochains form a group $C^q(\mathfrak{U}, \mathfrak{G})$. Define the coboundary homomorphism

$$\delta^q : C^q(\mathfrak{U}, \mathfrak{G}) \rightarrow C^{q+1}(\mathfrak{U}, \mathfrak{G})$$

by the formula:

$$(\delta^q f)(i_0, \ldots, i_{q+1}) = \sum_{k=0}^{q+1} (-1)^k r_W^{W_k}(f(i_0, \ldots, \hat{i}_k, \ldots, i_{q+1}))$$

for $f \in C^q(\mathfrak{U}, \mathfrak{S})$. Here the "roof" ($\hat{}$) over a symbol means that the symbol is to be omitted,

$$W = U_{i_0} \cap \cdots \cap U_{i_{q+1}} \text{ and } W_k = U_{i_0} \cap \cdots \cap \hat{U}_{i_k} \cap \cdots \cap U_{i_{q+1}}.$$

As usual $\delta^{q+1}\delta^q = 0$ and therefore cohomology groups can be defined:

$$H^q(\mathfrak{U}, \mathfrak{S}) = \text{kernel}(\delta^q)/\text{image}(\delta^{q-1}).$$

The cohomology groups $H^q(\mathfrak{U}, \mathfrak{S})$ with coefficients in a sheaf $\mathfrak{S}$ over X are then defined as the cohomology groups with coefficients in the canonical presheaf of $\mathfrak{S}$.

Cohomology groups $H^q(X, \mathfrak{S})$, $H^q(X, \mathfrak{S})$:

Let $\mathfrak{V} = \{V_j\}_{j \in J}$ be a refinement of the open covering $\mathfrak{U} = \{U_i\}_{i \in I}$. Choose a map $\tau: J \to I$ so that $V_j \subset U_{\tau j}$ for all $j \in J$. The map τ defines a homomorphism

$$\tau^*: C^q(\mathfrak{U}, \mathfrak{S}) \to C^q(\mathfrak{V}, \mathfrak{S})$$

by the formula

$$(\tau^* f)(j_0, \ldots, j_q) = r_W^{W'}(f(\tau j_0, \ldots, \tau j_q))$$

for $f \in C^q(\mathfrak{U}, \mathfrak{S})$. Here we have for the moment put $W = V_{j_0} \cap \cdots \cap V_{j_q}$ and $W' = U_{\tau j_0} \cap \cdots \cap U_{\tau j_q}$ so that $W \subset W'$.

For each $q \geqq 0$ there is a commutative diagram

$$\begin{array}{ccc} C^q(\mathfrak{U}, \mathfrak{S}) & \xrightarrow{\tau^*} & C^q(\mathfrak{V}, \mathfrak{S}) \\ \delta^q \downarrow & & \downarrow \delta^q \\ C^{q+1}(\mathfrak{U}, \mathfrak{S}) & \xrightarrow{\tau^*} & C^{q+1}(\mathfrak{V}, \mathfrak{S}). \end{array}$$

Therefore τ^* induces a homomorphism

$$t_{\mathfrak{V}}^{\mathfrak{U}}: H^q(\mathfrak{U}, \mathfrak{S}) \to H^q(\mathfrak{V}, \mathfrak{S}).$$

Lemma 2.6.1. *The homomorphism $t_{\mathfrak{V}}^{\mathfrak{U}}$ depends only on the open covering $\mathfrak{U}$ and the refinement $\mathfrak{V}$ of $\mathfrak{U}$, and not on the choice of refinement map $\tau: J \to I$. Furthermore $t_{\mathfrak{U}}^{\mathfrak{U}}$ is the identity, and if $\mathfrak{W}$ is a refinement of $\mathfrak{V}$ then $t_{\mathfrak{W}}^{\mathfrak{U}} = t_{\mathfrak{W}}^{\mathfrak{V}} t_{\mathfrak{V}}^{\mathfrak{U}}$.*

Proof: Let τ, τ' be two maps from J to I with $V_j \subset U_{\tau j} \cap U_{\tau' j}$. For each $q \geqq 1$ we define a homomorphism (homotopy operator)

$$k^q: C^q(\mathfrak{U}, \mathfrak{S}) \to C^{q-1}(\mathfrak{V}, \mathfrak{S})$$

by the formula

$$(k^q f)(j_0, \ldots, j_{q-1}) = \sum_{h=0}^{q-1} (-1)^h r_W^{W_h}(f(\tau j_0, \ldots, \tau j_h, \tau' j_h, \tau' j_{h+1}, \ldots, \tau' j_{q-1}))$$

for $f \in C^q(\mathfrak{U}, \mathfrak{S})$. Here we have for the moment put

$$W = V_{j_0} \cap \cdots \cap V_{j_{q-1}}$$

and

$$W_h = U_{\tau j_0} \cap \cdots \cap U_{\tau j_h} \cap U_{\tau' j_h} \cap U_{\tau' j_{h+1}} \cap \cdots \cap U_{\tau' j_{q-1}}$$

so that $W \subset W_h$. Then

$$k^1 \delta^0 = (\tau')^* - \tau^*$$

$$\delta^{q-1} k^q + k^{q+1} \delta^q = (\tau')^* - \tau^* \quad \text{for} \quad q \geqq 1 .$$

This proves the first part of the lemma. The second part follows immediately.

By Lemma 2.6.1, cofine coverings have naturally isomorphic cohomology groups. It is therefore possible, for the definition of the cohomology groups of the space X, to restrict attention to proper coverings of X (see the beginning of this paragraph for terminology).

Definition: The *cohomology group* $H^q(X, \mathfrak{S})$ *of the topological space* X *with coefficients in a presheaf* $\mathfrak{S}$ is the direct limit of the groups $H^q(\mathfrak{U}, \mathfrak{S})$ with respect to the homomorphisms $t^{\mathfrak{U}}_{\mathfrak{V}}$, where $\mathfrak{U}$ runs through all proper coverings of X.

The *cohomology groups* $H^q(X, \mathfrak{S})$ *with coefficients in a sheaf* $\mathfrak{S}$ *over* X are the cohomology groups with coefficients in the canonical presheaf of $\mathfrak{S}$.

The cohomology group $H^0(\mathfrak{U}, \mathfrak{S})$ is, by definition, the group of functions f which associate to each $i \in I$ a section f_i of $\mathfrak{S}|\mathfrak{U}_i$ such that $f_i = f_j$ on $U_i \cap U_j$. Therefore $H^0(\mathfrak{U}, \mathfrak{S}) = \Gamma(X, \mathfrak{S})$ which gives

Theorem 2.6.2. *The cohomology group* $H^0(X, \mathfrak{S})$ *is naturally isomorphic to the group* $\Gamma(X, \mathfrak{S})$ *of sections of* $\mathfrak{S}$ *over* X.

Now let $\mathfrak{S}$ be a sheaf over a closed subset Y of the topological space X and $\hat{\mathfrak{S}}$ the trivial extension of $\mathfrak{S}$ to X constructed in Theorem 2.4.3. With these notations we have

Theorem 2.6.3. *The cohomology groups* $H^q(Y, \mathfrak{S})$ *and* $H^q(X, \hat{\mathfrak{S}})$ *are naturally isomorphic.*

Proof: An open covering $\mathfrak{U} = \{U_i\}_{i \in I}$ of X defines an open covering $\mathfrak{U}|Y = \{U_i \cap Y\}_{i \in I}$ of Y. Every open covering of Y is obtained in this way. For each open set U of X the groups $\Gamma(U \cap Y, \mathfrak{S})$ and $\Gamma(U, \hat{\mathfrak{S}})$ are naturally isomorphic, and these isomorphisms are compatible with the restriction maps r^U_V when $V \subset U$. Therefore there is an isomorphism

$$C^q(\mathfrak{U}|Y, \mathfrak{S}) \cong C^q(\mathfrak{U}, \hat{\mathfrak{S}})$$

for each q which commutes with the coboundary homomorphisms in the cochain complexes $\{C^q(\mathfrak{U}|Y, \mathfrak{S})\}$ and $\{C^q(\mathfrak{U}, \hat{\mathfrak{S}})\}$. Therefore there is a natural isomorphism

$$H^q(\mathfrak{U}|Y, \mathfrak{S}) \simeq H^q(\mathfrak{U}, \hat{\mathfrak{S}})$$

and the statement of the theorem follows.

2.7. The exact cohomology sequence for presheaves

Let $\mathfrak{S}$, $\tilde{\mathfrak{S}}$ be two presheaves over the topological space X. A homomorphism $h = \{h_U\}$ from $\mathfrak{S}$ to $\tilde{\mathfrak{S}}$ (see 2.2) induces in a natural way a homomorphism h_* from $C^q(\mathfrak{U}, \mathfrak{S})$ to $C^q(\mathfrak{U}, \tilde{\mathfrak{S}})$. This homomorphism commutes with the coboundary homomorphisms and therefore defines a homomorphism

$$h_* : H^q(\mathfrak{U}, \mathfrak{S}) \to H^q(\mathfrak{U}, \tilde{\mathfrak{S}})$$

for each $q \geqq 0$. If $\mathfrak{V}$ is a refinement of $\mathfrak{U}$ there is a commutative diagram

$$\begin{array}{ccc} H^q(\mathfrak{U}, \mathfrak{S}) & \xrightarrow{h_*} & H^q(\mathfrak{U}, \tilde{\mathfrak{S}}) \\ t^{\mathfrak{U}}_{\mathfrak{V}} \downarrow & & \downarrow t^{\mathfrak{U}}_{\mathfrak{V}} \\ H^q(\mathfrak{V}, \mathfrak{S}) & \xrightarrow{h_*} & H^q(\mathfrak{V}, \tilde{\mathfrak{S}}) \end{array} \tag{12}$$

and hence in the direct limit a homomorphism

$$h_* : H^q(X, \mathfrak{S}) \to H^q(X, \tilde{\mathfrak{S}}) .$$

Now consider an exact sequence

$$0 \longrightarrow \mathfrak{S}' \xrightarrow{h'} \mathfrak{S} \xrightarrow{h} \mathfrak{S}'' \longrightarrow 0 \tag{13}$$

of presheaves over X (see 2.4). Here 0 denotes the *zero presheaf* which associates the zero group to each open set of X. Let S'_U, S_U, S''_U be the groups associated to the open set U by the presheaves $\mathfrak{S}'$, $\mathfrak{S}$, $\mathfrak{S}''$. Then S''_U is the quotient group S_U/S'_U. Therefore for each open covering $\mathfrak{U}$ of X the sequence

$$0 \longrightarrow C^q(\mathfrak{U}, \mathfrak{S}') \xrightarrow{h'_*} C^q(\mathfrak{U}, \mathfrak{S}) \xrightarrow{h_*} C^q(\mathfrak{U}, \mathfrak{S}'') \longrightarrow 0 \tag{14}$$

induced by (13) is exact.

The theory of cochain complexes implies that there is an exact cohomology sequence

$$\begin{gathered} 0 \longrightarrow H^0(\mathfrak{U}, \mathfrak{S}') \xrightarrow{h'_*} H^0(\mathfrak{U}, \mathfrak{S}) \xrightarrow{h_*} H^0(\mathfrak{U}, \mathfrak{S}'') \xrightarrow{\delta^0_*} H^1(\mathfrak{U}, \mathfrak{S}') \to \\ \cdots \to H^q(\mathfrak{U}, \mathfrak{S}') \xrightarrow{h'_*} H^q(\mathfrak{U}, \mathfrak{S}) \xrightarrow{h_*} H^q(\mathfrak{U}, \mathfrak{S}'') \xrightarrow{\delta^q_*} H^{q+1}(\mathfrak{U}, \mathfrak{S}') \to \cdots . \end{gathered} \tag{15}$$

The homomorphism δ^q_* is obtained in the following way: Represent the element $b \in H^q(\mathfrak{U}, \mathfrak{S}'')$ by a cochain $f \in C^q(\mathfrak{U}, \mathfrak{S}'')$ with $\delta^q f = 0$. By the exactness of (14) it is possible to choose a cochain $g \in C^q(\mathfrak{U}, \mathfrak{S})$ such that $h_*(g) = f$. Therefore $\delta^q g$ lies in the subgroup $C^{q+1}(\mathfrak{U}, \mathfrak{S}')$ of $C^{q+1}(\mathfrak{U}, \mathfrak{S})$ and $\delta^{q+1}(\delta^q g) = 0$. Then $\delta^q_* b \in H^{q+1}(\mathfrak{U}, \mathfrak{S}')$ is the element represented by the cochain $\delta^q g$.

Now let $\mathfrak{V}$ be a refinement of the open covering $\mathfrak{U}$ of X. There is an exact cohomology sequence for $\mathfrak{V}$, corresponding to (15), and the diagram

$$\begin{array}{ccc} H^q(\mathfrak{U}, \mathfrak{S}'') & \xrightarrow{\delta^q_*} & H^{q+1}(\mathfrak{U}, \mathfrak{S}') \\ t^{\mathfrak{U}}_{\mathfrak{V}} \downarrow & & \downarrow t^{\mathfrak{U}}_{\mathfrak{V}} \\ H^q(\mathfrak{V}, \mathfrak{S}'') & \xrightarrow{\delta^q_*} & H^{q+1}(\mathfrak{V}, \mathfrak{S}') \end{array} \tag{16}$$

is commutative. Therefore in the direct limit there is a homomorphism for each $q \geqq 0$

$$\delta^q_* : H^q(X, \mathfrak{S}'') \to H^{q+1}(X, \mathfrak{S}') .$$

The commutative diagram (16), and the commutative diagrams (12) given by h' and h, imply that $t^{\mathfrak{U}}_{\mathfrak{V}}$ is a homomorphism from the exact cohomology sequence (15) for $\mathfrak{U}$ to the corresponding exact cohomology sequence for $\mathfrak{V}$. There is a commutative diagram

$$\begin{array}{ccccccccc} \cdots \to H^q(\mathfrak{U}, \mathfrak{S}') & \xrightarrow{h'_*} & H^q(\mathfrak{U}, \mathfrak{S}) & \xrightarrow{h_*} & H^q(\mathfrak{U}, \mathfrak{S}'') & \xrightarrow{\delta^q_*} & H^{q+1}(\mathfrak{U}, \mathfrak{S}') \to \cdots \\ \downarrow t^{\mathfrak{U}}_{\mathfrak{V}} & & \downarrow t^{\mathfrak{U}}_{\mathfrak{V}} & & \downarrow t^{\mathfrak{U}}_{\mathfrak{V}} & & \downarrow t^{\mathfrak{U}}_{\mathfrak{V}} \\ \cdots \to H^q(\mathfrak{V}, \mathfrak{S}') & \xrightarrow{h'_*} & H^q(\mathfrak{V}, \mathfrak{S}) & \xrightarrow{h_*} & H^q(\mathfrak{V}, \mathfrak{S}'') & \xrightarrow{\delta^q_*} & H^{q+1}(\mathfrak{V}, \mathfrak{S}') \to \cdots . \end{array} \tag{17}$$

The direct limit of exact sequences is again an exact sequence and therefore (17) implies

Lemma 2.7.1. *An exact sequence* $0 \to \mathfrak{S}' \to \mathfrak{S} \to \mathfrak{S}'' \to 0$ *of presheaves over a topological space* X *gives a natural exact cohomology sequence*

$$\begin{array}{l} 0 \to H^0(X, \mathfrak{S}') \to H^0(X, \mathfrak{S}) \to H^0(X, \mathfrak{S}'') \to H^1(X, \mathfrak{S}') \to \cdots \\ \cdots \to H^q(X, \mathfrak{S}') \to H^q(X, \mathfrak{S}) \to H^q(X, \mathfrak{S}'') \xrightarrow{\delta^q_*} H^{q+1}(X, \mathfrak{S}') \to \cdots . \end{array} \tag{18}$$

Corollary: *Let* $0 \to \mathfrak{S}' \to \mathfrak{S} \xrightarrow{h} \tilde{\mathfrak{S}} \to \mathfrak{S}'' \to 0$ *be an exact sequence of presheaves over a topological space* X, *and suppose that* $H^q(X, \mathfrak{S}') = H^q(X, \mathfrak{S}'') = 0$ *for all* $q \geqq 0$. *Then* $h_* : H^q(X, \mathfrak{S}) \to H^q(X, \tilde{\mathfrak{S}})$ *is an isomorphism for all* $q \geqq 0$.

Proof: Let $h(\mathfrak{S})$ be the image of h. Then the exact sequences $0 \to \mathfrak{S}' \to \mathfrak{S} \to h(\mathfrak{S}) \to 0$ and $0 \to h(\mathfrak{S}) \to \tilde{\mathfrak{S}} \to \mathfrak{S}'' \to 0$ imply that h_* is the composition of isomorphisms $H^q(X, \mathfrak{S}) \to H^q(X, h(\mathfrak{S})) \to$ $\to H^q(X, \tilde{\mathfrak{S}})$ for all $q \geqq 0$.

2.8. Paracompact spaces

Certain results of sheaf theory can be proved only for paracompact spaces (see however the bibliographical note to Chapter One). In this section we collect the definitions and theorems on paracompact spaces which are needed. We follow the definitions given by Bourbaki (Topologie générale). Thus compact, locally compact and paracompact spaces are all Hausdorff spaces by definition.

Definition: An open covering $\mathfrak{U} = \{U_i\}_{i \in I}$ of a topological space X is *point finite* if each point of X is contained in U_i for only finitely many $i \in I$. The covering $\mathfrak{U}$ is *locally finite* if each point of X has an open neighbourhood which meets U_i for only finitely many $i \in I$.

Definition: The topological space X is *paracompact* if it is a HAUSDORFF space and if every open covering of X has a locally finite refinement.

Theorem 2.8.1 (DIEUDONNÉ [1], Théorème 1). *Every paracompact space is normal.*

Theorem 2.8.2 (DIEUDONNÉ [1], Théorème 3). *Every locally compact space, which is the union of a countable number of compact subsets, is paracompact. In particular, every locally compact space with a countable basis is paracompact.*

The manifolds which occur in this book are, by definition, HAUSDORFF spaces with a countable basis [see 2.5, 3) and 4)]. They are therefore paracompact by Theorem 2.8.2. It is also true that every metric space is paracompact and that every CW-complex is paracompact [MORITA, Math. Japon. 1, 60–68 (1948) and Proc. Japan Acad. 30, 711–717 (1954)].

Theorem 2.8.3 (Shrinking Theorem, DIEUDONNÉ [1], Théorème 6). *Let $\mathfrak{U} = \{U_i\}_{i \in I}$ be a point finite open covering of a normal space X. Then there is an open covering $\mathfrak{V} = \{V_i\}_{i \in I}$ with the same index set I such that $\overline{V}_i \subset U_i$ for all $i \in I$.*

If $\varphi : X \to \mathbf{R}$ is a continuous function, the support $\operatorname{supp}\varphi = \overline{\{x \in X;\ \varphi(x) \neq 0\}}$ is the smallest closed set outside which φ is zero.

Definition: Let $\mathfrak{U} = \{U_i\}_{i \in I}$ be an open covering of the topological space X. A system $\{\varphi_i\}_{i \in I}$ of real valued continuous functions defined on X is called a *partition of unity associated to* $\mathfrak{U}$ if

1) $\varphi_i(x) \geqq 0$ *for* $x \in X$,

2) $\operatorname{supp}\varphi_i \subset U_i$,

3) *each point* $x \in X$ *has an open neighbourhood which meets* $\operatorname{supp}\varphi_i$ *for only finitely many* $i \in I$,

4) $\sum_{i \in I} \varphi_i(x) = 1$ *for all* $x \in X$. [The sum can be formed because of 3).]

Theorem 2.8.4. *X is paracompact if and only if X is a* HAUSDORFF *space and every open covering of X admits an associated partition of unity.*

Proof: Suppose that X is paracompact and that $\mathfrak{U} = \{U_i\}_{i \in I}$ is an open covering of X. Then X is normal (2.8.1) and $\mathfrak{U}$ has a locally finite (and therefore point finite) refinement $\mathfrak{U}' = \{U'_i\}_{i \in I}$. By the shrinking theorem (2.8.3) there are open coverings $\mathfrak{V} = \{V_i\}_{i \in I}$ and $\mathfrak{W} = \{W_i\}_{i \in I}$ of X such that $\overline{W}_i \subset V_i$ and $\overline{V}_i \subset U'_i$. By the URYSOHN lemma there exists a real valued non-negative function φ'_i on X which is continuous, identically 1 on $\overline{W}_i$ and identically 0 outside V_i. Since $\mathfrak{V}$ and $\mathfrak{W}$ are locally finite coverings the sum $\psi = \sum_{i \in I} \varphi'_i$ is a never zero continuous function. The functions $\varphi_i = \varphi'_i/\psi$ satisfy properties 1), 2), 3), 4).

Conversely suppose any open covering $\mathfrak{U} = \{U_i\}_{i \in I}$ of X admits an associated partition of unity $\{\varphi_i\}_{i \in I}$. Let V_i be the interior of the closed set $\operatorname{supp}\varphi_i$. Then $\mathfrak{V} = \{V_i\}_{i \in I}$ is an open covering of X [by 4)], is a refinement of $\mathfrak{U}$ [by 2)], and is locally finite [by 3)]. Therefore X is paracompact.

2.9. Cohomology groups for paracompact spaces

Let $\mathfrak{S}$ be a presheaf over a topological space X and $\mathfrak{S}$ the corresponding sheaf (see 2.2). Let $\tilde{\mathfrak{S}}$ be the canonical presheaf of $\mathfrak{S}$ and $h : \mathfrak{S} \to \tilde{\mathfrak{S}}$ the natural homomorphism defined in 2.3. For each $q \geqq 0$ there is a cohomology homomorphism $h_* : H^q(X, \mathfrak{S}) \to H^q(X, \tilde{\mathfrak{S}}) = H^q(X, \mathfrak{S})$ defined by h.

Theorem 2.9.1. *Let $\mathfrak{S}$ be a presheaf over a paracompact space X and let $\mathfrak{S}$ be the corresponding sheaf. Then the natural homomorphism $h_* : H^q(X, \mathfrak{S}) \to H^q(X, \mathfrak{S})$ is an isomorphism.*

The above theorem shows that the cohomology groups of a paracompact space with coefficients in a presheaf $\mathfrak{S}$ depend only on the corresponding sheaf $\mathfrak{S}$. We first prove a preliminary lemma.

Lemma 2.9.2. *Let X be a paracompact space, and $\mathfrak{S}$ a presheaf over X with the zero sheaf as corresponding sheaf. Let $\mathfrak{U} = \{U_i\}_{i \in I}$ be an open covering of X and $f \in C^q(\mathfrak{U}, \mathfrak{S})$. Then there is a refinement $\mathfrak{V} = \{V_j\}_{j \in J}$ of $\mathfrak{U}$ and a map $\tau : J \to I$ with $V_j \subset U_{\tau j}$ for all $j \in J$ such that the cochain $\tau^* f \in C^q(\mathfrak{V}, \mathfrak{S})$ is zero.*

Proof (see SERRE [2], p. 218): Let $\mathfrak{S} = \{S_U, r^U_V\}$. We first make the following remark. Let U be any open neighbourhood of a point $x \in X$ and let $g \in S_U$. Then there exists an open neighbourhood V of x such that $r^U_V g = 0$. If $V \subset U$ and $g \in S_U$ the element $r^U_V g$ will be referred to as "g regarded as an element of S_V".

Now let $\mathfrak{U} = \{U_i\}_{i \in I}$ be an open covering of X. A cochain $f \in C^q(\mathfrak{U}, \mathfrak{S})$ associates to each $(q+1)$-ple $(i_0, \ldots, i_q)$ an element $f(i_0, \ldots, i_q)$ of $S_{(U_{i_0} \cap \cdots \cap U_{i_q})}$. We must now construct a refinement $\mathfrak{V} = \{V_j\}_{j \in J}$ of $\mathfrak{U}$ with the required properties.

Without loss of generality we may assume that $\mathfrak{U}$ is locally finite. By 2.8.1 and 2.8.3 there is an open covering $\mathfrak{W} = \{W_i\}_{i \in I}$ of X with $\overline{W}_i \subset U_i$. Let $J = X$ and let $\tau : X \to I$ be a map for which $x \in W_{\tau x}$. For each point $x \in X$ we choose an open neighbourhood V_x of x which fulfils the following conditions:

a) If $x \in U_i$ then $V_x \subset U_i$. If $x \in W_i$ then $V_x \subset W_i$.

b) If $V_x \cap W_i$ is non-empty then $V_x \subset U_i$.

c) If $x \in U_{i_0} \cap \cdots \cap U_{i_q}$ then $f(i_0, \ldots, i_q)$, regarded as an element of S_{V_x}, is zero. [By a) the set V_x is contained in $U_{i_0} \cap \cdots \cap U_{i_q}$.]

Conditions a) and b) can be fulfilled because $\mathfrak{U}$ and $\mathfrak{W}$ are locally finite coverings and $\overline{W}_i \subset U_i$. By the remark at the beginning of the proof, V_x can then be chosen sufficiently small so that c) holds. Let $\mathfrak{V} = \{V_x\}_{x \in X}$.

We shall now show that the cochain $\tau^* f \in C^q(\mathfrak{V}, \mathfrak{S})$ is zero, *i. e.* that, for all $(x_0, \ldots, x_q)$, the element $f(\tau x_0, \ldots, \tau x_q)$, regarded as an element of $S_{(V_{x_0} \cap \cdots \cap V_{x_q})}$ is zero. If $V_{x_0} \cap \cdots \cap V_{x_q}$ is empty there is nothing to prove. If not, then $V_{x_0} \cap V_{x_k}$ is non-empty for all k with $0 \leq k \leq q$. By a), $V_{x_0} \cap W_{\tau x_k}$ is non-empty and therefore, by b), $V_{x_0} \subset U_{\tau x_k}$ for all k with $0 \leq k \leq q$. It follows from c) that $f(\tau x_0, \ldots, \tau x_q)$, regarded as an element of $S_{V_{x_0}}$, is zero. This implies immediately the corresponding result for the smaller set $V_{x_0} \cap \cdots \cap V_{x_q}$. Q. E. D.

Remark 1: In the particular case $q = 0$, Lemma 2.9.2 holds for an arbitrary topological space X. It is sufficient to choose an open covering $\mathfrak{V} = \{V_x\}_{x \in X}$ and a map $\tau : X \to I$ such that V_x is an open neighbourhood of x, $V_x \subset U_{\tau x}$, and $f(\tau x)$ regarded as an element of S_{V_x} is zero.

We now come to the proof of Theorem 2.9.1. The method is due to SERRE.

Let $\tilde{\mathfrak{S}}$ be the canonical presheaf of $\mathfrak{S}$ and $h : \mathfrak{S} \to \tilde{\mathfrak{S}}$ the natural homomorphism. There is an exact sequence of presheaves

$$0 \to \mathfrak{S}' \to \mathfrak{S} \xrightarrow{h} \tilde{\mathfrak{S}} \to \mathfrak{S}'' \to 0 \tag{19}$$

in which both $\mathfrak{S}'$ and $\mathfrak{S}''$ have the zero sheaf as corresponding sheaf. Lemma 2.9.2 implies that $H^q(X, \mathfrak{S}') = H^q(X, \mathfrak{S}'') = 0$ for all $q \geqq 0$. Therefore $h_* : H^q(X, \mathfrak{S}) \to H^q(X, \tilde{\mathfrak{S}})$ is an isomorphism by the corollary to Lemma 2.7.1. Q. E. D.

Remark 2: In the particular case $q = 0$, Theorem 2.9.1 holds for an arbitrary topological space X provided that $\mathfrak{S}' = 0$ (that is, provided that h is a monomorphism).

2.10. The exact cohomology sequence for sheaves

Consider an exact sequence

$$0 \to \mathfrak{S}' \to \mathfrak{S} \to \mathfrak{S}'' \to 0 \tag{20}$$

of sheaves over the topological space X. For each open set U of X there is, in the notation of 2.4, an exact sequence

$$0 \to \Gamma(U, \mathfrak{S}') \to \Gamma(U, \mathfrak{S}) \to S''_U \to 0 \,. \tag{21}$$

Let $\mathfrak{S}$, $\mathfrak{S}'$ be the canonical presheaves of $\mathfrak{S}$, $\mathfrak{S}'$, but let $\mathfrak{S}''$ be the presheaf determined by the groups S''_U. Then there is an exact sequence

$$0 \to \mathfrak{S}' \to \mathfrak{S} \to \mathfrak{S}'' \to 0 \tag{22}$$

which, by Lemma 2.7.1, gives an exact cohomology sequence. By definition $H^q(X, \mathfrak{S}') = H^q(X, \mathfrak{S}')$ and $H^q(X, \mathfrak{S}) = H^q(X, \mathfrak{S})$. The sheaf constructed from $\mathfrak{S}''$ is $\mathfrak{S}''$. If X is paracompact then by Theorem 2.9.1 the natural homomorphism $H^q(X, \mathfrak{S}'') \to H^q(X, \mathfrak{S}'')$ is an isomorphism. The groups $H^q(X, \mathfrak{S}'')$ can therefore be replaced by $H^q(X, \mathfrak{S}'')$ in the exact cohomology sequence given by (22). Moreover the resulting homomorphism

$$\delta^q_* : H^q(X, \mathfrak{S}'') \to H^{q+1}(X, \mathfrak{S}')$$

is defined in a natural way. We therefore obtain

Theorem 2.10.1. *An exact sequence*

$$0 \to \mathfrak{S}' \xrightarrow{h'} \mathfrak{S} \xrightarrow{h} \mathfrak{S}'' \to 0 \tag{23}$$

of sheaves over a paracompact space X gives an exact cohomology sequence

$$0 \to H^0(X, \mathfrak{S}') \xrightarrow{h'_*} H^0(X, \mathfrak{S}) \xrightarrow{h_*} H^0(X, \mathfrak{S}'') \xrightarrow{\delta^0_*} H^1(X, \mathfrak{S}') \to \cdots$$

$$\cdots \to H^q(X, \mathfrak{S}') \xrightarrow{h'_*} H^q(X, \mathfrak{S}) \xrightarrow{h_*} H^q(X, \mathfrak{S}'') \xrightarrow{\delta^q_*} H^{q+1}(X, \mathfrak{S}') \to \cdots$$

in which all homomorphisms are defined in a natural way.

Remark: The remarks in the previous section imply that, for an arbitrary (not necessarily paracompact) topological space X, an exact sequence (23) of sheaves over X gives an exact cohomology sequence

$$0 \to H^0(X, \mathfrak{S}') \to H^0(X, \mathfrak{S}) \to H^0(X, \mathfrak{S}'') \to H^1(X, \mathfrak{S}') \to H^1(X, \mathfrak{S}) \to H^1(X, \mathfrak{S}'').$$

We now come to some applications of the exact cohomology sequence which are used in Chapter Four. Let $\mathbf{K}$ be a field and let $\mathfrak{S}$ be a sheaf of $\mathbf{K}$-modules over X (see 2.1). Then the cohomology groups $H^q(X, \mathfrak{S})$ are vector spaces over $\mathbf{K}$. Let $\dim H^q(X, \mathfrak{S})$ denote dimension over $\mathbf{K}$.

Definition: A sheaf $\mathfrak{S}$ of $\mathbf{K}$-modules over X is *of type* (F) if the cohomology groups $H^q(X, \mathfrak{S})$ are finite dimensional vector spaces over $\mathbf{K}$ and if $\dim H^q(X, \mathfrak{S}) = 0$ for all but a finite number of $q \geqq 0$.

If $\mathfrak{S}$ is of type (F) the EULER-POINCARÉ characteristic $\chi(X, \mathfrak{S})$ can be defined by the formula

$$\chi(X, \mathfrak{S}) = \sum_{q=0}^{\infty} (-1)^q \dim H^q(X, \mathfrak{S}) .$$

Theorem 2.10.2. *Let* $0 \to \mathfrak{S}' \to \mathfrak{S} \to \mathfrak{S}'' \to 0$ *be an exact sequence of sheaves over a paracompact space X. If two of the sheaves* $\mathfrak{S}', \mathfrak{S}, \mathfrak{S}''$ *are of type* (F) *then so is the third, and*

$$\chi(X, \mathfrak{S}) = \chi(X, \mathfrak{S}') + \chi(X, \mathfrak{S}'') .$$

Proof: By direct application of Theorem 2.10.1.

Theorem 2.10.3. *Let* $0 \to \mathfrak{S}_1 \to \mathfrak{S}_2 \to \mathfrak{S}_3 \to \cdots \to \mathfrak{S}_n \to 0$ *be an exact sequence of sheaves over a paracompact space* X *which are all of type* (F). *Then*

$$\sum_{i=1}^{n} (-1)^i \chi(X, \mathfrak{S}_i) = 0 .$$

Proof: Let $\mathfrak{R}_r$ be the kernel of the homomorphism from $\mathfrak{S}_r$ to $\mathfrak{S}_{r+1}$ and apply Theorem 2.10.2 to the exact sequences

$$0 \to \mathfrak{R}_r \to \mathfrak{S}_r \to \mathfrak{R}_{r+1} \to 0 .$$

2.11. Fine sheaves

In applications of sheaf theory results on the vanishing of cohomology groups are of particular importance.

Definition: Let $\mathfrak{S}$ be a sheaf over a paracompact space X. Then $\mathfrak{S}$ is a *fine sheaf* if for each locally finite open covering $\mathfrak{U} = \{U_i\}_{i \in I}$ of X there is a system $\{h_i\}_{i \in I}$ of homomorphisms $h_i : \mathfrak{S} \to \mathfrak{S}$ such that:

I) *For each* $i \in I$ *there is a closed set* A_i *of* X *such that* $A_i \subset U_i$ *and*

$$h_i(S_x) = 0 \textit{ for } x \notin A_i, \ (S_x = \text{stalk of } \mathfrak{S} \text{ at } x) .$$

II) $\sum_{i \in I} h_i$ *is the identity.* (The sum can be formed because $\mathfrak{U}$ is locally finite.)

Theorem 2.11.1. *Let* $\mathfrak{S}$ *be a fine sheaf over a paracompact space* X. *Then* $H^q(X, \mathfrak{S})$ *vanishes for* $q \geqq 1$.

Proof (see Cartan [4], Exposé XVII): Since X is paracompact it is sufficient to prove that $H^q(\mathfrak{U}, \mathfrak{S})$ vanishes for $q \geqq 1$ for any locally finite open covering $\mathfrak{U} = \{U_i\}_{i \in I}$. We define for $q \geqq 1$ a homomorphism (homotopy operator)

$$k^q : C^q(\mathfrak{U}, \mathfrak{S}) \to C^{q-1}(\mathfrak{U}, \mathfrak{S})$$

in the following way. Let $f \in C^q(\mathfrak{U}, \mathfrak{S})$. The cochain $k^q f$ associates to each q-ple $(i_0, \ldots, i_{q-1})$ a section $(k^q f)(i_0, \ldots, i_{q-1})$ of $\mathfrak{S}$ over $U_{i_0} \cap \cdots \cap U_{i_{q-1}}$. For each index $i \in I$ let $t(i, i_0, \ldots, i_{q-1})$ be the section of $\mathfrak{S}$ over $U_{i_0} \cap \cdots \cap U_{i_{q-1}}$ which is equal to $h_i(f(i, i_0, \ldots, i_{q-1}))$ over the smaller set $U_i \cap U_{i_0} \cap \cdots \cap U_{i_{q-1}}$ and is zero outside this smaller set. The h_i are homomorphisms with properties I) and II). We define

$$(k^q f)(i_0, \ldots, i_{q-1}) = \sum_{i \in I} t(i, i_0, \ldots, i_{q-1}) .$$

The sum can be formed because $\mathfrak{U}$ is locally finite. Let δ^q be the coboundary homomorphism $C^q(\mathfrak{U}, \mathfrak{S}) \to C^{q+1}(\mathfrak{U}, \mathfrak{S})$. It is easy to prove that, for $q \geqq 1$, the homomorphism $k^{q+1} \delta^q + \delta^{q-1} k^q$ is equal to the identity. This completes the proof.

The above proof is a generalisation of the cone construction which is used to prove that the cohomology groups of a simplex (with constant coefficients) are trivial.

Now consider the sheaf $\mathbf{C}_c$ over a paracompact space X (see 2.5, Example 2)). Let $\mathfrak{U}$ be a locally finite open covering of X. By 2.8.4, $\mathfrak{U}$ has an associated partition of unity $\{\varphi_i\}_{i\in I}$. The functions φ_i can be used to define homomorphisms $h_i: \mathbf{C}_c \to \mathbf{C}_c$ as follows. Let S_U be the $\mathbf{C}$-module of complex valued continuous functions defined on U. For $f \in S_U$ define $h_i(f) = \varphi_i f$. This defines a homomorphism h_i from the presheaf $\{S_U\}$ to itself and therefore also a homomorphism h_i from $\mathbf{C}_c$ to itself (see 2.2). The homomorphisms h_i satisfy properties I) and II) of the definition of a fine sheaf. This proves

Theorem 2.11.2. *The sheaf $\mathbf{C}_c$ of germs of local complex valued continuous functions over a paracompact space X is fine.*

Exactly the same proof shows that the sheaf of germs of local real valued continuous functions over a paracompact space X is fine. Theorem 2.11.2 should be regarded as a typical example of a whole class of similar theorems.

Now let X be a differentiable manifold [see 2.5, Example 3)] and $\mathfrak{U} = \{U_i\}_{i\in I}$ an open covering of X. Then it is possible to find an associated partition of unity $\{\varphi_i\}_{i\in I}$ in which the functions φ_i are differentiable (DE RHAM [1], § 2). With the help of such a differentiable partition of unity it is possible to prove that many sheaves over X are fine. For instance the sheaf $\mathbf{C}_\mathfrak{d}$ of germs of local complex valued differentiable functions is fine. Similarly the sheaf $\mathfrak{A}^p$ of germs of exterior differential forms of degree p with real (or complex) differentiable functions as coefficients is fine. The canonical presheaf of this sheaf is obtained by associating to each open set U of X the $\mathbf{R}$-module (or $\mathbf{C}$-module) of exterior differential forms of degree p defined on U (see DE RHAM [1], § 4).

2.12. Resolutions of sheaves. Theorem of DE RHAM

Consider an exact sequence

$$0 \to \mathfrak{S} \xrightarrow{h} \mathfrak{S}_0 \xrightarrow{h^0} \mathfrak{S}_1 \xrightarrow{h^1} \mathfrak{S}_2 \xrightarrow{h^2} \cdots \xrightarrow{h^{p-1}} \mathfrak{S}_p \xrightarrow{h^p} \cdots \tag{24}$$

of sheaves over a paracompact space X. The sequence is called a *resolution* of the sheaf $\mathfrak{S}$ if the cohomology groups $H^q(X, \mathfrak{S}_p)$ vanish for $q \geqq 1$ and $p \geqq 0$. By Theorem 2.11.1 this is the case if each $\mathfrak{S}_p$ is a fine sheaf. An exact sequence (24) with $\mathfrak{S}_p$ fine for all $p \geqq 0$ is called a *fine resolution* of $\mathfrak{S}$. The exact sequence (24) defines a sequence

$$0 \to \Gamma(X, \mathfrak{S}) \xrightarrow{h_*} \Gamma(X, \mathfrak{S}_0) \xrightarrow{h_*^0} \Gamma(X, \mathfrak{S}_1) \xrightarrow{h_*^1} \cdots \xrightarrow{h_*^{p-1}} \Gamma(X, \mathfrak{S}_p) \xrightarrow{h_*^p} \cdots \tag{25}$$

which in general is exact only at $\Gamma(X, \mathfrak{S})$ and $\Gamma(X, \mathfrak{S}_0)$. Since $h_*^{p+1} h_*^p = 0$ the groups $\Gamma(X, \mathfrak{S}_p)$, $p \geqq 0$, and the homomorphisms h_*^p form an abstract cochain complex.

Theorem 2.12.1. *Consider a resolution* (24) *of a sheaf* $\mathfrak{S}$ *over a paracompact space* X. *The q-th cohomology group of the abstract complex* $\{\Gamma(X, \mathfrak{S}_p), p \geqq 0\}$ *is naturally isomorphic to the cohomology group* $H^q(X, \mathfrak{S})$ *for* $q \geqq 0$. *In other words*

$H^q(X, \mathfrak{S}) \cong$ kernel (h_*^q)/image (h_*^{q-1}) *for* $q \geqq 1$,

$H^0(X, \mathfrak{S}) \cong$ kernel (h_*^0).

Proof: Clearly kernel $(h_*^0) = \Gamma(X, \mathfrak{S})$ which by Theorem 2.6.2 is the cohomology group $H^0(X, \mathfrak{S})$. This proves the statement for $q = 0$. Let $\mathfrak{R}_p$ be the kernel of the homomorphism $h^p : \mathfrak{S}_p \to \mathfrak{S}_{p+1}$. The sequence (24) gives an exact sequence of sheaves over X

$$0 \to \mathfrak{R}_p \to \mathfrak{S}_p \xrightarrow{h^p} \mathfrak{R}_{p+1} \to 0 \tag{26}$$

for each $p \geqq 0$. Since the cohomology groups $H^q(X, \mathfrak{S}_p)$ vanish for $q \geqq 1$ the exact cohomology sequence of (26) gives natural isomorphisms

$$H^{q-1}(X, \mathfrak{R}_{p+1}) \cong H^q(X, \mathfrak{R}_p) \text{ for } q \geqq 2\,. \tag{27}$$

Since $\mathfrak{R}_0 = \mathfrak{S}$ repeated application of (27) gives

$$H^1(X, \mathfrak{R}_{q-1}) \cong H^q(X, \mathfrak{S}) \text{ for } q \geqq 1\,. \tag{28}$$

The exact cohomology sequence of (26), with p replaced by $q - 1$, contains the exact sequence

$$H^0(X, \mathfrak{S}_{q-1}) \xrightarrow{h_*^{q-1}} H^0(X, \mathfrak{R}_q) \to H^1(X, \mathfrak{R}_{q-1}) \to 0\,. \tag{29}$$

Since $H^0(X, \mathfrak{R}_q)$ is the kernel of h_*^q, and $H^0(X, \mathfrak{S}_{q-1}) = \Gamma(X, \mathfrak{S}_{q-1})$ the theorem follows from (28) and (29).

Let X be a differentiable manifold [see 2.5, Example 3)] and let $\mathfrak{A}^p$ be the sheaf of germs of differentiable p-forms over X (see the end of 2.11). If U is an open set of X then $\Gamma(U, \mathfrak{A}^p)$ is the **R**-module of differentiable p-forms defined on U. The exterior derivative d is a homomorphism from $\Gamma(U, \mathfrak{A}^p)$ to $\Gamma(U, \mathfrak{A}^{p+1})$. In terms of local coordinates the derivative of a p-form

$$\omega = \sum_{i_1 < \cdots < i_p} f_{i_1, \ldots, i_p} \, dx_{i_1} \wedge \cdots \wedge dx_{i_p}$$

is the $(p + 1)$-form

$$d\omega = \sum_{i_1 < \cdots < i_p} df_{i_1, \ldots, i_p} \wedge dx_{i_1} \wedge \cdots \wedge dx_{i_p}\,.$$

Let **R** be the constant sheaf of real numbers, h the embedding of **R** in the sheaf $\mathfrak{A}^0$ of germs of real valued differentiable functions, and $h^p : \mathfrak{A}^p \to \mathfrak{A}^{p+1}$ the homomorphism defined by the exterior derivative.

Theorem 2.12.2 (POINCARÉ Lemma). *The sequence*

$$0 \to \mathbf{R} \xrightarrow{h} \mathfrak{A}^0 \xrightarrow{h^1} \mathfrak{A}^1 \xrightarrow{h^2} \cdots \xrightarrow{h^{p-1}} \mathfrak{A}^p \xrightarrow{h^p} \cdots$$

is exact.

Proof: Since $dd = 0$ it follows that $h^{p+1} h^p = 0$ for $p \geqq 0$. It is therefore sufficient to prove the following result. Let ω be a p-form ($p \geqq 1$) defined on an open set U. If $d\omega = 0$ then there exists a $(p-1)$-form α defined on an open set $V \subset U$ such that $\omega = d\alpha$ on V. This is a local result, so we may assume that X is n-dimensional euclidean space. The required result is then the classical form of the POINCARÉ Lemma. It can for instance be proved by induction.

Theorem 2.12.3 (DE RHAM). *Let X be a differentiable manifold and let A^p, $p \geqq 0$, be the* **R**-*module of differentiable p-forms defined on the whole of X. Let Z^p denote the kernel of the* **R**-*homomorphism $d: A^p \to A^{p+1}$. Then $dA^{p-1} \subset Z^p$ for $p \geqq 1$. There are isomorphisms*

$$H^0(X, \mathbf{R}) = Z^0 \text{ and } H^p(X, \mathbf{R}) \cong Z^p/dA^{p-1}, \; p \geqq 1\,.$$

Proof: The exact sequence of Theorem 2.12.2 is a fine resolution of the constant sheaf **R**. The homomorphisms h^p_* of Theorem 2.12.1 are in this case the exterior derivative of forms, so that the result follows from Theorem 2.12.1.

Remark: Exactly the same proof gives the corresponding result for complex valued differentiable p-forms. Let A^p be the **C**-module of p-forms defined on the whole of X with local complex valued differentiable functions as coefficients. Let Z^p be the kernel of the **C**-homomorphism $d: A^p \to A^{p+1}$. Then there are isomorphisms

$$H^0(X, \mathbf{C}) \cong Z^0 \text{ and } H^p(X, \mathbf{C}) \cong Z^p/dA^{p-1}, \; p \geqq 1\,.$$

§ 3. Fibre bundles

3.1. Let X be a topological space. A sheaf $\mathfrak{S} = (S, \pi, X)$ of (not necessarily abelian) groups over X is defined, as in 2.1, by properties I), II) together with property III) in the following slightly modified form:

III) *Every stalk has the structure of a group. The group operations associate, to points α, β in S_x, the elements $\alpha\,\beta$, $\alpha\,\beta^{-1}$ in S_x. $\alpha\,\beta^{-1}$ depends continuously on α and β.* (It then follows that the identity 1_x of the group S_x depends continuously on x and that $\alpha\,\beta$, $\alpha^{-1}\,\beta$ depend continuously on α, β.)

The definitions of presheaf, canonical presheaf, etc. carry over with similar modification. As in 2.3 the group $\Gamma(U, \mathfrak{S})$ of sections of $\mathfrak{S}$ over an open set U of X and the restriction homomorphisms r^U_V are defined. The identity of $\Gamma(U, \mathfrak{S})$ is the section $x \to 1_x$. [If U is empty then by definition $\Gamma(U, \mathfrak{S})$ consists only of the identity element.]

Cohomology groups $H^q(X, \mathfrak{S})$ cannot be defined in the non-abelian case. It is nevertheless possible, for $q = 1$, to define a cohomology set $H^1(X, \mathfrak{S})$ with a distinguished element **1**. If $\mathfrak{S}$ is a sheaf of abelian groups then the cohomology set $H^1(X, \mathfrak{S})$ agrees with the cohomology groups defined in 2.6. The distinguished element then corresponds to the zero element of the cohomology group. The cohomology set can again be defined with coefficients in an arbitrary presheaf. For convenience we formulate the definition only for the canonical presheaf, *i. e.* for the sheaf $\mathfrak{S}$ itself.

The cohomology set $H^1(\mathfrak{U}, \mathfrak{S})$.

Let $\mathfrak{U} = \{U_i\}_{i \in I}$ be an open covering of X. A $\mathfrak{U}$-cocycle is a function f which associates, to each ordered pair i, j of elements in I, an element $f_{ij} \in \Gamma(U_i \cap U_j, \mathfrak{S})$ such that

$$f_{ij} f_{jk} = f_{ik} \text{ in } U_i \cap U_j \cap U_k \text{ for all } i, j, k \in I .$$

Equations of this type are always to be understood as holding between the restrictions of sections to a common domain of definition. It follows from the definition that f_{ii} is equal to the identity element of $\Gamma(U_i, \mathfrak{S})$ and that $f_{ij} = f_{ji}^{-1}$.

The set of $\mathfrak{U}$-cocycles is denoted by $Z^1(\mathfrak{U}, \mathfrak{S})$. Cocycles f, f' are said to be equivalent if for each $i \in I$ there exists an element $g_i \in \Gamma(U_i, \mathfrak{S})$ such that

$$f'_{ij} = g_i^{-1} f_{ij} g_j \text{ in } U_i \cap U_j \text{ for all } i, j \in I .$$

The cohomology set $H^1(\mathfrak{U}, \mathfrak{S})$ is the set of equivalence classes of $\mathfrak{U}$-cocycles. Let $\mathfrak{V} = \{V_j\}_{j \in J}$ be a refinement of $\mathfrak{U}$ and let τ be a map from J to I with $V_r \subset U_{\tau r}$ for all $r \in J$. A $\mathfrak{U}$-cocycle f defines a $\mathfrak{V}$-cocycle $\tau^* f$ by

$$(\tau^* f)_{r,s} = f_{\tau r, \tau s} \text{ in } V_r \cap V_s \text{ for all } r, s \in J .$$

The map τ^* induces a natural map

$$t^{\mathfrak{U}}_{\mathfrak{V}} : H^1(\mathfrak{U}, \mathfrak{S}) \to H^1(\mathfrak{V}, \mathfrak{S})$$

with properties as in Lemma 2.6.1. If $'\tau$ is another map from J to I with $V_r \subset U_{'\tau r}$ then the sections $g_r = f_{\tau r, '\tau r} \in \Gamma(V_r, \mathfrak{S})$ define an equivalence

$$('\tau^* f)_{r,s} = g_r^{-1} (\tau^* f)_{r,s} g_s \text{ in } V_r \cap V_s \text{ for all } r, s \in J$$

between $'\tau^* f$ and $\tau^* f$. Therefore the map $t^{\mathfrak{U}}_{\mathfrak{V}}$ does not depend on the choice of the map τ.

The cohomology set $H^1(X, \mathfrak{S})$ is the direct limit of the sets $H^1(\mathfrak{U}, \mathfrak{S})$, with respect to the maps $t^{\mathfrak{U}}_{\mathfrak{V}}$, as $\mathfrak{U}$ runs through all proper open coverings of X (see 2.6 and the beginning of § 2). If $\mathfrak{V}$ is a refinement of $\mathfrak{U}$ it can be shown that the map $t^{\mathfrak{U}}_{\mathfrak{V}}$ is one-one, and so $H^1(\mathfrak{U}, \mathfrak{S})$ can be regarded as a subset of $H^1(\mathfrak{V}, \mathfrak{S})$. If $\mathfrak{U}$ and $\mathfrak{V}$ are cofine then $H^1(\mathfrak{U}, \mathfrak{S})$ is identified with $H^1(\mathfrak{V}, \mathfrak{S})$ in a natural way. It follows that $H^1(X, \mathfrak{S})$ can be regarded as the union of all sets $H^1(\mathfrak{U}, \mathfrak{S})$, as $\mathfrak{U}$ runs through all proper open coverings of X. The distinguished element $\mathbf{1} \in H^1(X, \mathfrak{S})$ is represented, for any open covering $\mathfrak{U} = \{U_i\}_{i \in I}$, by the cocycle $f_{ij} = 1 \in \Gamma(U_i \cap U_j, \mathfrak{S})$.

We now consider the particular case in which G is a group and $\mathfrak{S}$ is a sheaf of germs of functions with values in G. The functions may be continuous, differentiable or holomorphic depending on the structure of X and G. These cases are distinguished by the symbols $G_\mathfrak{c}$, $G_\mathfrak{d}$, G_ω which agree with those used in 2.5.

If X is a topological space and G is a topological group then $G_\mathfrak{c}$ is the sheaf for which $\Gamma(U, G_\mathfrak{c})$ is the group of continuous functions from U to G.

If X is a differentiable manifold (see 2.5) and G is a real LIE group then $G_\mathfrak{d}$ is the sheaf for which $\Gamma(U, G_\mathfrak{d})$ is the group of differentiable (*i. e.* C^∞-differentiable, see 2.5) functions from U to G.

If X is a complex manifold (see 2.5) and G is a complex LIE group then G_ω is the sheaf for which $\Gamma(U, G_\omega)$ is the group of holomorphic functions from U to G.

Convention: If the sheaf $G_\mathfrak{d}$ over X is mentioned it will be assumed implicitly that X is a differentiable manifold and G is a LIE group. If the sheaf G_ω over X is mentioned it will be assumed implicitly that X is a complex manifold and G is a complex LIE group.

The sheaf $G_\mathfrak{d}$ over X is a subsheaf of the sheaf $G_\mathfrak{c}$ over X. The sheaf G_ω over X is a subsheaf of the sheaf $G_\mathfrak{d}$ over X. There are natural maps

$$H^1(X, G_\mathfrak{d}) \to H^1(X, G_\mathfrak{c}), \quad H^1(X, G_\omega) \to H^1(X, G_\mathfrak{d}) \tag{1}$$

together with the composite map

$$H^1(X, G_\omega) \to H^1(X, G_\mathfrak{c}).$$

If $h: G' \to G$ is a continuous (or differentiable, or holomorphic) transformation of topological groups (or LIE groups, or complex LIE groups) there are sheaf homomorphisms

$$G'_\mathfrak{c} \to G_\mathfrak{c}, \ G'_\mathfrak{d} \to G_\mathfrak{d}, \ G'_\omega \to G_\omega$$

and natural maps

$$H^1(X, G'_\mathfrak{c}) \to H^1(X, G_\mathfrak{c}), \ H^1(X, G'_\mathfrak{d}) \to H^1(X, G_\mathfrak{d}), \ H^1(X, G'_\omega) \to H^1(X, G_\omega). \tag{2}$$

If $G' = G$, and h is the inner automorphism
$h(g) = a^{-1} g\, a \ (g \in G)$ determined by an element $a \in G$ (2*)
then the natural maps (2) are all equal to the identity.

3.2. a). Let X be a topological space, and let G be a topological group with identity element $e \in G$. Consider an effective continuous action of a group G on a topological space F. Here *continuous action* means a continuous map $G \times F \to F$ which maps $g \times f \in G \times F$ to $g f \in F$, such that $g_1(g_2 f) = (g_1 g_2) f$ and $e f = f$ for all $f \in F$. *Effective* means that if $g f = f$ for some g, and all $f \in F$, then $g = e$.

Definition: A topological space W, together with a continuous map (projection) $\pi : W \to X$, is called a *fibre bundle over X with structure group G and (typical) fibre F* if there exists a system of coordinate transformations, that is

I) *an open covering $\mathfrak{U} = \{U_i\}_{i \in I}$ of X and homeomorphisms $h_i : \pi^{-1}(U_i) \to U_i \times F$ which map the "fibre" $\pi^{-1}(u)$ onto $u \times F$*, and

II) *elements $g_{ij} \in \Gamma(U_i \cap U_j, G_c)$ for all $i, j \in I$ such that*

$$(h_i h_j^{-1})(u \times f) = u \times g_{ij}(u) f \text{ for all } u \in U_i \cap U_j,\ f \in F\,. \tag{3}$$

Remark: Since the action of G on F is effective, the element g_{ij} is determined uniquely by h_i and h_j. The g_{ij} clearly define a cocycle $g \in Z^1(\mathfrak{U}, G_c)$ and hence an element of the cohomology set $H^1(\mathfrak{U}, G_c)$. For example consider the trivial fibre bundle with $W = X \times F$ and π the product projection. Any open covering $\mathfrak{U} = \{U_i\}_{i \in I}$ satisfies I), and the functions $g_{ij} = 1 \in \Gamma(U_i \cap U_j, G_c)$ satisfy II). In this case g defines the distinguished element of the cohomology set $H^1(\mathfrak{U}, G_c)$.

The definition of a fibre bundle over X will be complete once we specify under what circumstances different systems of coordinate transformations define the same fibre bundle. A homeomorphism $h_U : \pi^{-1}(U) \to U \times F$, U open in X, is called an *admissible chart* for the system of coordinate transformations I), II) if there are elements $g_{U,i} \in \Gamma(U \cap U_i, G_c)$ for each $i \in I$ such that

$$(h_U h_i^{-1})(u \times f) = u \times g_{U,i}(u) f \text{ for all } u \in U \cap U_i,\ f \in F\,. \tag{3*}$$

Definition: Two systems of coordinate transformations make W (together with the projection π) the *same* fibre bundle W over X with structure group G and fibre F if and only if every admissible chart for one system is an admissible chart for the other system.

Definition: Let W (projection π) and W' (projection π') be fibre bundles over X with structure group G and fibre F. An *isomorphism* k from W to W' is a homeomorphism $k : W \to W'$ such that, for each point $x \in X$,

I) *the fibre $\pi^{-1}(x)$ maps onto the fibre $\pi'^{-1}(x)$, and*

II) *there is an open neighbourhood U of x, an element $g_U \in \Gamma(U, G_c)$, and admissible charts $h_U : \pi^{-1}(U) \to U \times F$ for W and $h'_U : \pi'^{-1}(U) \to U \times F$ for W' such that*

$$h'_U\, k\, h_U^{-1}(u \times f) = u \times g_U(u) f \text{ for all } u \in U,\ f \in F\,.$$

Given an open covering $\mathfrak{U} = \{U_i\}_{i \in I}$ of X and a $\mathfrak{U}$-cocycle $g = \{g_{ij}\} \in Z^1(\mathfrak{U}, G_\mathfrak{c})$, a fibre bundle W_g over X with structure group G and fibre F can be constructed. It is sufficient to form the disjoint union of the cartesian products $U_i \times F$ and to identify, for each $u \in U_i \cap U_j$, the points $u \times f \in U_j \times F$ and $u \times g_{ij}(u) f \in U_i \times F$. The identification space W_g is a fibre bundle with projection induced by the product projections $U_i \times F \to U_i$. If $g \in Z^1(\mathfrak{U}, G_\mathfrak{c})$ and $h \in Z^1(\mathfrak{V}, G_\mathfrak{c})$ then W_g is isomorphic to W_h if and only if g and h represent the same element of the cohomology set $H^1(X, G_\mathfrak{c})$. Every fibre bundle over X with structure group G and fibre F is isomorphic to a fibre bundle W_g for some g. We obtain

Theorem 3.2.1. *The isomorphism classes of fibre bundles over X with structure group G and fibre F (with a given effective continuous action of G on F) are in a natural one-one correspondence with the elements of the cohomology set $H^1(X, G_\mathfrak{c})$. The trivial fibre bundle $W = X \times F$ corresponds to the distinguished element $\mathbf{1} \in H^1(X, G_\mathfrak{c})$.*

Fibre bundles in the isomorphism class corresponding to $\xi \in H^1(X, G_\mathfrak{c})$ are said to be *associated to* ξ. If $F = G$ and the action of G on itself is left translation, then fibre bundles with structure group and fibre G are called *principal bundles.*

Convention: Elements of $H^1(X, G_\mathfrak{c})$ will be referred to as *G-bundles.* On the other hand the words *fibre bundle, principal bundle* will refer to a particular fibre bundle or principal bundle (as in the above definitions) and not to an isomorphism class.

3.2. b). The definitions and results of 3.2. a) carry over to the differentiable and holomorphic cases. Thus let X be a differentiable (complex) manifold and G a real (complex) Lie group. For further details on Lie groups see, for instance, Pontrjagin [1]. Consider an effective differentiable (holomorphic) action $G \times F \to F$ of G on a differentiable (complex) manifold F. In the remaining definitions it is only necessary to replace $G_\mathfrak{c}$ throughout by the sheaf $G_\mathfrak{b}$ (G_ω) over X. A fibre bundle W is then automatically a differentiable (complex) manifold. The projection π is a differentiable (holomorphic) map. An isomorphism between two fibre bundles is a differentiable (holomorphic) homeomorphism.

We speak of continuous, or differentiable, or complex analytic, fibre bundles and G-bundles according as the sheaf $G_\mathfrak{c}$, or $G_\mathfrak{b}$, or G_ω, is used in the definition. Let W be a continuous, or differentiable, or complex analytic, fibre bundle over X with projection π. A *section* of W over an open set U of X is a continuous, or differentiable, or holomorphic, function $s: U \to W$ for which πs is the identity. If a section over the whole of X exists, we also say simply that W has a section.

Remark: The pattern of 3.2 a) can be used to define many other sorts of fibre bundle (*e.g.* real analytic, algebraic). One has only to replace $G_\mathfrak{c}$ by another sheaf. In general one speaks of fibre bundles with structure sheaf (see GROTHENDIECK [1] and HOLMANN [1]). The sheaves $G_{\mathfrak{c}}$, $G_{\mathfrak{d}}$ and G_{ω} suffice for the purposes of the present work.

3.2. c). Consider a continuous action of the topological group G on the topological space F which is not effective. The elements h of G which act trivially on F (that is $h f = f$ for all $f \in F$) form a closed normal subgroup N of G. There is an effective continuous action of the topological group G/N on F.

If G is a real LIE group then so is any closed subgroup N of G. Therefore a differentiable action of G on the differentiable manifold F defines an effective differentiable action of the real LIE group G/N on F.

If G is a complex LIE group then a closed subgroup of G need not be a complex LIE group. It is, however, easy to prove that the closed normal subgroup N, defined by a holomorphic action of G on a complex manifold F, is a complex LIE group. There is then an effective holomorphic action of the complex LIE group G/N on F.

There are natural maps [see **3.1** (2)]

$t: H^1(X, G_{\mathfrak{c}}) \to H^1(X, (G/N)_{\mathfrak{c}})$, X a topological space,

$t: H^1(X, G_{\mathfrak{d}}) \to H^1(X, (G/N)_{\mathfrak{d}})$, X a differentiable manifold,

$t: H^1(X, G_{\omega}) \to H^1(X, (G/N)_{\omega})$, X a complex manifold.

Let W be a fibre bundle with structure group G/N and fibre F which is associated to $t\,\xi$, $\xi \in H^1(X, G_{\mathfrak{c}})$. In this case we also speak of W as a fibre bundle with structure group G and fibre F associated to ξ. Similarly for $G_{\mathfrak{d}}$ and G_{ω}.

3.2. d). The following remarks apply to the continuous, differentiable, and also to the complex analytic, cases.

Let E be a principal bundle over X with structure group and fibre G. There is an effective action of G on E defined by right translation on each fibre. With respect to the local product structure $U \times G$ of E (admissible chart) the action of an element $a \in G$ is given by $(u \times g)\, a = u \times g\, a$. This operation of $a \in G$ on E does not depend on the choice of admissible chart because the coordinate transformations (3), (3*) are defined by left translation.

Consider an action (not necessarily effective) of G on F. We now show how to construct, from the principal bundle E, a fibre bundle W over X with fibre F. Form the cartesian product $E \times F$ and identify $e\, a \times f$ with $e \times a\, f$ for each $a \in G$, $e \in E$, $f \in F$. The identification space W can be regarded in a natural way as a fibre bundle over X with structure group G and fibre F. The fibre bundles W and E are associated to the same G-bundle.

3.3. Let Y, X be topological spaces, $\varphi: Y \to X$ a continuous map, and G a topological group. There is a natural map

$$\varphi^*: H^1(X, G_{\mathfrak{c}}) \to H^1(Y, G_{\mathfrak{c}})\,. \tag{4}$$

If ξ is represented with respect to an open covering $\mathfrak{U} = \{U_i\}_{i \in I}$ of X by a $\mathfrak{U}$-cocycle $\{g_{ij}\}$, then $\varphi^* \xi$ is represented with respect to the open covering $\varphi^{-1} \mathfrak{U} = \{\varphi^{-1} U_i\}_{i \in I}$ of Y by the $\varphi^{-1} \mathfrak{U}$-cocycle $\{g_{ij} \varphi\}$. $\varphi^* \xi$ is called the G-bundle *induced* from the G-bundle ξ by the map φ.

Let W (projection π) be a fibre bundle over X with structure group G and fibre F which is associated to ξ. The following construction gives a fibre bundle $\varphi^* W$ over Y which is associated to $\varphi^* \xi$. Let $\varphi^* W$ be the subspace of $Y \times W$ consisting of all points $y \times w \in Y \times W$ with $\varphi(y) = \pi(w)$. The projection of the fibre bundle $\varphi^* W$ is induced by the product projection $Y \times W \to Y$.

Let $\varphi : Y \to X$ be a differentiable, or holomorphic, map of differentiable, or complex, manifolds X, Y and let G be a real, or complex, Lie group. There is a natural map

$$\varphi^* : H^1(X, G_{\mathfrak{d}}) \to H^1(Y, G_{\mathfrak{d}}) \text{ or } \varphi^* : H^1(X, G_{\omega}) \to H^1(Y, G_{\omega}) . \quad (4')$$

The definition of φ^* and the construction of the fibre bundle $\varphi^* W$ follows just as in the continuous case.

3.4. a). Let G' be a closed subgroup of the topological group G. Consider the space G/G' of left cosets $x G'$, $x \in G$, and the map σ from G to G/G'. Let $e \in G$ be the identity element. The statement

$$\sigma : G \to G/G' \textit{ admits a local section} \quad (5)$$

means that there is an open neighbourhood U of $\sigma(e)$ in G/G' and a continuous map $s : U \to G$ for which σs is the identity.

Theorem 3.4.1 (see Steenrod [1], 7.4). *If* (5) *holds then G can be regarded in a natural way as a principal bundle over G/G' with structure group and fibre G' and projection σ.*

Theorem 3.4.2. *Let G' be a closed subgroup of the real* Lie *group G. Then G' is a real* Lie *group and $G \xrightarrow{\sigma} G/G'$ admits a local differentiable (in fact, real analytic) section. G can be regarded in a natural way as a differentiable principal bundle over G/G' with structure group and fibre G' and projection σ.*

Theorem 3.4.3. *Let G' be a closed complex* Lie *subgroup of the complex* Lie *group G. Then $G \xrightarrow{\sigma} G/G'$ admits a local holomorphic section. G can be regarded in a natural way as a complex analytic principal bundle over G/G' with structure group and fibre G' and projection σ.*

The existence of the local differentiable (holomorphic) section s which is asserted in Theorem 3.4.2 (Theorem 3.4.3) can be proved by means of canonical coordinates in an open neighbourhood of $e \in G$. In the special cases which arise in this book it is actually easy to construct s directly.

3.4. b). The following exposition is valid in the continuous, differentiable or complex analytic cases. X will denote a topological space, differentiable manifold or complex manifold and G a topological, real LIE, or complex LIE group according to the case considered. Let G' be a closed subgroup of G. In the continuous case it will be assumed that (5) holds. In the complex analytic case it will be assumed that G' is a complex LIE subgroup of G.

Convention: Let W be a fibre bundle with structure group G and fibre F which is associated to a G-bundle ξ over X [see 3.2 a) and 3.2. c)]. Let h denote the natural embedding of the set of G'-bundles over X in the set of G-bundles over X induced by the embedding of G' in G (see 3.1). If there exists a G'-bundle $\check{\xi}$ over X with $h\,\check{\xi} = \xi$ we say *"the structure group of W can be reduced to G'"*. If such a G'-bundle arises naturally from the context we say that the structure group can be reduced to G' in a natural way.

Let E (projection π) be a principal bundle with fibre G which is associated to a G-bundle ξ over X. Let E/G' be the identification space obtained by identifying, in each fibre of E, points which correspond under right multiplication by elements of G' [see 3.2. d)]. Consider the commutative diagram

$$\begin{array}{ccc} E & \xrightarrow{\sigma} & E/G' \\ & {}_{\pi}\searrow \quad \swarrow_{\varrho} & \\ & X & \end{array}$$

Theorem 3.4.4. *E can be regarded in a natural way as a principal bundle over E/G' with structure group and fibre G' and projection σ. Let $\check{\xi}$ denote the corresponding G'-bundle over E/G'.*

E/G' can be regarded in a natural way as a fibre bundle over X with structure group G, fibre G/G' and projection ϱ (G acts on G/G' by left translation; see 3.2. c)). E/G' is associated to the G-bundle ξ.

Let h be the map from the set of G'-bundles over E/G' to the set of G-bundles over E/G'. Then

$$h\,\check{\xi} = \varrho^*\,\xi\,. \tag{6}$$

(After "lifting" by ϱ the structure group of ξ can be reduced to G' in a natural way.)

The proof follows from Theorem 3.4.1, Theorem 3.4.2 or Theorem 3.4.3 according to the case considered. We leave the first parts to the reader and show only how to obtain equation (6). Let W be the subspace of $E/G' \times E$ consisting of all points $c \times d$ in $E/G' \times E$ with $\varrho(c) = \pi(d)$. By 3.3, W is a principal bundle over E/G' with fibre G which is associated to $\varrho^*\,\xi$. By 3.2. d) there is a fibre bundle $\tilde{W}$ over E/G' which is constructed from $E \times G$ by the identifications $d\,a \times a^{-1}\,g = d \times g$ for all $a \in G'$,

$d \in E$, $g \in G$. $\tilde{W}$ has structure group G' and fibre G. The action of G' on G is left translation and therefore $\tilde{W}$ can be regarded as a principal bundle over E/G' with structure group and fibre G which is associated to $h\check{\xi}$. The rule $k(d \times g) = \sigma(d) \times dg$ for $d \in E$, $g \in G$ gives a well defined map $k: \tilde{W} \to W$ which is an isomorphism of principal bundles. This completes the proof of (6).

In the following theorem the notations of Theorem 3.4.4 are used to state conditions under which the structure group of ξ can be reduced to G'. We also use the terminology of 3.2. b), so that a section is assumed continuous, differentiable or holomorphic according to the case considered.

Theorem 3.4.5. *The structure group of ξ can be reduced to G' if and only if the fibre bundle E/G' over X has a section s.*

If a section s of E/G' is given, then the G'-bundle

$$\eta = s^*(\check{\xi})$$

is mapped to ξ by the embedding $G' \to G$. In this case there is an open covering $\mathfrak{U} = \{U_i\}_{i \in I}$ of X and a system of admissible charts $U_i \times G$ for E such that the coordinate transformations

$$g_{ij}: U_i \cap U_j \to G$$

map $U_i \cap U_j$ to the subgroup G' of G and such that, with respect to every chart $U_i \times G$, the section s associates to $u \in U_i$ the point of E/G' represented by $u \times e$ (here $e \in G$ is the identity element). The cocycle $\{g_{ij}\}$ represents the G-bundle ξ if the g_{ij} are regarded as maps to G, and represents the G'-bundle η if the g_{ij} are regarded as maps to G'.

Proofs of the theorems in this section can be found in STEENROD [1] and HOLMANN [1]. The essential fact in the continuous case is the assumption (5) that G/G' admits a local section. In the other two cases the analogous assumption is not necessary, because a local section always exists.

3.5. The action of the complex LIE group $\mathbf{GL}(q, \mathbf{C})$ on the complex vector space $\mathbf{C}_q$ (see 0.9) is continuous and effective. A *vector bundle* over X is a fibre bundle W over X with structure group $\mathbf{GL}(q, \mathbf{C})$ and fibre $\mathbf{C}_q$. This defines continuous vector bundles over a topological space X, differentiable vector bundles over a differentiable manifold X, and complex analytic vector bundles over a complex manifold X [see 3.2. b)]. If $q = 1$, W is called a *line bundle.*

The coordinate transformations between two admissible charts of W preserve the vector space structure on each fibre of W. Addition of points on a fibre, and multiplication of a point by a complex number, are therefore defined. Every fibre is a complex vector space. It follows that addition of sections over an open set U, and multiplication of a section over U by complex number, are defined. These operations remain within the domain of continuous, differentiable or holomorphic sections over U according to the case considered [see 3.2. b)]. Therefore the following sheaves over X can be defined:

I) $\mathfrak{C}(W)$ = *sheaf of germs of local continuous sections of a continuous vector bundle W over a topological space X.*

The canonical presheaf of $\mathfrak{C}(W)$ associates to each open set U of X the $\mathbf{C}$-module of all continuous sections of W over U. Similarly:

II) $\mathfrak{A}(W)$ = *sheaf of germs of local differentiable sections of a differentiable vector bundle W over a differentiable manifold X.*

III) $\Omega(W)$ = *sheaf of germs of local holomorphic sections of a complex analytic vector bundle W over a complex manifold X.*

The sheaf $\mathfrak{C}(W)$ is fine if X is paracompact. The sheaf $\mathfrak{A}(W)$ is fine. In both cases local sections can be multiplied by the (continuous or differentiable) functions φ_i of a partition of unity to define sheaf homomorphisms h_i (see 2.11).

Let W be a vector bundle associated to a (continuous, differentiable or complex analytic) $\mathbf{GL}(q, \mathbf{C})$-bundle ξ over X. The following construction gives a principal bundle E over X with structure group and fibre $\mathbf{GL}(q, \mathbf{C})$ which is associated to L:

The fibre of E over $x \in X$ is the set of all isomorphisms between the fixed vector space $\mathbf{C}_q$ and the fibre W_x of W over x.

Vector bundles W with $\mathbf{GL}(q, \mathbf{R})$ or $\mathbf{GL}^+(q, \mathbf{R})$ as structure group and $\mathbf{R}^q$ as fibre (see 0.9) are defined similarly. The construction of a principal bundle E from W follows just as for vector bundles with fibre $\mathbf{C}_q$.

3.6. a). Let A, B be arbitrary finite dimensional vector spaces over a field $\mathbf{K}$. The direct sum $A \oplus B$ and the tensor product $A \otimes B$ are again vector spaces over $\mathbf{K}$ of dimension $\dim(A \oplus B) = \dim A + \dim B$ and $\dim(A \otimes B) = \dim A \dim B$. Vectors $a \in A$, $b \in B$ define vectors $a \oplus b \in A \oplus B$ and $a \otimes b \in A \otimes B$. The product $a \otimes b$ is linear in each factor, and the vector space $A \otimes B$ is generated by the elements of the form $a \otimes b$. There is also a vector space $\mathrm{Hom}(A, B)$ over $\mathbf{K}$, whose elements are the homomorphisms (linear maps) from A to B. For each finite dimensional vector space A over $\mathbf{K}$ the dual vector space A^* of linear forms is defined. $A^* = \mathrm{Hom}(A, \mathbf{K})$ by definition and $\dim(A^*) = \dim(A)$. The vector space $\lambda^p A$ of p-vectors is also defined. Vectors $a_1, a_2, \ldots, a_p \in A$ define a vector $a_1 \wedge a_2 \wedge \cdots \wedge a_p \in \lambda^p A$, which depends linearly on each factor. A permutation of the factors $a_1, a_2, \ldots, a_p$ multiplies $a_1 \wedge a_2 \wedge \cdots \wedge a_p$ by the sign of the permutation and $a_1 \wedge a_2 \wedge \cdots \wedge a_p = 0$ if two factors agree. The elements of the form $a_1 \wedge a_2 \wedge \cdots \wedge a_p$ generate $\lambda^p A$. If $\dim(A) = q$ then $\dim(\lambda^p A) = \binom{q}{p}$. (For full details of these definitions from multilinear algebra see Bourbaki, Algèbre, Chap. II.)

3.6. b). Let W be a vector bundle over X. The fibre W_x over the point $x \in X$ is a complex vector space isomorphic to the typical fibre $\mathbf{C}_q$. Let W' be another vector bundle over X with fibre W'_x and typical fibre $\mathbf{C}_{q'}$ (see 3.5).

It is possible to define in a natural way the vector bundles $W \oplus W'$ (WHITNEY sum of W and W'), $W \otimes W'$ (tensor product), $\operatorname{Hom}(W, W')$, W^* (dual bundle) and $\lambda^p W$ (bundle of p-vectors). The fibres of these vector bundles over the point $x \in X$ are respectively the complex vector spaces $W_x \oplus W'_x$, $W_x \otimes W'_x$, $\operatorname{Hom}(W_x, W'_x)$, W^*_x and $\lambda^p W_x$. The vector bundle $\lambda^p(W^*)$ is called the bundle of p-forms of W.

In terms of admissible charts $U \times \mathbf{C}_q$ for W and $U \times \mathbf{C}_{q'}$ for W' the product $U \times (\mathbf{C}_q \otimes \mathbf{C}_{q'})$ is an admissible chart for $W \otimes W'$. Coordinate transformations of W, W' induce coordinate transformations of $W \otimes W'$ in a natural way. Similarly in the other cases. This is a general principle formulated by MILNOR (compare LANG [1], Chap. III, § 4 or MILNOR, Der Ring der Vektorraumbündel eines topologischen Raumes, Bonn 1959, lecture notes by P. DOMBROWSKI).

If W and W' are both continuous, differentiable, or complex analytic then so are the new vector bundles defined above. The following theorem holds in the continuous, in the differentiable, and in the complex analytic case.

Theorem 3.6.1. *Let W, W', W'' be vector bundles over X. There are isomorphisms*

$$(W \oplus W') \oplus W'' \cong W \oplus (W' \oplus W''), \qquad W \oplus W' \cong W' \oplus W,$$
$$(W \otimes W') \otimes W'' \cong W \otimes (W' \otimes W''), \qquad W \otimes W' \cong W' \otimes W,$$
$$(W \oplus W') \otimes W'' \cong (W \otimes W'') \oplus (W' \otimes W''),$$
$$(W \oplus W')^* \cong W^* \oplus (W')^*, \qquad (W \otimes W')^* \cong W^* \otimes (W')^*,$$
$$\operatorname{Hom}(W, W') \cong W^* \otimes W', \qquad (W^*)^* \cong W.$$

If W has typical fibre $\mathbf{C}_n$ then for all $0 \leqq p \leqq n$,

$$(\lambda^p W)^* \cong \lambda^p(W^*), \quad \lambda^n(W^*) \otimes \lambda^p W \cong \lambda^{n-p}(W^*).$$

For the proof of Theorem 3.6.1 see the index of BOURBAKI, Algèbre, Chap. III under the heading *Isomorphisme canonique.*

The operations of WHITNEY sum, tensor product, etc., defined in this section for vector bundles with a complex vector space as fibre, can be defined in exactly the same way for vector bundles with a real vector space as fibre. Theorem 3.6.1 holds similarly.

3.6. c). Let ξ be a continuous, differentiable or complex analytic $\mathbf{GL}(q, \mathbf{C})$-bundle over X and ξ' a corresponding $\mathbf{GL}(q', \mathbf{C})$-bundle over X. We now define a $\mathbf{GL}(q + q', \mathbf{C})$-bundle $\xi \oplus \xi'$ (WHITNEY sum of ξ and ξ'), and a $\mathbf{GL}(q\,q', \mathbf{C})$-bundle $\xi \otimes \xi'$ (tensor product of ξ and ξ'). These bundles are again continuous, differentiable or complex analytic according to the case considered.

Let W, W' be vector bundles associated to ξ, ξ'. Then $\xi \oplus \xi'$ is defined as the $\mathbf{GL}(q+q', \mathbf{C})$-bundle determined by $W \oplus W'$. It depends only on ξ and ξ'. Let $\mathfrak{U} = \{U_i\}_{i \in I}$ be an open covering of X for which ξ, ξ' can be represented by $\mathfrak{U}$-cocycles $\{g_{ij}\}$, $\{g'_{ij}\}$,

$$g_{ij}: U_i \cap U_j \to \mathbf{GL}(q, \mathbf{C}), \quad g'_{ij}: U_i \cap U_j \to \mathbf{GL}(q', \mathbf{C}).$$

Then the $\mathbf{GL}(q+q', \mathbf{C})$-bundle $\xi \oplus \xi'$ is represented by a $\mathfrak{U}$-cocycle $\{h_{ij}\}$ where

$$h_{ij}(x) = \begin{pmatrix} g_{ij}(x) & 0 \\ 0 & g'_{ij}(x) \end{pmatrix} \in \mathbf{GL}(q+q', \mathbf{C}) \quad \text{for} \quad x \in U_i \cap U_j.$$

Similarly $\xi \otimes \xi'$ is defined as the $\mathbf{GL}(q\,q', \mathbf{C})$-bundle determined by $W \otimes W'$. It is represented by a $\mathfrak{U}$-cocycle $\{h_{ij}\}$ where

$$h_{ij}(x) = g_{ij}(x) \otimes g'_{ij}(x) \in \mathbf{GL}(q\,q', \mathbf{C}) \quad \text{for} \quad x \in U_i \cap U_j$$

and where $\otimes$ denotes KRONECKER product of matrices.

For each continuous, differentiable or complex analytic $\mathbf{GL}(q, \mathbf{C})$-bundle ξ over X the dual $\mathbf{GL}(q, \mathbf{C})$-bundle ξ^* and the $\mathbf{GL}\left(\binom{q}{p}, \mathbf{C}\right)$-bundle $\lambda^p\,\xi$ are defined. These are again continuous, differentiable or complex analytic according to the case considered. Let W be a vector bundle associated to ξ. Then ξ^* is defined as the $\mathbf{GL}(q, \mathbf{C})$-bundle determined by W^*. If ξ is represented by a $\mathfrak{U}$-cocycle $\{g_{ij}\}$ then ξ^* is represented by the $\mathfrak{U}$-cocycle $\{g^*_{ij}\}$, where

$$g^*_{ij}(x) = (g^{-1}_{ij}(x))^\dagger \in \mathbf{GL}(q, \mathbf{C}) \quad \text{for} \quad x \in U_i \cap U_j$$

is the transpose of the inverse of the matrix $g_{ij}(x)$.

Similarly $\lambda^p\,\xi$ is defined as the $\mathbf{GL}\left(\binom{q}{p}, \mathbf{C}\right)$-bundle determined by $\lambda^p\,W$. It is represented by the $\mathfrak{U}$-cocycle $\{g^{(p)}_{ij}\}$ where

$$g^{(p)}_{ij}(x) = g_{ij}(x)^{(p)} \in \mathbf{GL}\left(\binom{q}{p}, \mathbf{C}\right) \quad \text{for} \quad x \in U_i \cap U_j$$

is the p-th compound matrix (matrix of $p \times p$ minors) of the matrix $g_{ij}(x)$.

A suitable $\mathfrak{U}$-cocycle can also be obtained as follows. Choose a definite isomorphism which identifies the vector space $\lambda^p\,\mathbf{C}_q$ with the vector space $\mathbf{C}_{\binom{q}{p}}$. [Which isomorphism is chosen will be immaterial by 3.1 (2*).] The group $\mathbf{GL}(q, \mathbf{C})$ operates on $\mathbf{C}_q$ and hence on $\mathbf{C}_{\binom{q}{p}}$, giving a holomorphic homomorphism ψ_p from $\mathbf{GL}(q, \mathbf{C})$ to $\mathbf{GL}\left(\binom{q}{p}, \mathbf{C}\right)$. Then $\lambda^p\,\xi$ is represented by the cocycle $\psi_p(g_{ij})$.

We write $\mathbf{C}^* = \mathbf{GL}(1, \mathbf{C})$. Then $\lambda^0\,\xi$ is the trivial $\mathbf{C}^*$-bundle. For a $\mathbf{GL}(q, \mathbf{C})$-bundle ξ the $\mathbf{C}^*$-bundle $\lambda^q\,\xi$ is represented by the $\mathfrak{U}$-cocycle $\{g^{(q)}_{ij}\}$ where $g^{(q)}_{ij}(x)$ is the determinant of $g_{ij}(x)$ for all $x \in U_i \cap U_j$.

The definitions in this section carry over immediately for $\mathbf{GL}(q, \mathbf{R})$- and $\mathbf{GL}^+(q, \mathbf{R})$-bundles.

3.7. In the case $q = 1$ the group $\mathbf{GL}(1, \mathbf{C}) = \mathbf{C}^*$ is the multiplicative group of non-zero complex numbers. The tensor product $\xi \otimes \xi'$ of two $\mathbf{C}^*$-bundles ξ, ξ' is again a $\mathbf{C}^*$-bundle. If ξ, ξ' are represented by $\mathfrak{U}$-cocycles $\{g_{ij}\}$, $\{g'_{ij}\}$ then $\xi \otimes \xi'$ is represented by the $\mathfrak{U}$-cocycle $\{g_{ij}\, g'_{ij}\}$. (The complex valued never zero functions g_{ij}, g'_{ij} defined on $U_i \cap U_j$ are continuous, differentiable, or holomorphic according to the case considered.)

The group operation in $H^1(X, \mathbf{C}^*_{\mathfrak{c}})$, $H^1(X, \mathbf{C}^*_{\mathfrak{d}})$ and $H^1(X, \mathbf{C}^*_{\omega})$ in the sense of sheaf theory (see **2.5** and **2.6**) is therefore the tensor product. If ξ is represented by $\{g_{ij}\}$ then the inverse ξ^{-1} is the $\mathbf{C}^*$-bundle represented by $\{g_{ij}^{-1}\}$. In fact $\xi^{-1} = \xi^*$ so that $\xi \otimes \xi^* = \mathbf{1}$.

3.8. We collect here some further remarks about the $\mathbf{C}^*$-bundles considered in **2.5**. If X is paracompact there is an exact cohomology sequence

$$\cdots \to H^1(X, \mathbf{C}_{\mathfrak{c}}) \to H^1(X, \mathbf{C}^*_{\mathfrak{c}}) \xrightarrow{\delta^1_*} H^2(X, \mathbf{Z}) \to H^2(X, \mathbf{C}_{\mathfrak{c}}) \to \cdots .$$

By **2.11** the sheaf $\mathbf{C}_{\mathfrak{c}}$ is fine and the groups $H^1(X, \mathbf{C}_{\mathfrak{c}})$ and $H^2(X, \mathbf{C}_{\mathfrak{c}})$ are zero. Therefore δ^1_* is an isomorphism between the group of continuous $\mathbf{C}^*$-bundles over X and the second integer cohomology group of X.

If X is a differentiable manifold there is again an exact sequence

$$\to H^1(X, \mathbf{C}_{\mathfrak{d}}) \to H^1(X, \mathbf{C}^*_{\mathfrak{d}}) \xrightarrow{\delta^1_*} H^2(X, \mathbf{Z}) \to H^2(X, \mathbf{C}_{\mathfrak{d}}) .$$

By **2.11** the sheaf $\mathbf{C}_{\mathfrak{d}}$ is fine and therefore δ^1_* is an isomorphism between the group of differentiable $\mathbf{C}^*$-bundles and $H^2(X, \mathbf{Z})$. It then follows that the natural homomorphism

$$H^1(X, \mathbf{C}^*_{\mathfrak{d}}) \to H^1(X, \mathbf{C}^*_{\mathfrak{c}})$$

of **3.1** (1) is an isomorphism.

If X is a complex manifold there is an exact sequence

$$\to H^1(X, \mathbf{C}_{\omega}) \to H^1(X, \mathbf{C}^*_{\omega}) \xrightarrow{\delta^1_*} H^2(X, \mathbf{Z}) \to H^2(X, \mathbf{C}_{\omega}) .$$

This sequence is discussed further in **15.9**.

§ 4. Characteristic classes

Important special cases of the reduction of the structure group of a fibre bundle are discussed in **4.1**. The definition of Chern classes of a continuous $\mathbf{U}(q)$-bundle in **4.2** depends on a fundamental theorem of Borel [2] on the cohomology of classifying spaces. The Pontrjagin classes of a continuous $\mathbf{O}(q)$-bundle are defined in **4.5**.

4.1. a). The following notations are used in addition to those of 0.9. Let L_r be the r-dimensional linear subspace of $\mathbf{C}_q$ defined (with respect to the coordinates $z_1, z_2, \ldots, z_q$) by $z_{r+1} = z_{r+2} = \cdots = z_q = 0$. The invertible $q \times q$ matrices which map L_r to itself form a subgroup $\mathbf{GL}(r, q-r; \mathbf{C})$ of $\mathbf{GL}(q, \mathbf{C})$. Matrices $A \in \mathbf{GL}(r, q-r; \mathbf{C})$ have the form

$$A = \begin{pmatrix} A' & B \\ 0 & A'' \end{pmatrix}$$

where $A' \in \mathbf{GL}(r, \mathbf{C})$, $A'' \in \mathbf{GL}(q-r, \mathbf{C})$ and B is an arbitrary complex matrix with r rows and $q-r$ columns.

The subgroup $\mathbf{GL}(r, q-r; \mathbf{R})$ of $\mathbf{GL}(q, \mathbf{R})$ is defined similarly. Matrices $A \in \mathbf{GL}(r, q-r; \mathbf{R})$ have the above form with $A' \in \mathbf{GL}(r, \mathbf{R})$, $A'' \in \mathbf{GL}(q-r, \mathbf{R})$ and B an arbitrary real $r \times (q-r)$ matrix. Let $\mathbf{GL}^+(r, q-r; \mathbf{R})$ be the subgroup of those $A \in \mathbf{GL}(r, q-r; \mathbf{R})$ with $A' \in \mathbf{GL}^+(r, \mathbf{R})$ and $A'' \in \mathbf{GL}^+(q-r, \mathbf{R})$.

$$\mathfrak{G}(r, q-r; \mathbf{C}) = \mathbf{GL}(q, \mathbf{C})/\mathbf{GL}(r, q-r; \mathbf{C}) = \mathbf{U}(q)/\mathbf{U}(r) \times \mathbf{U}(q-r)$$

is the GRASSMANN manifold of r-dimensional linear subspaces of $\mathbf{C}_q$. Similarly the real GRASSMANN manifolds

$$\mathfrak{G}(r, q-r; \mathbf{R}) = \mathbf{GL}(q, \mathbf{R})/\mathbf{GL}(r, q-r; \mathbf{R}) = \mathbf{O}(q)/\mathbf{O}(r) \times \mathbf{O}(q-r)$$

$$\mathfrak{G}^+(r, q-r; \mathbf{R}) = \mathbf{GL}^+(q, \mathbf{R})/\mathbf{GL}^+(r, q-r; \mathbf{R}) = \mathbf{SO}(q)/\mathbf{SO}(r) \times \mathbf{SO}(q-r)$$

represent the r-dimensional linear subspaces of $\mathbf{R}^q$ and the r-dimensional oriented linear subspaces of $\mathbf{R}^q$ respectively.

The invertible $q \times q$ complex matrices which map L_r to itself for each r form a subgroup $\Delta(q, \mathbf{C})$ of $\mathbf{GL}(q, \mathbf{C})$. Clearly $\Delta(q, \mathbf{C})$ is the subgroup of matrices in $\mathbf{GL}(q, \mathbf{C})$ which are triangular (all coefficients below the diagonal are zero).

The group $\mathbf{T}^q = \Delta(q, \mathbf{C}) \cap \mathbf{U}(q)$ of unitary diagonal matrices is a q-dimensional torus. $\mathbf{F}(q) = \mathbf{GL}(q, \mathbf{C})/\Delta(q, \mathbf{C}) = \mathbf{U}(q)/\mathbf{T}^q$ is the manifold of "flags" in $\mathbf{C}_q$. Each such flag is a sequence $0 = E_0 \subset E_1 \subset \cdots \subset E_q = \mathbf{C}_q$ of linear subspaces ($\dim E_k = k$) of $\mathbf{C}_q$. Note that these descriptions of the GRASSMANN manifolds and the flag manifold refer to linear subspaces (*i. e.* subspace through the origin of $\mathbf{C}_q$).

4.1. b). Certain results on fibre bundles over a topological space X depend on the assumption that X is paracompact (see 2.8).

Let I denote the unit interval $0 \leqq t \leqq 1$. Two continuous maps $f_0, f_1 : X \to Y$ are *homotopic* if there exists a continuous map $F : X \times I \to Y$ such that $F(x, 0) = f_0(x)$ and $F(x, 1) = f_1(x)$ for all $x \in X$. A *cell* is a space homeomorphic to $\mathbf{R}^N$ for some N.

I) *Let X be a paracompact space, W a continuous fibre bundle over a space Y, and $f_0, f_1 : X \to Y$ homotopic maps. Then the induced bundles $f_0^* W$, $f_1^* W$* (see 3.3) *are isomorphic.*

Proofs of I) can be found in DOLD [3], 7.10; in HOLMANN [1], VI. 2.3; in CARTAN [1], Exp. VIII, for X locally compact and paracompact; in STEENROD [1], 11.5, for X locally compact with a countable basis; and in ATIYAH-BOTT [1], Prop. 1.3, for X compact and W a vector bundle.

II) *Let X be a paracompact space and A a closed subspace (possibly empty) of X. If W is a fibre bundle over X with fibre a cell then every section s of W over A can be extended to a section over X.*

If it is assumed that the section s can already be extended to an open neighbourhood of A (which is the case in most applications) then II) is a special case of a theorem of DOLD [3], 2.8. Other proofs can be found in HOLMANN [1], VI. 3.1; in CARTAN [1], Exp. VIII, for X locally compact and paracompact; in STEENROD [1], 12.2, for X normal with a countable basis; and in ATIYAH-BOTT [1], Lemma 1.1, for W a vector bundle.

Now let G be a real LIE group and G^0 a closed (LIE) subgroup for which G/G^0 is a cell. The embedding $G^0 \subset G$ induces [3.1 (2)] a map

$$H^1(X, G^0_{\mathfrak{c}}) \to H^1(X, G_{\mathfrak{c}}) \ . \tag{1}$$

III) *If X is paracompact the map* (1) *is bijective.*

Proof (STEENROD [1], 12.7): By 3.4.2 and 3.4.5 the section extension property II) implies that every fibre bundle over X with structure group G is isomorphic to a fibre bundle W with structure group G^0. Therefore (1) is surjective. Now suppose that W, W' are fibre bundles over X with structure group G^0 which are isomorphic as bundles with structure group G. There is an open covering $\{U_i\}_{i\in J}$ of X such that W, W' are given by coordinate transformations $g_{ij}: U_i \cap U_j \to G^0$, $g'_{ij}: U_i \cap U_j \to G^0$ and, for some continuous functions $h_i: U_i \to G$,

$$g'_{ij} = h_i^{-1} g_{ij} h_j \quad \text{in} \quad U_i \cap U_j \quad \text{for all} \quad i, j \in J \ .$$

Now let I be the unit interval $0 \leqq t \leqq 1$, and U_i^0, U_i^1 the open sets in $X \times I$ defined by

$$U_i^0 = \{(x, t) \in X \times I;\ x \in U_i,\ 0 \leqq t < 1\} \ ,$$
$$U_i^1 = \{(x, t) \in X \times I;\ x \in U_i,\ 0 < t \leqq 1\} \ .$$

Construct a fibre bundle $\tilde{W}$ over $X \times I$ with structure group G and coordinate transformations

$$g_{ij}^{00}: U_i^0 \cap U_j^0 \to G^0$$
$$g_{ij}^{11}: U_i^1 \cap U^1 \to G^0$$
$$g_{ij}^{01}: U_i^0 \cap U_j^1 \to G$$

by $g_{ij}^{00}(x, t) = g'_{ij}(x)$, $g_{ij}^{11}(x, t) = g_{ij}(x)$, $g_{ij}^{01}(x, t) = h_i(x)\, g'_{ij}(x) = g_{ij}(x)\, h_j(x)$.

Then W has structure group G, reduced to G^0 over the closed set $A = X \times \{0\} \cup X \times \{1\}$ of $X \times I$. By 3.4.5, and II) applied to the paracompact space $X \times I$ and the closed subspace A, the fibre bundle $\tilde{W}$ is isomorphic to a fibre bundle with structure group G^0 whose restriction to $X \times \{0\}$, $X \times \{1\}$ is W', W. Consider the maps $f_0, f_1 : X \to X \times I$ with $f_0(x) = x \times \{0\}$, $f_1(x) = x \times \{1\}$. By I), W is isomorphic to W'. Therefore (1) is injective and the proof of III) is complete.

If X is a differentiable manifold then X is paracompact (2.8.2). Let G be a real LIE group and G^0 a closed (LIE) subgroup for which G/G^0 is a cell. There is [3.1 (1), (2)] a commutative diagram of maps

$$\begin{array}{ccc} H^1(X, G^0_{\mathfrak{d}}) & \to & H^1(X, G_{\mathfrak{d}}) \\ \downarrow & & \downarrow \\ H^1(X, G^0_{\mathfrak{c}}) & \to & H^1(X, G_{\mathfrak{c}}) \,. \end{array} \tag{1*}$$

IV) *Each map in* (1*) *is bijective.*

Proof: The lower horizontal arrow is a bijective map by III). A direct proof that the vertical arrows are bijective maps is given in HOLMANN [1], VI. 1.1.

When G^0 is a compact subgroup of G an alternative proof that the arrows are bijective maps can be given using the STEENROD approximation theorem *(every continuous section of a differentiable fibre bundle over X can be approximated arbitrarily closely by a differentiable section,* STEENROD [1], 6.7*)* and the classification theorem for fibre bundles with structure group G^0 (for references see the bibliographical note to Chapter One). The general case then follows by application of the theorem that the quotient space of a connected LIE group modulo a maximal compact subgroup is a cell (see STEENROD [1], 12.14).

Properties III) and IV) allow the sets in (1) and (1*) to be identified in a natural way. They can be applied in particular for

$G^0 = \mathbf{U}(q)$, $G = \mathbf{GL}(q, \mathbf{C})$

$G^0 = \mathbf{U}(r) \times \mathbf{U}(q-r)$, $G = \mathbf{GL}(r, q-r; \mathbf{C})$ or $\mathbf{GL}(r, \mathbf{C}) \times \mathbf{GL}(q-r, \mathbf{C})$

$G^0 = \mathbf{T}^q$, $G = \Delta(q, \mathbf{C})$ or $\mathbf{G} = \mathbf{C}^* \times \cdots \times \mathbf{C}^*$ q times

$G^0 = \mathbf{O}(q)$, $G = \mathbf{GL}(q, \mathbf{R})$

$G^0 = \mathbf{SO}(q)$, $G = \mathbf{GL}^+(q, \mathbf{R})$

$G^0 = \mathbf{O}(r) \times \mathbf{O}(q-r)$, $G = \mathbf{GL}(r, q-r; \mathbf{R})$ or $\mathbf{GL}(r, \mathbf{R}) \times \mathbf{GL}(q-r, \mathbf{R})$

$G^0 = \mathbf{SO}(r) \times \mathbf{SO}(q-r)$, $G = \mathbf{GL}^+(r, q-r; \mathbf{R})$
or $\mathbf{GL}^+(r, \mathbf{R}) \times \mathbf{GL}^+(q-r, \mathbf{R})$.

4.1. c). The following results hold in the continuous, differentiable and complex analytic cases.

Let $h : \mathbf{GL}(r, q-r; \mathbf{C}) \to \mathbf{GL}(r, \mathbf{C}) \times \mathbf{GL}(q-r, \mathbf{C})$ be the homomorphism defined by $h(A) = A' \times A''$ [see 4.1. a)]. The kernel of h

consists of matrices of the form $\begin{pmatrix} 1 & B \\ 0 & 1 \end{pmatrix}$, where B is a matrix with r rows and $q-r$ columns, and can therefore be identified with a complex vector space of dimension $r(q-r)$. This gives an exact sequence

$$0 \to \mathbf{C}_{r(q-r)} \to \mathbf{GL}(r, q-r; \mathbf{C}) \xrightarrow{h} \mathbf{GL}(r, \mathbf{C}) \times \mathbf{GL}(q-r, \mathbf{C}) \to 0\,. \quad (2)$$

By 3.1 (2) the homomorphism h associates, to each $\mathbf{GL}(r, q-r; \mathbf{C})$-bundle ξ a $\mathbf{GL}(r, \mathbf{C}) \times \mathbf{GL}(q-r, \mathbf{C})$-bundle: that is, a pair (ξ', ξ'') where ξ' is a $\mathbf{GL}(r, \mathbf{C})$-bundle, called a *subbundle* of ξ, and ξ'' is a $\mathbf{GL}(q-r, \mathbf{C})$-bundle, called a *quotient bundle* of ξ.

Convention: The statement "*the* $\mathbf{GL}(q, \mathbf{C})$*-bundle* ξ *has subbundle* ξ' *and quotient bundle* ξ''" means that there exists a $\mathbf{GL}(r, q-r; \mathbf{C})$-bundle which is mapped to ξ by the embedding $\mathbf{GL}(r, q-r; \mathbf{C}) \subset \mathbf{GL}(q, \mathbf{C})$ and which has subbundle ξ' and quotient bundle ξ''.

Let $\varphi_k: \boldsymbol{\Delta}(q, \mathbf{C}) \to \mathbf{C}^*$ be the homomorphism which picks out the k-th diagonal coefficient a_{kk} of the triangular matrix $A \in \boldsymbol{\Delta}(q, \mathbf{C})$. By 3.1 (2) the homomorphism φ_k associates to each $\boldsymbol{\Delta}(q, \mathbf{C})$-bundle ξ a $\mathbf{C}^*$-bundle ξ_k. The ordered set $\xi_1, \xi_2, \ldots, \xi_q$ is the set of *diagonal* $\mathbf{C}^*$-bundles of ξ.

Convention: The statement "*the* $\mathbf{GL}(q, \mathbf{C})$*-bundle* ξ *has diagonal* $\mathbf{C}^*$*-bundles* $\xi_1, \ldots, \xi_q$" means that there exists a $\boldsymbol{\Delta}(q, \mathbf{C})$-bundle which is mapped to ξ by the embedding $\boldsymbol{\Delta}(q, \mathbf{C}) \subset \mathbf{GL}(q, \mathbf{C})$ and whose ordered set of diagonal $\mathbf{C}^*$-bundles is $\xi_1, \ldots, \xi_q$.

Theorem 4.1.1. *Suppose that the* $\mathbf{GL}(q, \mathbf{C})$*-bundle* ξ *has diagonal* $\mathbf{C}^*$*-bundles* $\xi_1, \ldots, \xi_q$ *and that the* $\mathbf{GL}(q', \mathbf{C})$*-bundle* ξ' *has diagonal* $\mathbf{C}^*$*-bundles* $\xi'_1, \ldots, \xi'_{q'}$. *Then*

ξ^* *has* q *diagonal* $\mathbf{C}^*$*-bundles* $\xi_q^{-1}, \ldots, \xi_1^{-1}$,

$\xi \oplus \xi'$ *has* $q + q'$ *diagonal* $\mathbf{C}^*$*-bundles* $\xi_1, \ldots, \xi_q, \xi'_1, \ldots, \xi'_{q'}$,

$\xi \otimes \xi'$ *has* $q\,q'$ *diagonal* $\mathbf{C}^*$*-bundles* $\xi_i \otimes \xi'_j$,

$\lambda^p \xi$ *has* $\binom{q}{p}$ *diagonal* $\mathbf{C}^*$*-bundles* $\xi_{i_1} \otimes \cdots \otimes \xi_{i_p}$ $(1 \leqq i_1 < \cdots < i_p \leqq q)$.

Proof: Apply 3.6. c) and 3.1 (2*).

4.1. d). The following discussion is again valid in the continuous, differentiable and complex analytic cases.

Let W be a vector bundle (fibre $\mathbf{C}_q$) over X and let E be the principal bundle [fibre $\mathbf{GL}(q, \mathbf{C})$] of isomorphisms from $\mathbf{C}_q$ to W constructed in 3.5. By Theorem 3.4.4 there is a fibre bundle ${}^{[r]}W = E/\mathbf{GL}(r, q-r; \mathbf{C})$ over X with fibre the GRASSMANN manifold $\mathfrak{G}(r, q-r; \mathbf{C})$. The fibre ${}^{[r]}W_x$ is the GRASSMANN manifold of r-dimensional subspaces of the complex vector space W_x. The fibre bundles $W, E, {}^{[r]}W$ are all associated to the same $\mathbf{GL}(q, \mathbf{C})$-bundle ξ.

Now suppose that ${}^{[r]}W$ has a section s. Then s associates to each $x \in X$ a r-dimensional linear subspace W'_x of W_x which depends continuously (or differentiably, or complex analytically) on x. By Theorem 3.4.5 the section s determines a $\mathbf{GL}(r, q-r; \mathbf{C})$-bundle with subbundle ξ' and quotient bundle ξ''. The union of all the W'_x is a vector bundle W' over X which is associated to the $\mathbf{GL}(r, \mathbf{C})$-bundle ξ'. The union of all the $W''_x = W_x/W'_x$ is a vector bundle W'' over X which is associated to the $\mathbf{GL}(q-r, \mathbf{C})$-bundle ξ''.

Remark 1: *Every point $x \in X$ has an open neighbourhood U over which W is isomorphic to the product $U \times \mathbf{C}_q$. The isomorphism can be chosen so that W' is defined in $U \times \mathbf{C}_q$ by the equations $z_{r+1} = \cdots = z_q = 0$.* Here $\mathbf{C}_q$ is the vector space of q-ples $z_1, \ldots, z_q$ and the form of the isomorphism follows from Theorem 3.4.5.

Let W, $\tilde{W}$ be vector bundles over X. A *homomorphism* $W \to \tilde{W}$ is a continuous (or differentiable, or holomorphic) map from W to $\tilde{W}$ which maps each fibre W_x linearly to $\tilde{W}_x$. A sequence of vector bundles and homomorphisms

$$0 \to W' \to W \to W'' \to 0 \tag{3}$$

is *exact* if for each $x \in X$ the corresponding sequence

$$0 \to W'_x \to W_x \to W''_x \to 0 \tag{3*}$$

is exact. In this case we write $W'' = W/W'$ and call W' a *subbundle*, W'' a *quotient bundle*, of W.

Let W be a vector bundle (fibre $\mathbf{C}_q$) over X. A section s of the fibre bundle ${}^{[r]}W$ defines in a natural way an exact sequence (3) with a subbundle W' (fibre $\mathbf{C}_r$) and a quotient bundle W'' (fibre $\mathbf{C}_{q-r}$). Conversely any such exact sequence determines a section of ${}^{[r]}W$. If W', W and W'' are associated to the $\mathbf{GL}(r, \mathbf{C})$-bundle ξ', the $\mathbf{GL}(q, \mathbf{C})$-bundle ξ and the $\mathbf{GL}(q-r, \mathbf{C})$-bundle ξ'' respectively then there exists an exact sequence (3) if and only if ξ has subbundle ξ' and quotient bundle ξ''.

Remark 2: *By Remark* 1 *an exact sequence* (3) *satisfies the condition: every point $x \in X$ has an open neighbourhood U over which W', W and W'' are isomorphic to $U \times \mathbf{C}_r$, $U \times \mathbf{C}_q$ and $U \times \mathbf{C}_{q-r}$ respectively and over which the exact sequence* (3) *corresponds to the exact sequence*

$$0 \to \mathbf{C}_r \to \mathbf{C}_r \oplus \mathbf{C}_{q-r} \to \mathbf{C}_{q-r} \to 0\,.$$

The proofs of the following theorems are left partly to the reader (see 3.6):

Theorem 4.1.2. *Consider an exact sequence*

$$0 \to W' \to W \to W'' \to 0 \tag{3}$$

of vector bundles over X, and let $\tilde{W}$ be another vector bundle over X. Then there are exact sequences

$$0 \to \mathrm{Hom}(\tilde{W}, W') \to \mathrm{Hom}(\tilde{W}, W) \to \mathrm{Hom}(\tilde{W}, W'') \to 0 \tag{4}$$

$$0 \to W' \otimes \tilde{W} \to W \otimes \tilde{W} \to W'' \otimes \tilde{W} \to 0 \tag{5}$$

obtained in a natural way from (3). *In addition there is an exact sequence obtained by "dualising"* (3)

$$0 \to (W'')^* \to W^* \to (W')^* \to 0 . \tag{6}$$

Theorem 4.1.3. *An exact sequence*

$$0 \to F \to W \to W'' \to 0$$

of vector bundles over X with F a line bundle determines in a natural way an exact sequence

$$0 \to \lambda^{p-1} W'' \otimes F \to \lambda^p W \to \lambda^p W'' \to 0 . \tag{7}$$

Proof: There are natural homomorphisms $\lambda^p W \to \lambda^p W''$ and $\lambda^{p-1} W \otimes F \to \lambda^p W$. The latter is zero on the kernel of the natural homomorphism from $\lambda^{p-1} W \otimes F$ onto $\lambda^{p-1} W'' \otimes F$, and therefore induces a homomorphism $\lambda^{p-1} W'' \otimes F \to \lambda^p W$. This defines the homomorphisms in (7) in a natural way. It is easy to check that (7) is exact.

By dualising the above theorem one obtains

Theorem 4.1.3*. *An exact sequence*

$$0 \to W' \to W \to F \to 0$$

of vector bundles over X with F a line bundle determines in a natural way an exact sequence

$$0 \to \lambda^p W' \to \lambda^p W \to \lambda^{p-1} W' \otimes F \to 0 . \tag{7*}$$

4.1. e). The following results are once again valid in the continuous, differentiable and complex analytic cases.

We consider the situation discussed at the beginning of 4.1. d) and construct from the vector bundle W (fibre $\mathbf{C}_q$) a fibre bundle ${}^{\Delta}W = E/\boldsymbol{\Delta}(q, \mathbf{C})$ over X with structure group $\mathbf{GL}(q, \mathbf{C})$ and fibre the flag manifold

$$\mathbf{F}(q) = \mathbf{GL}(q, \mathbf{C})/\boldsymbol{\Delta}(q, \mathbf{C}) .$$

The fibre ${}^{\Delta}W_x$ is the manifold of flags in the complex vector space W_x. The fibre bundles W and ${}^{\Delta}W$ are associated to the same $\mathbf{GL}(q, \mathbf{C})$-bundle ξ.

Now suppose that ${}^{\Delta}W$ has a section s. Then s associates to each $x \in X$ a flag $s(x)$ in W_x which depends continuously (or differentiably, or complex analytically) on x. The flag $s(x)$ is an increasing sequence ${}_xL_0 \subset {}_xL_1 \subset \cdots \subset {}_xL_q = W_x$ of subspaces of W_x with $\dim {}_xL_r = r$ [see 4.1. a)]. For each r the union $\bigcup_{x \in X} {}_xL_r$ is by 4.1. d) a vector bundle $W_{(r)}$ over X with fibre $\mathbf{C}_r$. There are exact sequences

$$0 \to W_{(r)} \to W_{(r+1)} \to A_{r+1} \to 0$$

with A_r a line bundle and $A_1 = W_{(1)}$. We call $A_1, \ldots, A_q$ the diagonal line bundles determined by the section s. By **3.4.5** the section s determines a $\boldsymbol{\Delta}(q, \mathbf{C})$-bundle which is mapped to ξ by the embedding $\boldsymbol{\Delta}(q, \mathbf{C}) \subset \subset \mathbf{GL}(q, \mathbf{C})$. The line bundles $A_1, \ldots, A_q$ are associated to the diagonal $\mathbf{C}^*$-bundles of this $\boldsymbol{\Delta}(q, \mathbf{C})$-bundle.

Remark: *Every point $x \in X$ has an open neighbourhood U over which W is isomorphic to the product $U \times \mathbf{C}_q$ and over which $W_{(r)}$ is defined in $U \times \mathbf{C}_q$ by $z_{r+1} = \cdots = z_q = 0$* [see **3.4.5** and **4.1. d)**].

4.1. f). The following two theorems hold only in the continuous and differentiable cases. It is assumed that X is paracompact (in the differentiable case this is no restriction, since every differentiable manifold is paracompact by **2.8.2**).

Theorem 4.1.4. *If the $\mathbf{GL}(q, \mathbf{C})$-bundle ξ over X has the $\mathbf{GL}(r, \mathbf{C})$-bundle ξ' as subbundle and the $\mathbf{GL}(q-r, \mathbf{C})$-bundle ξ'' as quotient bundle then ξ is equal to the* Whitney *sum of ξ' and ξ''*:

$$\xi = \xi' \oplus \xi'' .$$

Theorem 4.1.5. *If the $\mathbf{GL}(q, \mathbf{C})$-bundle ξ over X has diagonal $\mathbf{C}^*$-bundles $\xi_1, \xi_2, \ldots, \xi_q$ then*

$$\xi = \xi_1 \oplus \xi_2 \oplus \cdots \oplus \xi_q .$$

Proofs: Both theorems follow from properties III) and IV) of **4.1 b)**. The set of $\mathbf{GL}(r, q-r; \mathbf{C})$-bundles can be identified with the set of $\mathbf{U}(r) \times \mathbf{U}(q-r)$-bundles, and hence with the set of $\mathbf{GL}(r, \mathbf{C}) \times \mathbf{GL}(q-r, \mathbf{C})$-bundles. This proves 4.1.4. The set of $\boldsymbol{\Delta}(q, \mathbf{C})$-bundles can be identified with the set of $\mathbf{T}^q$-bundles and hence with the set of $\mathbf{C}^* \times \mathbf{C}^* \times \cdots \times \mathbf{C}^*$-bundles ($q$ factors). This proves 4.1.5.

Remark: The following alternative proof makes it clearer why Theorems 4.1.4 and 4.1.5 are false in the complex analytic case (Atiyah [3]). Consider an exact sequence (3) of continuous, or differentiable, or complex analytic vector bundles over X. The exact sequence (4), with $\tilde{W} = W''$, defines a corresponding exact sequence of sheaves of germs of local sections (see **3.5** and **16.1**). Denote the corresponding cohomology exact sequence by

$$H^0(X, \operatorname{Hom}(W'', W')) \to H^0(X, \operatorname{Hom}(W'', W)) \to$$
$$\to H^0(X, \operatorname{Hom}(W'', W'')) \xrightarrow{\delta^0_*} H^1(X, \operatorname{Hom}(W'', W')) .$$

The identity homomorphism $W'' \to W''$ defines an element $I \in H^0(X, \operatorname{Hom}(W'', W''))$ and hence an element $\delta^0_*(I) \in H^1(X, \operatorname{Hom}(W'', W'))$. The exactness shows that there is a splitting homomorphism $W'' \to W$ of (3) if and only if $\delta^0_*(I) = 0$. Therefore W is isomorphic to $W' \oplus W''$ if $\delta^0_*(I) = 0$. In the continuous and differentiable cases the sheaves

$\mathfrak{C}(\mathrm{Hom}(W'', W'))$ and $\mathfrak{A}(\mathrm{Hom}(W'', W'))$ defined in 3.5 are fine and therefore $H^1(X, \mathrm{Hom}(W'', W')) = 0$. This proves 4.1.4. Repeated application of the same result proves Theorem 4.1.5.

4.1. g). The results of 4.1 d), together with Theorem 4.1.4, also hold in the real case and are summarised in

Theorem 4.1.6. *Let ξ be a $\mathbf{GL}(q, \mathbf{R})$-bundle over X and W an associated vector bundle with fibre $\mathbf{R}^q$. Consider the principal bundle E (fibre $\mathbf{GL}(q, \mathbf{R})$) of isomorphisms from $\mathbf{R}^q$ to W. The fibre bundle ${}^{[r]}W = E/\mathbf{GL}(r, q-r; \mathbf{R})$ has fibre ${}^{[r]}W_x$ over $x \in X$ the* GRASSMANN *manifold of (unoriented) linear subspaces of W_x. If ${}^{[r]}W$ has a section s then the union of subspaces $s(x)$ of W_x is a vector bundle W' over X associated to a $\mathbf{GL}(r, \mathbf{R})$-bundle ξ'. The union of the $W_x/s(x)$ is a vector bundle W'' over X associated to a $\mathbf{GL}(q-r, \mathbf{R})$-bundle ξ''. Moreover ξ is equal to the* WHITNEY *sum $\xi' \oplus \xi''$; in other words W is isomorphic to $W' \oplus W''$.*

The theorem remains true if throughout $\mathbf{GL}$ is replaced by $\mathbf{GL}^+$ and unoriented is replaced by oriented. It is also true in the differentiable case.

4.2. In this section we define the CHERN cohomology classes of a continuous $\mathbf{U}(q)$-bundle over an "admissible" space. A space X will be called *admissible* if it is locally compact, the union of a countable number of compact subsets, and finite dimensional. The first two conditions imply (2.8.2) that X is paracompact. In the third condition we use the following definition of dimension: the space X is *of dimension* $\leqq n$ if every open covering $\mathfrak{U}$ of X has a refinement $\mathfrak{V}$ such that each point of X lies in at most $n+1$ open sets of $\mathfrak{V}$. Under this definition a n-dimensional differentiable manifold (see 2.5) is of dimension n. In the sequel *it will be assumed that all bundles considered are defined over admissible spaces.*

The CHERN classes will be defined as integral cohomology classes of X. Unless otherwise stated the cohomology groups of X with coefficients in an additive group A are to be understood as the cohomology groups of X with coefficients in the constant sheaf A [see 2.5, Example 1)]. Then $H^i(X, A)$ is the i-th ČECH cohomology group (with arbitrary supports) of X with coefficients in A. The direct sum $H^*(X, A) = \sum_i H^i(X, A)$ is a graded ring with respect to cup product if A is a commutative ring. The cohomology groups of X with coefficients in a sheaf $\mathfrak{S}$ can also be defined by "alternating cochains" (SERRE [2]) and hence $H^i(X, \mathfrak{S}) = 0$ for $i > n = \dim X$. In particular $H^i(X, A) = 0$ for $i > n$. For X a locally finite polyhedron, and in particular for X a differentiable manifold, $H^i(X, A)$ is naturally isomorphic to the corresponding simplicial cohomology group (EILENBERG-STEENROD [1], p. 250).

The unitary group $\mathbf{U}(N) = 1 \times \mathbf{U}(N)$ is a normal subgroup of $\mathbf{U}(q) \times \mathbf{U}(N)$. Therefore $\mathbf{U}(q+N)/\mathbf{U}(N)$ is a principal bundle with structure group $\mathbf{U}(q)$ over the GRASSMANN manifold $\mathfrak{G}(q, N; \mathbf{C})$. The homogeneous space $\mathbf{U}(q+N)/\mathbf{U}(N)$ is the STIEFEL manifold of unitary-orthogonal q-frames at the origin of $\mathbf{C}_{q+N}$. The homotopy groups $\pi_i(\mathbf{U}(q+N)/\mathbf{U}(N))$ are zero for $1 \leqq i \leqq 2N$ (STEENROD [1], 25.7). The fibre bundle $\mathbf{U}(q+N)/\mathbf{U}(N)$ is associated to a $\mathbf{U}(q)$-bundle over $\mathfrak{G}(q, N; \mathbf{C})$ which is called the universal $\mathbf{U}(q)$-bundle.

Let X be an admissible space with $\dim X \leqq 2N$. Then the classification theorem (STEENROD [1], 19.4; CARTAN [1], Exposé VIII) implies that the $\mathbf{U}(q)$-bundles over X are in one-one correspondence with homotopy classes of continuous maps from X to $\mathfrak{G}(q, N; \mathbf{C})$. More precisely, every $\mathbf{U}(q)$-bundle over X can be induced by such a map from the universal $\mathbf{U}(q)$-bundle, and two such maps are homotopic if and only if they induce the same $\mathbf{U}(q)$-bundle. In order to define the CHERN classes of a $\mathbf{U}(q)$-bundle over X it is sufficient to define the CHERN classes of the universal $\mathbf{U}(q)$-bundle over $\mathfrak{G}(q, N; \mathbf{C})$. We adopt a slightly different approach which gives "axioms" for the CHERN classes together with a proof of uniqueness and existence. This approach avoids any confusion over signs (a comparison with other definitions of CHERN classes can be found in BOREL-HIRZEBRUCH [1]).

Axioms for the CHERN classes:

Axiom I: *For every continuous $\mathbf{U}(q)$-bundle ξ over an admissible space X and every integer $i \geqq 0$ there is a* CHERN *class $c_i(\xi) \in H^{2i}(X, \mathbf{Z})$. The class $c_0(\xi) = 1$ is the unit element.*

We write $c(\xi) = \sum_{i=0}^{\infty} c_i(\xi)$. Since X is finite dimensional this is a finite sum. The element $c(\xi)$ of the cohomology ring $H^*(X, \mathbf{Z})$ is called the (total) CHERN class of ξ. A continuous map $f: Y \to X$ induces a map $f^*: H^1(X, \mathbf{U}(q)_c) \to H^1(Y, \mathbf{U}(q)_c)$ and a homomorphism

$$f^*: H^*(X, \mathbf{Z}) \to H^*(Y, \mathbf{Z}).$$

Axiom II *(Naturality)*: $c(f^* \xi) = f^* c(\xi)$.

Axiom III: *If $\xi_1, \ldots, \xi_q$ are continuous $\mathbf{U}(1)$-bundles over X then*

$$c(\xi_1 \oplus \cdots \oplus \xi_q) = c(\xi_1) \ldots c(\xi_q).$$

Let $(z_0 : \cdots : z_n)$ be homogeneous coordinates for the complex projective space $\mathbf{P}_n(\mathbf{C})$. The open sets U_i defined by $z_i \neq 0$ form an open covering of $\mathbf{P}_n(\mathbf{C})$. Let η_n be the $\mathbf{C}^*$-bundle defined by the cocycle $\{g_{ij}\} = \{z_j z_i^{-1}\}$. Then η_n is complex analytic but can be regarded as a continuous $\mathbf{C}^*$-bundle and hence as a $\mathbf{U}(1)$-bundle over $\mathbf{P}_n(\mathbf{C})$. The hyperplane $z_0 = 0$, with the induced orientation, is a $\mathbf{P}_{n-1}(\mathbf{C})$ and represents a $(2n-2)$-dimensional integral homology class of $\mathbf{P}_n(\mathbf{C})$. The corresponding cohomology class [with respect to the natural orientation of $\mathbf{P}_n(\mathbf{C})$] is denoted by h_n. The class h_n is a generator of $H^2(\mathbf{P}_n(\mathbf{C}), \mathbf{Z}) = \mathbf{Z}$.

Axiom IV *(Normalisation)*: $c(\eta_n) = 1 + h_n$.

Remarks: Let $j: \mathbf{P}_{n-1}(\mathbf{C}) \to \mathbf{P}_n(\mathbf{C})$ be the embedding of a hyperplane. Then $j^* h_n = h_{n-1}$ and $j^* \eta_n = \eta_{n-1}$ in agreement with Axiom II. We give two geometrical interpretations of the $\mathbf{U}(1)$-bundle η_n. Let $\mathbf{P}_n(\mathbf{C})$ be embedded in $\mathbf{P}_{n+1}(\mathbf{C})$ as the hyperplane $z_{n+1} = 0$ and let $x_0 \in \mathbf{P}_{n+1}(\mathbf{C})$ be the point $(0 : \cdots : 0 : 1)$. There is a continuous map $\pi: \mathbf{P}_{n+1}(\mathbf{C}) - \{x_0\} \to \mathbf{P}_n(\mathbf{C})$ defined by $\pi(z_0 : \cdots : z_n : z_{n+1}) = (z_0 : \cdots : z_n)$. Define a homeomorphism $h_i: \pi^{-1}(U_i) \to U_i \times \mathbf{C}$ by

$$h_i(z_0 : \cdots : z_n : z_{n+1}) = (z_0 : \cdots : z_n : 0) \times \frac{z_{n+1}}{z_i}.$$

Then $h_i h_j^{-1}((z_0 : \cdots : z_n : 0) \times w) = (z_0 : \cdots : z_n : 0) \times \frac{z_j}{z_i} w$.

Therefore $\mathbf{P}_{n+1}(\mathbf{C}) - \{x_0\}$ is a vector bundle H over $\mathbf{P}_n(\mathbf{C})$ with structure group $\mathbf{C}^*$ and fibre $\mathbf{C}$ which is associated to the $\mathbf{U}(1)$-bundle η_n.

The second interpretation involves the continuous map $\pi: \mathbf{C}_{n+1} - \{0\} \to \mathbf{P}_n(\mathbf{C})$ defined by $\pi(z_0, \ldots, z_n) = (z_0 : \cdots : z_n)$. Define a homeomorphism $h_i: \pi^{-1}(U_i) \to U_i \times \mathbf{C}^*$ by $h_i(z_0, \ldots, z_n) = (z_0 : \cdots : z_n) \times z_i$. Then $h_i h_j^{-1}((z_0 : \cdots : z_n) \times w) = (z_0 : \cdots : z_n) \times \frac{z_i}{z_j} w$.

Therefore $\mathbf{C}_{n+1} - \{0\}$ is a principal bundle E with structure group $\mathbf{C}^*$ which is associated to the $\mathbf{U}(1)$-bundle η_n^{-1}. It follows that the principal bundle $\mathbf{U}(n+1)/\mathbf{U}(n)$ over the GRASSMANN manifold $\mathfrak{G}(1, n; \mathbf{C}) = \mathbf{P}_n(\mathbf{C})$ is associated to η_n^{-1}. Thus η_n^{-1} is the universal bundle over $\mathbf{P}_n(\mathbf{C})$.

Convention: By 4.1. b) (1) the continuous $\mathbf{U}(q)$-bundles over X are in one-one correspondence with the continuous $\mathbf{GL}(q, \mathbf{C})$-bundles. Differentiable $\mathbf{U}(q)$- and $\mathbf{GL}(q, \mathbf{C})$-bundles and complex analytic $\mathbf{GL}(q, \mathbf{C})$-bundles can all be regarded as continuous bundles [see 3.1 (1)]. Therefore CHERN classes are defined in these cases also. If W is a vector bundle over X with fibre $\mathbf{C}_q$ associated to a $\mathbf{GL}(q, \mathbf{C})$-bundle ξ, we call $c(\xi)$ the (total) CHERN class of W and write $c(W) = c(\xi)$.

Uniqueness of CHERN classes:

a) If $\xi \in H^1(X, \mathbf{U}(1)_c)$ there is, for n sufficiently large, a continuous map $f: X \to \mathbf{P}_n(\mathbf{C})$ such that $\xi = f^* \eta_n$. By Axioms II and IV, $c(\xi) = f^*(1 + h_n)$ is determined uniquely. In particular $c_i(\xi) = 0$ for $i > 1$.

b) Now let $\xi \in H^1(X, \mathbf{U}(q)_c)$. Construct a fibre bundle $Y_\xi \xrightarrow{\varrho} X$ with fibre $\mathbf{F}(q) = \mathbf{U}(q)/\mathbf{T}^q$ which is associated to ξ. The space Y_ξ is again admissible. By 3.4.4 and 4.1.5 the $\mathbf{U}(q)$-bundle $\varrho^* \xi$ is equal to the WHITNEY sum of q diagonal $\mathbf{U}(1)$-bundles $\xi_1, \ldots, \xi_q$ over Y_ξ whose CHERN classes $c(\xi_i) = 1 + \gamma_i$, where $\gamma_i \in H^2(Y_\xi, \mathbf{Z})$, are determined uniquely by a). Axioms II and III imply

$$\varrho^* c(\xi) = c(\varrho^* \xi) = \prod_{i=1}^{q} (1 + \gamma_i). \tag{8}$$

A spectral sequence argument shows that $\varrho^*: H^*(X, \mathbf{Z}) \to H^*(Y_\xi, \mathbf{Z})$ is a monomorphism (BOREL [2]; see also ROTHENBERG-STEENROD [1]). Therefore $c(\xi)$ is determined uniquely. In particular we have shown that if ξ is a $\mathbf{U}(q)$-bundle $c_i(\xi) = 0$ for $i > q$.

Remark: By the induction argument used later in 18.3 it is actually sufficient to know that $\varrho^*: H^*(X, \mathbf{Z}) \to H^*(Y, \mathbf{Z})$ is a monomorphism when $\varrho: Y \to X$ is a fibre bundle with fibre $\mathbf{P}_{q-1}(\mathbf{C})$ associated to a $\mathbf{U}(q)$-bundle (see GROTHENDIECK [4]).

Existence of CHERN classes:

The proof follows the same pattern as the proof of uniqueness. The CHERN classes of a $\mathbf{U}(1)$-bundle ξ are defined by a). It must be proved (from the classification theorem and the first of the remarks after Axiom IV) that $c(\xi) = f^*(1 + h_n)$ depends only on ξ and not on the choice of f and n. It is clear that $c(\xi)$ satisfies Axiom II for $\mathbf{U}(1)$-bundles ξ. The definition of $c(\xi)$ for a $\mathbf{U}(q)$-bundle ξ follows with the help of (8):

Let E be a principal bundle with $\mathbf{U}(q)$ as fibre which is associated to ξ and let $Y_\xi = E/\mathbf{T}^q$. By 3.4.4 there is a $\mathbf{T}^q$-bundle $\check{\xi}$ over Y_ξ which is mapped to $\varrho^* \xi$ by the embedding $\mathbf{T}^q \subset \mathbf{U}(q)$. We denote the diagonal $\mathbf{U}(1)$-bundles of $\check{\xi}$ by $\xi_1, \ldots, \xi_q$ and write $c(\xi_i) = 1 + \gamma_i$. Since $\varrho^*: H^*(X, \mathbf{Z}) \to H^*(Y_\xi, \mathbf{Z})$ is a monomorphism, $c(\xi)$ can be defined by (8) once it is shown that the elementary symmetric functions σ_j in the γ_i lie in the image of ϱ^*.

Let N be the normaliser of $\mathbf{T} = \mathbf{T}^q$ in $\mathbf{U}(q)$. Thus $N = \{a \in \mathbf{U}(q);\ a^{-1}\,\mathbf{T}\,a = \mathbf{T}\}$. It is known that $N/\mathbf{T}$ is a finite group Φ isomorphic to the group of permutations of q objects. Each element $\alpha \in \Phi$ (represented by $a \in N$) defines a fibre preserving homeomorphism $\tilde{\alpha}: Y_\xi \to Y_\xi$. With respect to a chart $V \times (\mathbf{U}(q)/\mathbf{T})$, where V is an open set of X, $\tilde{\alpha}$ is given by right translation:

$$\tilde{\alpha}(v \times g\,\mathbf{T}) = v \times g\,a\,\mathbf{T} = v \times g\,\mathbf{T}\,a \quad \text{for } v \in V,\ g \in \mathbf{U}(q),\ g\,\mathbf{T} \in \mathbf{U}(q)/\mathbf{T}\,.$$

Since $\tilde{\alpha}$ is fibre preserving it defines an automorphism $\tilde{\alpha}^*$ of the ring $H^*(Y_\xi, \mathbf{Z})$ whose restriction to $\varrho^* H^*(X, \mathbf{Z})$ is the identity. In addition there is an outer automorphism $t \to a^{-1}\,t\,a$ of $\mathbf{T}$ which depends only on α and induces an automorphism $\alpha^\#$ of $H^1(Y_\xi, \mathbf{T}_c)$ [see 3.1 (2)]. Since the outer automorphism is a permutation of the diagonal coefficients of the diagonal matrices $t \in \mathbf{T}$, the diagonal $\mathbf{U}(1)$-bundles of $\alpha^\# \check{\xi}$ are obtained from $\xi_1, \ldots, \xi_q$ by the same permutation. It can be shown that $\alpha^\# \check{\xi} = \tilde{\alpha}^* \check{\xi}$ (where $\tilde{\alpha}^*$ is induced from $\tilde{\alpha}$ as in 3.3). Therefore $\tilde{\alpha}^*$ permutes the diagonal $\mathbf{U}(1)$-bundles ξ_i, and the cohomology automorphism $\tilde{\alpha}^*$ permutes the γ_i [by Axiom II for $\mathbf{U}(1)$-bundles which is already established]. In this way Φ acts as the full group of permutations of $\gamma_1, \ldots, \gamma_q$. A necessary condition for an element $x \in H^*(Y_\xi, \mathbf{Z})$ to lie in $\varrho^* H^*(X, \mathbf{Z})$ is that x remains invariant under all operations of Φ. By a fundamental theorem of BOREL [2], which depends on a spectral sequence argument, the elementary symmetric functions σ_j in the γ_i actually occur as images under ϱ^*. The above condition is therefore also sufficient. The CHERN classes of ξ can now be defined by $\sigma_j = \varrho^* c_j(\xi)$. Clearly they do not depend on the choice of E and do satisfy Axioms I, II and IV.

It remains to prove Axiom III. Let ξ be a $\mathbf{U}(q)$-bundle over X which is the WHITNEY sum of $\mathbf{U}(1)$-bundles $\xi'_1, \ldots, \xi'_q$ over X. Let ξ_i be the i-th diagonal $\mathbf{U}(1)$-bundle of $\check{\xi}$. Then the fibre bundle Y_ξ has a section $s: X \to Y_\xi$ such that $s^* \xi_i = \xi'_i$ for $i = 1, \ldots, q$. Therefore

$$c(\xi) = s^* \varrho^* c(\xi) = s^* \prod_{i=1}^{q} c(\xi_i) = \prod_{i=1}^{q} c(\xi'_i) .$$

Remark: For the universal $\mathbf{U}(q)$-bundle ξ the spaces $X = \mathfrak{G}(q, N; \mathbf{C})$ and Y_ξ are triangulable. Therefore if classes $c(\xi)$ are defined for continuous $\mathbf{U}(q)$-bundles over triangulable spaces X which satisfy Axioms I—IV these classes must agree with the CHERN classes. If X is triangulable, characteristic classes $c_i(\xi) \in H^{2i}(X, \mathbf{Z})$ can be defined for a $\mathbf{U}(q)$-bundle ξ over X by obstruction theory (see STEENROD [1]). One constructs a fibre bundle $E/\mathbf{U}(i-1)$ associated to ξ with the STIEFEL manifold $\mathfrak{S}_{q,i} = \mathbf{U}(q)/\mathbf{U}(i-1)$ of unitary $(q-i+1)$-frames in $\mathbf{C}_q$ as fibre. The first non-zero homotopy group of $\mathfrak{S}_{q,i}$ is $\pi_{2i-1}(\mathfrak{S}_{q,i})$ which is infinite cyclic. This defines a first obstruction

$$c_i(\xi) \in H^{2i}(X, \pi_{2i-1}(\mathfrak{S}_{q,i}))$$

to the existence of a section of $E/\mathbf{U}(i-1)$ over the $2i$-dimensional skeleton of X. In order to represent $c_i(\xi)$ as an element of $H^{2i}(X, \mathbf{Z})$ it is necessary to choose an isomorphism between $\pi_{2i-1}(\mathfrak{S}_{q,i})$ and $\mathbf{Z}$. The generator of $\pi_{2i-1}(\mathfrak{S}_{q,i})$ which will correspond to $1 \in \mathbf{Z}$ is defined as follows. Choose a fixed $(q-i)$-frame in $\mathbf{C}_q$. The complementary subspace is a complex vector space $\mathbf{C}_i$ which is oriented. The sphere $\mathbf{S}^{2i-1}$ of unit vectors in $\mathbf{C}_i$ is therefore oriented. Each point of this sphere can be added to the fixed $(q-i)$-frame to define a $(q-i+1)$-frame in $\mathbf{C}_q$ and hence a point of $\mathfrak{S}_{q,i}$. The map from the *oriented* $\mathbf{S}^{2i-1}$ to $\mathfrak{S}_{q,i}$ defined in this way is the required generator of $\pi_{2i-1}(\mathfrak{S}_{q,i})$. This defines the characteristic classes of obstruction theory as elements of $H^{2i}(X, \mathbf{Z})$. A detailed discussion shows that they satisfy Axioms I—IV and hence agree with the CHERN classes.

4.3. Axioms I, II, III determine the CHERN classes uniquely once $c_1(\xi)$ is defined for ξ a $\mathbf{U}(1)$- or $\mathbf{C}^*$-bundle (Axiom IV). In this section we assume that the base space X is admissible and give two alternative definitions for $c_1(\xi)$.

Theorem 4.3.1. *Let $\xi \in H^1(X, \mathbf{C}_c^*)$ be a continuous $\mathbf{C}^*$-bundle over X. If $\delta_*^1 : H^1(X, \mathbf{C}_c^*) \to H^2(X, \mathbf{Z})$ is the isomorphism defined in 3.8 then $c_1(\xi) = \delta_*^1(\xi)$.*

Proof: Since δ_*^1 commutes with maps it is sufficient to prove the result $\delta_*^1(\eta_n) = h_n$ for the bundle η_n of Axiom IV. For $n \geqq 2$ the embedding $j : \mathbf{P}_{n-1}(\mathbf{C}) \to \mathbf{P}_n(\mathbf{C})$ induces an isomorphism $j^* : H^2(\mathbf{P}_n(\mathbf{C}), \mathbf{Z}) \to \to H^2(\mathbf{P}_{n-1}(\mathbf{C}), \mathbf{Z})$. Since $j^* \delta_*^1(\eta_n) = \delta_*^1(j^* \eta_n)$ and $j^* h_n = h_{n-1}$ it is sufficient to prove that $\delta_*^1(\eta_1) = h_1$ for the RIEMANN sphere $\mathbf{S}^2 = \mathbf{P}_1(\mathbf{C})$.

The cohomology class h_1 is by definition dual to the homology class represented by a single point. In simplicial cohomology h_1 is therefore represented by a cochain which associates the value 1 to one 2-simplex (oriented by the natural orientation of $\mathbf{S}^2$) and the value 0 to all other 2-simplexes. There is a natural identification between simplicial cohomology and ČECH cohomology. $\mathbf{S}^2$ can be regarded as a complex plane (closed by the point ∞) parametrised by $z = z_1/z_0$. Triangulate $\mathbf{S}^2$ as a tetrahedron so that $z = 0$ is a vertex and ∞ is in the interior of the face opposite 0. Name the other three vertices A, B, C in positive direction round the origin. The open stars S_0, S_A, S_B, S_C of the vertices of the tetrahedron form an open covering of $\mathbf{S}^2$ whose nerve is isomorphic to the tetrahedron. This isomorphism induces the natural identification between ČECH and simplicial cohomology. The $\mathbf{C}^*$-bundle η_1 can be defined by maps f_{rs} from $S_r \cap S_s$ to $\mathbf{C}^*$:

$$f_{0A} = f_{0B} = f_{0C} = z; \; f_{A0} = f_{B0} = f_{C0} = z^{-1}; \; \text{all other } f_{rs} = 1.$$

$\delta_*^1(\eta_1)$ can by definition be represented by the cocycle

$$c_{rst} = \frac{1}{2\pi i} (\log f_{rs} + \log f_{st} + \log f_{tr})$$

where for each r, s we choose a branch log of the logarithm in the simply connected domain $S_r \cap S_s$. For example choose $\log f_{0A}$ arbitrarily and choose $\log f_{0B}$, $\log f_{0C}$ by analytic continuation of $\log f_{0A}$ in a positive direction round the origin ($\log f_{A0} = -\log f_{0A}, \ldots$). For r and s both non-zero, $\log f_{rs} = 0$. Therefore $c_{0CA} = 1$, $c_{rst} = +1, (-1)$ for r, s, t an even, (odd) permutation of 0, C, A and $c_{rst} = 0$ otherwise. But $0CA$ is a positive ordering of a 2-simplex with respect to the natural orientation of $\mathbf{S}^2$, and therefore c_{0CA} represents the cohomology class h_1. This completes the proof of Theorem 4.3.1.

Let ξ be a $\mathbf{U}(1)$-bundle over an oriented compact manifold X. Consider an associated fibre bundle $B \to X$ with fibre the unit disc $|z| \leqq 1$, $z \in \mathbf{C}$. An element $e^{2\pi i \phi} \in \mathbf{U}(1)$ operates on B by $z \to e^{2\pi i \phi} z$. The unit disc is oriented in a natural way. B is a manifold with boundary with an orientation induced from those of X and of the fibre. Let S denote the boundary of B. Then $S \to X$ is a fibre bundle with fibre $\mathbf{S}^1$ which is associated to ξ. Let $s : X \to B - S$ be the embedding of the manifold X as the zero section of B. Following THOM [1] consider the GYSIN homomorphism

$$s_* : H^i(X, \mathbf{Z}) \to H^{i+2}_{\mathrm{cp}}(B - S, \mathbf{Z}) , \quad i \geqq 0 .$$

The second group is cohomology with compact supports. If D_X denotes the POINCARÉ isomorphism from the cohomology groups with compact supports to the homology groups of dual dimension in X, and D_{B-S} denotes the corresponding isomorphism for cohomology and homology with compact supports in $B - S$, then $s_*(a) = D^{-1}_{B-S}(s_* D_X(a))$ for $a \in H^i(X, \mathbf{Z})$. Let $\tilde{B}$ be the compact space obtained from B by collapsing S to a point. There is a natural isomorphism g^* from $H^j_{\mathrm{cp}}(B - S, \mathbf{Z})$ to $H^j(\tilde{B}, \mathbf{Z})$ for $j > 0$. The bundle ξ over B is trivial over $B - s(X)$ and can therefore be regarded as a bundle $\tilde{\xi}$ over $\tilde{B}$. In the above notations we have

Theorem 4.3.2. *Let* $1 \in H^0(X, \mathbf{Z})$ *be the unit element. Then, under the above assumptions,*

$$g^* s_*(1) = c_1(\tilde{\xi}) \quad \textit{and} \quad s^* s_*(1) = c_1(\xi) .$$

The second equation states that the CHERN *class* $c_1(\xi)$ *is the restriction to* $s(X)$ *of the cohomology class (compact supports) corresponding to the homology class (compact supports) of* $B - S$ *given by the oriented submanifold* $s(X)$.

Proof: The second equation follows from the first. The definition given by THOM [1] shows immediately that s_* commutes with maps. The first equation need therefore be proved only for the bundle η_n over $\mathbf{P}_n(\mathbf{C})$. In this case (see the remarks after Axiom IV) $\tilde{B} = \mathbf{P}_{n+1}(\mathbf{C})$, $S = \mathbf{S}^{2n+1}$ and $\tilde{\eta}_n = \eta_{n+1}$. The orientation of B induces the natural orientation on $\mathbf{P}_{n+1}(\mathbf{C})$. Since X is the naturally oriented hyperplane $\mathbf{P}_n(\mathbf{C})$ of $\mathbf{P}_{n+1}(\mathbf{C})$ it follows that $g^* s_*(1) = h_{n+1} = c_1(\eta_{n+1}) = c_1(\tilde{\eta}_n)$.

Q. E. D.

4.4. In this section we show how to calculate the CHERN classes of the bundles ξ^*, $\xi \oplus \xi'$, $\xi \otimes \xi'$, $\lambda^p \xi$ (see 3.6) from those of ξ and ξ'. For this purpose we prove a lemma which allows all such calculations to be reduced to the case in which every bundle involved is a WHITNEY sum of $\mathbf{U}(1)$-bundles.

Lemma 4.4.1. *Let* ξ_i *be a continuous* $\mathbf{U}(q_i)$*-bundle over an admissible space* X (see 4.2) *for a finite number of values* $i = 1, \ldots, N$. *There is an admissible space* Y *and a continuous map* $\varphi : Y \to X$ *such that*

I) $\varphi^* : H^*(X, \mathbf{Z}) \to H^*(Y, \mathbf{Z})$ *is a monomorphism,*

II) $\varphi^* \xi_i$ *is a sum of* $\mathbf{U}(1)$*-bundles for all* $i = 1, \ldots, N$.

Proof: By repeated application of the construction in part b) of the proof of uniqueness of CHERN classes (4.2).

Lemma 4.4.2. *Let* ξ_1, ξ_2 *be two* $\mathbf{U}(1)$*-bundles over an admissible space* X. *Then* $c_1(\xi_1 \otimes \xi_2) = c_1(\xi_1) + c_1(\xi_2)$.

Proof: By 3.7 and Theorem 4.3.1.

We adopt the following convention. Let $a_i, b_i, c_i, \ldots$ $(i = 1, 2, \ldots)$ be commutative indeterminates. We put $a_0 = b_0 = c_0 = \cdots = 1$ and consider formal factorisations

$$\sum_{i=0}^{k} a_i x^i = \prod_{j=1}^{k} (1 + \alpha_j x), \quad \sum_{i=0}^{m} b_i x^i = \prod_{j=1}^{m} (1 + \beta_j x), \text{ etc.}$$

Every polynomial which is symmetric in each of the sets of variables $\alpha_j, \beta_j, \gamma_j, \ldots$ can be written in a unique way as a polynomial in the elementary symmetric functions $a_i, b_i, c_i, \ldots$. If particular values are substituted for $a_i, b_i, c_i, \ldots$ then the polynomial takes a well defined value. In applications the particular values will always be even dimensional elements of a cohomology ring.

Theorem 4.4.3. *Let ξ be a $\mathbf{U}(q)$-bundle and ξ' a $\mathbf{U}(q')$-bundle over an admissible space X. Consider formal factorisations*

$$\sum_{i=0}^{q} c_i(\xi) x^i = \prod_{j=1}^{q} (1 + \gamma_j x), \quad \sum_{i=0}^{q'} c_i(\xi') x^i = \prod_{k=1}^{q'} (1 + \delta_k x).$$

Then, subject to the above conventions,

$$\text{I)} \sum_{i=0}^{q} c_i(\xi^*) x^i = \prod_{j=1}^{q} (1 - \gamma_j x), \text{ i.e. } c_i(\xi^*) = (-1)^i c_i(\xi).$$

$$\text{II)} \sum_{i=0}^{q+q'} c_i(\xi \oplus \xi') x^i = \prod_{j=1}^{q} (1 + \gamma_j x) \prod_{k=1}^{q'} (1 + \delta_k x),$$

$$\text{i.e. } c(\xi \oplus \xi') = c(\xi) c(\xi').$$

$$\text{III)} \sum_{i=0}^{qq'} c_i(\xi \otimes \xi') x^i = \prod_{j,k} (1 + (\gamma_j + \delta_k) x), \quad (1 \leqq j \leqq q, 1 \leqq k \leqq q').$$

$$\text{IV)} \sum_{i} c_i(\lambda^p \xi) x^i = \prod (1 + (\gamma_{j_1} + \cdots + \gamma_{j_p}) x)$$

where the product is over all $\binom{q}{p}$ combinations with $1 \leqq j_1 < \cdots < j_p \leqq q$.

Proof: By Theorem 4.1.1, Lemma 4.4.2 and Axiom III of 4.2 the above formulae hold if ξ, ξ' are sums of $\mathbf{U}(1)$-bundles. Therefore by Lemma 4.4.1 they hold in the general case.

Remark: Formula II) is the Whitney multiplication formula (also called the "duality formula"; see for instance Chern [2]). Formula III) with $q' = 1$ implies a formula of Kundert (Ann. of Math. **54**, 215–246 (1951)). If ξ is a fixed $\mathbf{U}(q)$-bundle over X, and ξ' runs through the group of $\mathbf{U}(1)$-bundles over X, then $\xi \otimes \xi'$ runs through the set of all $\mathbf{U}(q)$-bundles over X which are identical to ξ as $\mathbf{PU}(q)$-bundles [$\mathbf{PU}(q)$ = projective unitary group]. Hence the Chern classes of all these $\mathbf{U}(q)$-bundles can be calculated. But this is precisely the content of the formula of Kundert.

4.5. The PONTRJAGIN classes of a $\mathbf{O}(q)$ bundle ξ over an admissible space X (see 4.2) are defined in this section in terms of the CHERN classes of unitary bundles. By 4.1. b). IV) this defines also the PONTRJAGIN classes of $\mathbf{GL}(q, \mathbf{R})$-bundles over X. If W is a vector bundle over X with fibre $\mathbf{R}^q$ which is associated to ξ, the PONTRJAGIN classes of W are by definition the PONTRJAGIN classes of ξ.

We use the following commutative diagrams of embeddings

$$\begin{array}{ccc} \mathbf{U}(q) & \rightarrow & \mathbf{O}(2q) \\ \downarrow & & \downarrow \\ \mathbf{GL}(q, \mathbf{C}) & \rightarrow & \mathbf{GL}(2q, \mathbf{R}), \end{array} \quad \begin{array}{ccc} \mathbf{O}(q) & \rightarrow & \mathbf{U}(q) \\ \downarrow & & \downarrow \\ \mathbf{GL}(q, \mathbf{R}) & \rightarrow & \mathbf{GL}(q, \mathbf{C}) . \end{array} \tag{9}$$

In the first diagram the horizontal arrows denote the embeddings obtained if every linear map of $\mathbf{C}_q$ (coordinates $z_1, \ldots, z_q$) is regarded as a linear map of $\mathbf{R}^{2q}$ (coordinates $x_1, \ldots, x_{2q}$) by writing $z = x_{2k-1} + i\, x_{2k}$. In the second diagram the horizontal arrows denote the embeddings obtained if every matrix with real coefficients is regarded as a matrix with complex coefficients.

The second diagram of (9) defines a map ψ from $H^1(X, \mathbf{O}(q)_c)$ to $H^1(X, \mathbf{U}(q)_c)$ [see 3.1 (2)]. If ξ is a $\mathbf{O}(q)$-bundle over X we define

$$\breve{p}(\xi) = c(\psi(\xi)) = \sum_{i=0}^{\infty} c_i(\psi(\xi)) \in H^*(X, \mathbf{Z}) \text{ and } p_i(\xi) = (-1)^i c_{2i}(\psi(\xi)) \quad .$$

It can be proved, by considering the classifying space of $\mathbf{O}(q)$, that $2c_{2i+1}(\psi(\xi)) = 0$ (BOREL [2], ROTHENBERG-STEENROD [1]). The element $p_i(\xi) \in H^{4i}(X, \mathbf{Z})$ is called the i-th PONTRJAGIN class of ξ. The sum $p(\xi) = \sum_{i=0}^{\infty} p_i(\xi)$ is called the (total) PONTRJAGIN class of ξ. The properties of the CHERN classes imply immediately that

I) $p_0(\xi) = 1$.

II) $\breve{p}(f^* \xi) = f^* \breve{p}(\xi)$ *for any continuous map* $f: Y \rightarrow X$ *and* $\mathbf{O}(q)$*-bundle* ξ *over* X.

III) $\breve{p}(\xi_1 \oplus \xi_2) = \breve{p}(\xi_1)\, \breve{p}(\xi_2)$ for $\xi_1 \in H^1(X, \mathbf{O}(q_1)_c)$ and $\xi_2 \in H^1(X, \mathbf{O}(q_2)_c)$, where $\xi_1 \oplus \xi_2$ is the WHITNEY sum of ξ_1 and ξ_2.

Remark: The PONTRJAGIN class $p(\xi)$ does not satisfy the multiplication formula III). It is however true that

$p(\xi_1 \oplus \xi_2) = p(\xi_1)\, p(\xi_2)$ modulo elements of order 2 in $H^*(X, \mathbf{Z})$.

The first diagram of (9) defines a map ϱ from $H^1(X, \mathbf{U}(q)_c)$ to $H^1(X, \mathbf{O}(2q)_c)$. If ξ is a $\mathbf{U}(q)$-bundle over X then $\varrho(\xi)$ is an $\mathbf{O}(2q)$-bundle.

Theorem 4.5.1. *Let* ξ *be a* $\mathbf{U}(q)$*-bundle over* X. *Then*

$$\begin{aligned} \breve{p}(\varrho(\xi)) &= 1 - p_1(\varrho(\xi)) + p_2(\varrho(\xi)) - p_3(\varrho(\xi)) + \cdots \\ &= (1 + c_1(\xi) + c_2(\xi) + \cdots)(1 - c_1(\xi) + c_2(\xi) - \cdots) . \end{aligned}$$

If the c_i are regarded formally as elementary symmetric functions in the γ_i, then the $p_i(\varrho(\xi))$ are the elementary symmetric functions in the γ_i^2 (see 1.3).

Proof: Consider the composite embedding $\mathbf{U}(q) \to \mathbf{O}(2q) \to \mathbf{U}(2q)$. An element $A \in \mathbf{U}(q)$ defines an element of $\mathbf{U}(2q)$ which, under a well known automorphism of $\mathbf{U}(2q)$ independent of A, is mapped to $\begin{pmatrix} A & 0 \\ 0 & \bar{A} \end{pmatrix}$. Since A is unitary the complex conjugate matrix $\bar{A}$ is equal to the transpose of the inverse of A. Therefore $\psi(\varrho(\xi))$ is equal to the WHITNEY sum of ξ and ξ^* [see 3.1 (2*)]. The result now follows from the WHITNEY multiplication formula (Theorem 4.4.3).

Remark: If ξ is a $\mathbf{O}(q)$-bundle the same argument implies that $\varrho(\psi(\xi)) = \xi \oplus \xi$. If however ξ is *oriented* it is easy to check that the natural orientations on $\varrho(\psi(\xi))$ and $\xi \oplus \xi$ differ by a factor $(-1)^{\frac{1}{2}q(q-1)}$.

4.6. Let X be a (not necessarily orientable) m-dimensional differentiable manifold [see 2.5, Example 3)]. Let $\mathfrak{U} = \{U_i\}_{i \in I}$ be an open covering of X such that each U_i admits differentiable coordinates $x_1^{(i)}, \ldots, x_m^{(i)}$. The contravariant tangent $\mathbf{GL}(m, \mathbf{R})$-bundle ${}_{\mathbf{R}}\theta$ of X is the differentiable bundle represented by the $\mathfrak{U}$-cocycle $f = \{f_{ij}\}$, where

$$f_{ij} = \left(\frac{\partial x_r^{(i)}}{\partial x_s^{(j)}}\right) : U_i \cap U_j \to \mathbf{GL}(m, \mathbf{R})\,. \tag{10}$$

f_{ij} is the jacobian matrix of the coordinate transformation from U_j to U_i. The bundle ${}_{\mathbf{R}}\theta$ is an element of the cohomology set $H^1(X, \mathbf{GL}(m, \mathbf{R})_{\mathfrak{b}})$ and is called simply the tangent bundle of X.

An admissible chart $\varkappa$ of X is a differentiable homeomorphism from an open set $U_\varkappa$ of X to an open set $V_\varkappa$ of $\mathbf{R}^m$. Differentiable coordinates are defined on $U_\varkappa$ by $\varkappa$. In particular one can consider the open covering $\overline{\mathfrak{U}} = \{U_\varkappa\}_{\varkappa \in K}$, where K is the set of all admissible charts of X, and the $\overline{\mathfrak{U}}$-cocycle $f = \{f_{ij}\}$ can be defined by (10).

The cocycle f can be used to construct, by 3.2. a), a vector bundle ${}_{\mathbf{R}}\mathfrak{T}$ over X with structure group $\mathbf{GL}(m, \mathbf{R})$ and fibre $\mathbf{R}^m$. ${}_{\mathbf{R}}\mathfrak{T}$ is the vector bundle of contravariant tangent vectors of X. By 4.5 (9) the f_{ij} can be regarded as maps from $U_i \cap U_j$ to $\mathbf{GL}(m, \mathbf{C})$. The cocycle f then defines a vector bundle ${}_{\mathbf{R}}\mathfrak{T}_{\mathbf{C}}$ with fibre $\mathbf{C}_m$ called the complexification of ${}_{\mathbf{R}}\mathfrak{T}$.

Definition: The PONTRJAGIN *classes* $p_i(X) \in H^{4i}(X, \mathbf{Z})$ of a differentiable manifold X are the PONTRJAGIN classes of the tangent bundle ${}_{\mathbf{R}}\theta$ of X.

An oriented m-dimensional differentiable manifold X can be covered by open sets U_i which admit a differentiable coordinate system $x_1^{(i)}, \ldots, x_m^{(i)}$ consistent with the orientation. (The orientation is associated with the ordering $x_1^{(i)}, \ldots, x_m^{(i)}$.) The maps f_{ij} defined by (10) for such a covering give a cocycle

$$f_{ij} : U_i \cap U_j \to \mathbf{GL}^+(m, \mathbf{R})$$

which represents the contravariant tangent $\mathbf{GL}^+(m, \mathbf{R})$-bundle of the oriented manifold X. When regarded as a $\mathbf{GL}(m, \mathbf{R})$-bundle this bundle coincides with ${}_{\mathbf{R}}\theta$.

Now assume that $m = 2n$ is even and that X is again oriented.

Definition: An *almost complex structure* on the oriented differentiable manifold X is a differentiable $\mathbf{GL}(n, \mathbf{C})$-bundle θ over X which is mapped to the tangent $\mathbf{GL}^+(m, \mathbf{R})$-bundle over X by the embedding $\mathbf{GL}(n, \mathbf{C}) \to \mathbf{GL}^+(2n, \mathbf{R})$. If an almost complex structure on the oriented manifold X exists and is specified then X is called an *almost complex manifold* with tangent $\mathbf{GL}(m, \mathbf{C})$-bundle θ. The CHERN classes $c_i(X) \in H^{2i}(X, \mathbf{Z})$ of X are defined to be the CHERN classes of θ.

Note that an almost complex manifold is by definition oriented in a particular way. Definitions of almost complex structure in the literature vary slightly from that given here (e. g. STEENROD [1]). The above definition is sufficient for the purposes of the present work. Theorem 4.5.1 implies immediately

Theorem 4.6.1. *The* CHERN *classes c_i of the almost complex manifold X are related to the* PONTRJAGIN *classes p_i of X (regarded as a differentiable manifold) by the equation*

$$\breve{p} = \sum_{i=0}^{\infty} (-1)^i p_i = \sum_{i=0}^{\infty} c_i \sum_{j=0}^{\infty} (-1)^j c_j .$$

4.7. Now let X be a complex manifold of complex dimension n [see 2.5, Example 4)]. An admissible chart $\varkappa$ of X is a holomorphic homeomorphism from an open set $U_\varkappa$ of X to an open set $V_\varkappa$ of $\mathbf{C}_n$. The chart $\varkappa$ defines complex coordinates $z_1^{(\varkappa)}, \ldots, z_n^{(\varkappa)}$ on $U_\varkappa$. Let $\mathfrak{U}$ be the open covering $\{U_\varkappa\}_{\varkappa \in K}$, where K is the set of all admissible charts of X. The contravariant tangent $\mathbf{GL}(n, \mathbf{C})$-bundle θ of X is the complex analytic bundle represented by the $\mathfrak{U}$-cocycle $f = \{f_{ij}\}$, where

$$f_{ij} = \left(\frac{\partial z_r^{(i)}}{\partial z_s^{(j)}}\right) : U_i \cap U_j \to \mathbf{GL}(n, \mathbf{C}) .$$

As in 4.6 (10), f_{ij} is the jacobian matrix of the holomorphic coordinate transformation from U_j to U_i.

By 3.2. a) the cocycle f can be used to construct a vector bundle $\mathfrak{T}$ over X with fibre $\mathbf{C}_n$ which is associated to θ. $\mathfrak{T}$ is the (complex analytic) vector bundle of contravariant tangent vectors of X. Similarly the cocycle $\bar{f} = \{\bar{f}_{ij}\}$ of conjugate matrices can be used to construct a (differentiable) vector bundle $\bar{\mathfrak{T}}$ over X with fibre $\mathbf{C}_n$. The vector bundles dual to $\mathfrak{T}$ and $\bar{\mathfrak{T}}$ [see 3.6. b)] are denoted by $\boldsymbol{T}$ and $\bar{\boldsymbol{T}}$. Here $\boldsymbol{T}$ is the (complex analytic) vector bundle of covariant tangent vectors of X. Note that $\bar{\mathfrak{T}}$ and $\bar{\boldsymbol{T}}$ are not complex analytic.

The complex manifold X is oriented in a natural way (see the remark in 0.2). Therefore X can be regarded as an oriented differentiable manifold with an almost complex structure given by θ.

Definition: The CHERN *classes* $c_i(X) \in H^{2i}(X, \mathbf{Z})$ of a complex manifold X are the CHERN classes of the tangent bundle θ of X.

For X regarded as a differentiable manifold the vector bundle ${}_{\mathbf{R}}\mathfrak{T}_{\mathbf{C}}$ over X with fibre $\mathbf{C}_{2n}$ is defined as in 4.6. There are differentiable isomorphisms

$$ {}_{\mathbf{R}}\mathfrak{T}_{\mathbf{C}} = \mathfrak{T} \oplus \overline{\mathfrak{T}} \tag{11}$$

$$ {}_{\mathbf{R}}\mathfrak{T}^*_{\mathbf{C}} = \boldsymbol{T} \oplus \overline{\boldsymbol{T}} \tag{12}$$

$$ \lambda^r\, {}_{\mathbf{R}}\mathfrak{T}^*_{\mathbf{C}} = \sum_{p+q=r} \lambda^p \boldsymbol{T} \otimes \lambda^p \overline{\boldsymbol{T}}\,. \tag{13}$$

Here $\lambda^p \boldsymbol{T}$ is the (complex analytic) vector bundle of covariant p-vectors of X and $\lambda^q \overline{\boldsymbol{T}} = \overline{\lambda^q \boldsymbol{T}}$. The sum in (13) is in the sense of WHITNEY sum.

A differentiable section of the vector bundle $\lambda^r\, {}_{\mathbf{R}}\mathfrak{T}^*_{\mathbf{C}}$ is a differential form of degree r with differentiable complex valued coefficients. The WHITNEY sum (13) corresponds to the unique representation of such a form as a sum of forms of degree r and type (p, q), where $p, q \geqq 0$ and $p + q = r$.

Finally we mention the (complex analytic) principal tangent bundle of the complex manifold X. It is associated to the tangent $\mathbf{GL}(n, \mathbf{C})$-bundle θ of X and is constructed by the method of 3.5. The fibre of the principal tangent bundle at $x \in X$ is the set of all isomorphisms between the fixed vector space $\mathbf{C}_n$ and the complex vector space $\mathfrak{T}_x$ of contravariant tangent vectors to X at x.

4.8. Let X be a k-dimensional differentiable submanifold of an m-dimensional differentiable manifold Y. Then by definition X is a closed subset of Y with the property: each point $x \in X$ has an open neighbourhood U in Y with differentiable coordinates $u_1, u_2, \ldots, u_m$ for which $U \cap X$ is given by the equations $u_{k+1} = \cdots = u_m = 0$.

Let $j: X \to Y$ be the embedding and consider the contravariant tangent vector bundle ${}_{\mathbf{R}}\mathfrak{T}$ of Y. Let L be the associated fibre bundle over Y with fibre $\mathfrak{G}(k, m-k; \mathbf{R})$ constructed in 4.1. g). The field of tangent k-planes to X defines a differentiable section of $j^* L$. Therefore by Theorem 4.1.6 the restriction $j^*\, {}_{\mathbf{R}}\theta(Y)$ to X of the tangent bundle ${}_{\mathbf{R}}\theta(Y)$ of Y admits a subbundle and quotient bundle in a natural way. The subbundle is precisely the tangent bundle ${}_{\mathbf{R}}\theta(X)$ of X. The quotient bundle ${}_{\mathbf{R}}\nu$ is called the normal bundle of X in Y. By Theorem 4.1.4

$$ j^*\, {}_{\mathbf{R}}\theta(Y) = {}_{\mathbf{R}}\theta(X) \oplus {}_{\mathbf{R}}\nu\,. \tag{14}$$

The corresponding result holds if X and Y are oriented. The normal bundle is then a $\mathbf{GL}^+(m-k, \mathbf{R})$-bundle. In the special case $m - k = 2$ the normal bundle can be regarded as a $\mathbf{U}(1)$-bundle by applying 4.1. b) IV) to the embedding $\mathbf{U}(1) = \mathbf{SO}(2) \subset \mathbf{GL}^+(2, \mathbf{R})$ [see 4.5 (9)]. The CHERN class of the normal bundle ${}_{\mathbf{R}}\nu$ is therefore defined.

Theorem 4.8.1. *Let $j: X \to Y$ be the embedding of an oriented compact $(m-2)$-dimensional differentiable manifold X in an oriented compact m-dimensional differentiable manifold Y. Let $h \in H^2(Y, \mathbf{Z})$ be the cohomology class defined, with respect to the given orientations, by the $(m-2)$-dimensional homology class represented by X. Let ${}_{\mathbf{R}}\nu$ be the normal bundle of X in Y. Then*

$$c_1({}_{\mathbf{R}}\nu) = j^* h\,. \tag{15}$$

Proof: By 4.1.b) IV) applied to the embedding $\mathbf{SO}(m) \subset \mathbf{GL}^+(m, \mathbf{R})$, the tangent $\mathbf{GL}^+(m, \mathbf{R})$-bundle of Y can be regarded as a $\mathbf{SO}(m)$-bundle. Hence Y admits a Riemann metric. This metric can be used to construct a closed tubular neighbourhood B of X in Y. B is a fibre bundle with fibre the unit disc $|z| \leqq 1$, $z \in \mathbf{C}$, which is associated to the $\mathbf{U}(1)$-bundle ${}_{\mathbf{R}}\nu$ (Thom [2]). Let $\check{B}$ be the compact space obtained from B by collapsing the boundary S of B to a point. Equivalently $\check{B}$ is obtained from Y by collapsing the closed subset $Y - (B - S)$ to a point. The map $r: Y \to \check{B}$ defines a cohomology homomorphism $r^*: H^*(\check{B}, \mathbf{Z}) \to H^*(Y, \mathbf{Z})$. Then, in the notation of Theorem 4.3.2,

$$j^* h = j^* r^* g^* s_*(1) = s^* s_*(1) = c_1({}_{\mathbf{R}}\nu)\,.$$

4.9. Let $X = X_k$ be a complex submanifold of the complex manifold $Y = Y_n$ $(k \leqq n)$. Then by definition X is a closed subset of Y. Each point $x \in X$ has an open neighbourhood U in Y with complex coordinates $z_1, z_2, \ldots, z_n$ for which $U \cap X$ is given by the equations $z_{k+1} = \cdots = z_n = 0$. Let $j: X \to Y$ be the embedding and consider the tangent $\mathbf{GL}(n, \mathbf{C})$-bundle $\theta(Y)$ of Y. As in 4.8 [see the discussion in 4.1. d)] the restriction $j^* \theta(Y)$ of $\theta(Y)$ to X admits a subbundle and quotient bundle. The subbundle is the tangent bundle $\theta(X)$ of X. The quotient bundle ν is the (complex analytic) normal bundle of X in Y. If all the bundles are regarded as differentiable bundles then $j^* \theta(Y)$ is the Whitney sum of $\theta(X)$ and ν.

Now consider the special case in which $X = X_{n-1}$ is a complex submanifold of $Y = Y_n$ of complex codimension 1. In this case X is called a non-singular divisor of Y. There is a covering of Y by open sets U_i such that $X \cap U_i$ is given by an equation $f_i = 0$. Here f_i is a holomorphic function defined on U_i with non-zero partial derivatives at each point $y \in U_i \cap X$. The functions $f_{ij} = f_i f_j^{-1}$ are holomorphic and never zero on $U_i \cap U_j$. The cocycle $\{f_{ij}\}$ determines a complex analytic $\mathbf{C}^*$-bundle $[X]$ over Y which depends only on the divisor X. For example the bundle η_n of **4.2** is determined by the non-singular divisor $\mathbf{P}_{n-1}(\mathbf{C})$ of $\mathbf{P}_n(\mathbf{C})$. Clearly $j^*[X]$ is the (complex analytic) normal bundle of X in Y.

Theorem 4.9.1. *Let X be a non-singular divisor of the compact complex manifold Y, and let $h \in H^2(Y, \mathbf{Z})$ be the cohomology class represented by the oriented $(2n-2)$-cycle X. Then $c_1([X]) = h$.*

Proof: We use the notations of the proof of Theorem 4.8.1. The bundle $[X]$ is trivial over $Y - X$ and therefore there is a bundle $[\tilde{X}]$ over $\tilde{B}$ such that $[X] = r^* [\tilde{X}]$. As in Theorem 4.3.2,

$$c_1([X]) = r^* c_1([\tilde{X}]) = r^*(g^* s_*(1)) = h .$$

Finally let $X = X^{2k}$ be an oriented differentiable submanifold of an almost complex manifold $Y = Y_n$ $(2k < 2n)$ and suppose that an almost complex structure on X is given. Let $j: X \to Y$ be the embedding.

Definition: X is an *almost complex submanifold* of Y if there exists a differentiable $\mathbf{GL}(n-k, \mathbf{C})$-bundle ν over X such that

I) ν is mapped to the normal bundle of X in Y (see 4.8) by the embedding $\mathbf{GL}(n-k, \mathbf{C}) \to \mathbf{GL}^+(2n-2k, \mathbf{R})$,

II) $j^* \theta(Y) = \theta(X) \oplus \nu$.

This definition of almost complex submanifold is somewhat crude but sufficient for our purposes. By 4.8, condition I) is always satisfied in the case $n - k = 1$. Clearly a complex submanifold X of a complex manifold Y is also an almost complex submanifold of Y.

4.10. The definition of the CHERN classes by obstruction theory referred to at the end of 4.2 gives the following theorem (STEENROD[1], 39.7 and 41.8). Another proof is outlined in 4.11.

Theorem 4.10.1. *Let V_n be a compact almost complex manifold and $c_n \in H^{2n}(V_n, \mathbf{Z})$ the n-th* CHERN *class of V_n. The natural orientation of V_n defines an integer $c_n[V_n]$* (see 0.3) *which is equal to the* EULER-POINCARÉ *characteristic of V_n.*

The EULER-POINCARÉ characteristic of $\mathbf{P}_n(\mathbf{C})$ is equal to $n+1$. This fact can be used to calculate the CHERN and PONTRJAGIN classes of $\mathbf{P}_n(\mathbf{C})$.

Theorem 4.10.2. *Let $h_n \in H^2(\mathbf{P}_n(\mathbf{C}), \mathbf{Z})$ be the generator defined in* 4.2. *The* CHERN *class of the complex manifold $\mathbf{P}_n(\mathbf{C})$ is $(1+h_n)^{n+1} = \sum_{i=0}^{n} \binom{n+1}{i} h_n^i$. The* PONTRJAGIN *class of the differentiable manifold $\mathbf{P}_n(\mathbf{C})$ is $(1+h_n^2)^{n+1}$.*

Proof: By Theorem 4.10.1 the formula for the CHERN class is correct for $n = 1$. Now suppose the formula is proved for $\mathbf{P}_{n-1}(\mathbf{C})$ and consider the embedding $j: \mathbf{P}_{n-1}(\mathbf{C}) \to \mathbf{P}_n(\mathbf{C})$. By Theorem 4.9.1, the WHITNEY multiplication formula, and the fact that $j^* h_n = h_{n-1}$ (see 4.2), $j^* c(\mathbf{P}_n(\mathbf{C})) = c(\mathbf{P}_{n-1}(\mathbf{C}))$. $j^*(1+h_n) = j^*(1+h_n)^{n+1}$. But $j^*: H^{2i}(\mathbf{P}_n(\mathbf{C}), \mathbf{Z}) \to H^{2i}(\mathbf{P}_{n-1}(\mathbf{C}), \mathbf{Z})$ is an isomorphism for $i \leq n-1$, and therefore

$$c(\mathbf{P}_n(\mathbf{C})) = (1+h_n)^{n+1} \text{ modulo } H^{2n}(\mathbf{P}_n(\mathbf{C}), \mathbf{Z}) .$$

By Theorem 4.10.1, $c_n(\mathbf{P}_n(\mathbf{C})) = (n+1) h_n^n$. This completes the proof by induction of the formula for the CHERN class. The formula for the PONTRJAGIN class follows immediately from Theorem 4.6.1.

4.11. Let X be a compact oriented manifold and ξ a $\mathbf{SO}(q)$-bundle over X. The construction of Theorem 4.3.2 can be used to define the EULER class $e(\xi) \in H^q(X, \mathbf{Z})$ of ξ. Let $B \to X$ be a fibre bundle associated to ξ with the disc $\mathbf{D}^q = \left\{(x_1, \ldots, x_q) \in \mathbf{R}^q;\ \sum_{i=1}^{q} x_i^2 \leq 1\right\}$ as fibre. B is a manifold with boundary with an orientation induced from the orientations of X and $\mathbf{R}^q$. The boundary S of B is a fibre bundle over X with fibre $\mathbf{S}^{q-1}$. Let $s: X \to B - S$ be the embedding of X as the zero section of B. There is a GYSIN homomorphism

$$s_* : H^i(X, \mathbf{Z}) \to H^{i+q}_{\mathbf{cp}}(B - S, \mathbf{Z}), \quad i \geqq 0$$

defined as in 4.3. Let X' be another compact oriented manifold and $f: X' \to X$ a continuous map. Then B', S', s' can be constructed as above from the $\mathbf{SO}(q)$-bundle $f^* \xi$ and there is a natural map $f: B' - S' \to B - S$. With these notations we have

Theorem 4.11.1 (THOM [1]). *The* GYSIN *homomorphism* s_* *is an isomorphism for* $i \geqq 0$ *and the diagram*

$$\begin{array}{ccc} H^i(X, \mathbf{Z}) & \xrightarrow{f^*} & H^i(X', \mathbf{Z}) \\ \downarrow s_* & & \downarrow s'_* \\ H^{i+q}_{\mathbf{cp}}(B - S, \mathbf{Z}) & \xrightarrow{f^*} & H^{i+q}_{\mathbf{cp}}(B' - S', \mathbf{Z}) \end{array}$$

is commutative.

Let $1 \in H^0(X, \mathbf{Z})$ be the unit element. The EULER class $e(\xi)$ of ξ is defined by $e(\xi) = s^* s_* 1$. By 4.1. b) the EULER class is defined also for any $\mathbf{GL}^+(q, \mathbf{R})$-bundle ξ over X.

Theorem 4.11.2. *Let* X Y *be compact oriented manifolds,* $f: Y \to X$ *a continuous map,* ξ *a* $\mathbf{SO}(q)$*-bundle over* X *and* ξ' *a* $\mathbf{SO}(q')$*-bundle over* X*. Then*

I) $2e(\xi) = 0$ *if* q *is odd,*

II) $e(f^* \xi) = f^* e(\xi)$,

III) $e(\xi \oplus \xi') = e(\xi)\, e(\xi')$,

IV) $e(\xi) = c_1(\xi)$ *if* $q = 2$.

Proof: The definition of s_* implies that $s_*(s^* b \cdot c) = b \cdot s_* c$ for $b \in H^i_{\mathbf{cp}}(B - S, \mathbf{Z})$, $c \in H^j(X, \mathbf{Z})$. Therefore $s_*(2e(\xi)) = 2s_*(s^* s_* 1) = 2s_* 1 \cdot s_* 1 = 0$ for q odd since cup product is anticommutative. Since s_* is an isomorphism, $2e(\xi) = 0$ for q odd. This proves I). II) follows from Theorem 4.11.1. To prove III) let B, B' be the unit disc bundles of ξ, ξ' and C the unit disc bundle of $\xi \oplus \xi'$. Let $t: B \to C$, $t': B' \to C$ be the embeddings defined by the direct sum $\xi \oplus \xi'$ and $u = t\, s = t'\, s'$ the embedding of X in C induced by the zero section. By Theorem 4.11.1,

$s_* s'^* 1 = t^* t'_* 1$ and therefore $u_* 1 = t_* s_* 1 = t_* (s_* s'^* 1) = t_* t^* (t'_* 1) = t'_* 1 \cdot t_* 1$.

Hence $u^* u_* 1 = s^* t^* (t'_* 1) \cdot s'^* t'^* (t_* 1)$
$= s^* s_* (s'^* 1) \cdot s'^* s'_* (s^* 1)$ by Theorem 4.11.1.

Therefore $e(\xi \oplus \xi') = e(\xi) e(\xi')$ as required. IV) follows from the isomorphism $\mathbf{SO}(2) \simeq \mathbf{U}(1)$ and Theorem 4.3.2.

Now let η be a $\mathbf{U}(q)$-bundle over X. The embedding $\mathbf{U}(q) \to \mathbf{SO}(2q)$ of 4.5 (9) defines a $\mathbf{SO}(2q)$-bundle $\varrho(\eta)$ over X. It follows from properties II), III), IV) of 4.11.2 and from the splitting method (compare the proof of uniqueness of CHERN classes in 4.2) that

$$e(\varrho(\eta)) = c_q(\eta) . \tag{16}$$

Theorem 4.11.3. *Let $j: X \to Y$ be the embedding of an oriented compact k-dimensional differentiable submanifold X in an oriented compact m-dimensional manifold Y. Let $h \in H^{m-k}(Y, \mathbf{Z})$ be the cohomology class which corresponds to the oriented cycle X and ${}_{\mathbf{R}}\nu$ the normal $\mathbf{GL}^+(m-k, \mathbf{R})$-bundle of X in Y. Then*

$$e({}_{\mathbf{R}}\nu) = j^* h . \tag{17}$$

Proof: The definition of the EULER class shows that the proof of Theorem 4.8.1 also applies to give (17).

Consider the following particular case of Theorem 4.11.3. Y is the product manifold $X \times X$, $j: X \to X \times X$ is the diagonal embedding and ${}_{\mathbf{R}}\nu$ is equal to the tangent bundle ${}_{\mathbf{R}}\theta$ of X. An algebraic calculation, due to LEFSCHETZ, shows that

$$(h \cup h)\,[X \times X] = \sum_{i=0}^{k} (-1)^i b_i(X)$$

is the alternating sum of the BETTI numbers of X. Theorem 4.11.3 therefore implies that

$$e({}_{\mathbf{R}}\theta)\,[X] = j^* h\,[X] = (h \cup h)\,[X \times X] = E(X)$$

is the EULER-POINCARÉ characteristic of X. This proves

Theorem 4.11.4. *Let X be a compact oriented differentiable manifold with tangent bundle ${}_{\mathbf{R}}\theta$. Then $e({}_{\mathbf{R}}\theta)\,[X]$ is equal to the* EULER-POINCARÉ *characteristic $E(X)$ of X.*

Theorem 4.11.4, with (16), gives a proof of Theorem 4.10.1. Theorem 4.11.3, with (16), gives the following generalisation of Theorem 4.9.1:

Let $j: X \to Y$ be the embedding of a compact complex submanifold X in a compact complex manifold Y of complex codimension q. Let $h \in H^{2q}(Y, \mathbf{Z})$ be the cohomology class represented by the oriented cycle X and ν the complex normal bundle of X in Y. Then

$$c_q(\nu) = j^* h . \tag{18}$$

Remarks: 1). The definition of the EULER class, Theorem 4.11.1, and Theorem 4.11.2, actually hold for a $\mathbf{SO}(q)$-bundle ξ over an arbitrary admissible space X (see 4.2). Therefore (16) is valid in this case also.

2). If ξ is a $\mathbf{O}(q)$-bundle the definition of s_* fails because the disc bundle B is no longer oriented in a natural way. If all cohomology groups are taken with coefficients $\mathbf{Z}_2$ then Theorem 4.11.1 remains true in this case and $s^* s_* 1 \in H^q(X, \mathbf{Z}_2)$ is the q-th WHITNEY class $w_q(\xi)$ of ξ. The total WHITNEY class $w(\xi) = \sum_{i=0}^{q} w_i(\xi)$ can be defined. It satisfies

I) *For every continuous* $\mathbf{O}(q)$*-bundle* ξ *over an admissible space* X *and every integer* $i \geqq 0$ *there is a* WHITNEY *class* $w_i(\xi) \in H^i(X, \mathbf{Z}_2)$. $w_0(\xi) = 1$ *is the unit element.*

II) $w(f^* \xi) = f^* w(\xi)$.

III) $w(\xi \oplus \xi') = w(\xi)\, w(\xi')$.

IV) $w(\eta_n) = 1 + h_n$, *where* η_n *is the* $\mathbf{O}(1)$*-bundle over* n*-dimensional real projective space* $\mathbf{P}^n(\mathbf{R})$ *defined similarly to the* $\mathbf{U}(1)$*-bundle* η_n *of* 4.2, *and* h_n *is the non-zero element of* $H^1(\mathbf{P}^n(\mathbf{R}), \mathbf{Z}_2)$.

If X is a differentiable manifold with tangent bundle ${}_{\mathbf{R}}\theta$ the WHITNEY class $w(X) = w({}_{\mathbf{R}}\theta)$ is sometimes called the STIEFEL-WHITNEY class.

The proofs of existence and uniqueness of WHITNEY classes are precisely analogous to those for CHERN classes in 4.2. There is also a definition of w_1 like that of c_1 in Theorem 4.3.1. The exact sequence

$$1 \to \mathbf{SO}(q) \to \mathbf{O}(q) \xrightarrow{\varrho} \mathbf{Z}_2 \to 1$$

defines a homomorphism $\varrho_* : H^1(X, \mathbf{O}(q)_c) \to H^1(X, \mathbf{Z}_2)$ such that $\varrho_*(\xi) = w_1(\xi)$. Hence a differentiable manifold X is orientable if and only if $w_1(X) = 0$.

3). The embedding $\mathbf{SO}(q) \to \mathbf{O}(q)$ defines the WHITNEY class and PONTRJAGIN class for a $\mathbf{SO}(q)$-bundle ξ. In this case $w_q(\xi)$ is $e(\xi)$ reduced mod 2. If now ξ is a $\mathbf{SO}(2q)$-bundle then (see 4.5) the $\mathbf{SO}(4q)$-bundle $\varrho(\psi(\xi))$ differs from $\xi \oplus \xi$ by a change in orientation $(-1)^q$ and therefore

$$p_q(\xi) = (-1)^q c_{2q}(\psi(\xi)) = (-1)^{2q} e(\xi \oplus \xi) = (e(\xi))^2 .$$

Finally, if ξ is a $\mathbf{U}(q)$-bundle over X then $\varrho(\xi)$ is a $\mathbf{SO}(2q)$-bundle. In this case $w_{2i}(\varrho(\xi))$ is the reduction mod 2 of $c_i(\xi)$ and $w_{2i+1}(\varrho(\xi)) = 0$.

Bibliographical note

Proofs of Lemma 1.5.2 and Lemma 1.7.3, together with applications of multiplicative sequences to cohomology operations can be found in ATIYAH-HIRZEBRUCH [4].

The treatment of sheaf cohomology in § 2 is entirely in terms of ČECH cohomology theory, and the exact cohomology sequence is established only for paracompact spaces X (Theorem 2.10.1). The first definition of sheaf cohomology groups which satisfy the exact cohomology sequence for arbitrary spaces X was given by GROTHENDIECK [2]. These groups are defined by homological algebra or, equivalently, by flabby resolutions (GODEMENT [1]). For X paracompact the GROTHENDIECK cohomology groups are isomorphic to the ČECH groups. For general X the two cohomology theories are related by a spectral sequence (GODEMENT [1], Chap. II, 5.9.1).

The books by STEENROD [1] and HOLMANN [1] give fuller accounts of the theory of fibre bundles. It is very convenient to replace all conditions on the base space (paracompact, admissible etc.) by suitable conditions on the bundle. Such an exposition, in terms of *numerable* bundles, has been given by DOLD [3]; moreover fibre bundles are treated as a special case of more general (not necessarily locally trivial) fibrations. The results of § 3 have been generalised in other directions by GROTHENDIECK [1], FRENKEL [1], and HOLMANN [2].

Let G be a topological group, E a principal bundle associated to a G-bundle η over a paracompact space Y, and $[X, Y]$ the set of homotopy classes [see 4.1. b)] of continuous maps $X \to Y$. Consider the property:

(*) *the map* $T: [X, Y] \to H^1(X, G_c)$ *given by* $T(f) = f^* \eta$ *is a natural equivalence.*

Then (*) holds for all paracompact spaces X if and only if E is contractible (DOLD [3], 7.5). In this case Y is unique up to homotopy equivalence and called the *classifying space* $B(G)$ of G. Such spaces always exist (MILNOR [1], DOLD [3], 8.1). E is called the universal principal bundle. In general the classifying space has infinite dimension. For example

$$B(\mathbf{U}(q)) = \lim_{N\to\infty} \mathfrak{G}(q, N; \mathbf{C}) \quad\text{and}\quad B(\mathbf{O}(k)) = \lim_{N\to\infty} \mathfrak{G}(k, N; \mathbf{R}).$$

Suppose that E is arcwise connected and the homotopy groups $\pi_i(E)$ vanish for $1 \leqq i \leqq n$. In this case proofs that (*) holds for X have been given by DOLD [3], 7.6, for X paracompact and locally the retract of a CW-complex of dimension $\leqq n$; CARTAN [1], Exp. VIII, for X locally compact, paracompact and of dimension $\leqq n$; and STEENROD [1], 19.4, for X a finite cell complex of dimension $\leqq n$. The principal bundle E is then said to be n-universal. If G is a compact LIE group such bundles always exist with a finite dimensional differentiable manifold as base space (STEENROD [1], 19.6). For example the bundle $\mathbf{U}(q + N)/\mathbf{U}(N)$ over $\mathfrak{G}(q, N; \mathbf{C})$ is $2N$-universal (see 4.2) and the bundle $\mathbf{O}(k + N)\mathbf{O}/(N)$ over $\mathfrak{G}(k, N; \mathbf{R})$ is $(N - 1)$-universal.

The basic theorems on WHITNEY classes and CHERN classes are contained in STEENROD [1]. The WHITNEY classes of a manifold can be defined, without reference to any differentiable structure, by means of STEENROD operations and are therefore topological invariants (THOM [1]). The PONTRJAGIN classes are not topological invariants (MILNOR [6]). However NOVIKOV [1] has recently proved that the rational PONTRJAGIN classes in $H^*(X, \mathbf{Q})$ are topological invariants. Definitions of the rational PONTRJAGIN classes of a combinatorial (not necessarily differentiable) manifold X have been given by THOM [3] and ROHLIN-ŠVARC [1]. For applications to algebraic geometry over more general fields it is important to avoid homotopy theory and classifying spaces as was done to some extent by the axiomatic approach in 4.2. The exposition by GROTHENDIECK [4] also defines $c_i(\xi)$ in terms of $c_1(\xi)$ by means of splitting methods. A $\mathbf{GL}(q, \mathbf{C})$-bundle ξ determines, in the notation of 13.1 c), a fibre bundle $\psi: \overline{X} \to X$ with fibre $\mathbf{P}_{q-1}(\mathbf{C})$ and an exact sequence

$$0 \to \eta \to \psi^* \xi \to \overline{\xi} \to 0$$

of bundles over $\overline{X}$. Since $\eta^* \otimes \overline{\xi}$ is a $\mathbf{GL}(q - 1, \mathbf{C})$-bundle we have

$$0 = c_q(\eta^* \otimes \overline{\xi}) = c_q(\eta^* \otimes \psi^* \xi) = y^q + y^{q-1} \psi^* c_1(\xi) + \cdots + \psi^* c_q(\xi)$$

where $y = -c_1(\eta)$. Since ψ^* is a monomorphism this formula (the "formula of HIRSCH") may be taken as the definition of the $c_i(\xi)$ for $i > 1$. The same method applies to WHITNEY classes and to other characteristic classes which occur in algebraic geometry (GROTHENDIECK [4]).

An excellent presentation of characteristic classes, which is formulated throughout in terms of singular cohomology theory and includes the combinatorial PONTRJAGIN classes has been given by MILNOR (Lectures on characteristic classes. Notes by J. STASHEFF. Princeton University 1957).

Chapter Two

The cobordism ring

In this chapter all manifolds are compact, orientable and differentiable of class C^∞. Several results from the cobordism theory of THOM [2] are stated. They are used to express the index of a manifold M^{4k} as a polynomial in the PONTRJAGIN classes of M^{4k} (Theorem 8.2.2). This result is needed in 19.5 to provide an essential step in the proof of the RIEMANN-ROCH theorem.

§ 5. PONTRJAGIN numbers

5.1. Let V^n be an oriented compact differentiable manifold. The value of an n-dimensional cohomology class x on the fundamental cycle of V^n is denoted by $x[V^n]$. If A is a (constant) additive group, and $x \in H^n(V^n, A)$ then $x[V^n] \in A$. This definition extends naturally to give $x[V^n] \in A \otimes B$ whenever $x \in H^n(V^n, A) \otimes B$ for some additive group B. The value of $x[V^n]$ depends on x and on the orientation of V^n; if V^n is connected it is determined by x up to sign.

Now let $n = 4k$ be divisible by 4 and let $p_i \in H^{4i}(V^{4k}, \mathbf{Z})$ be the PONTRJAGIN classes of V^{4k} defined in 4.6. Every product $p_{j_1} p_{j_2} \dots p_{j_r}$ of weight $k = j_1 + j_2 + \dots + j_r$ defines an integer $p_{j_1} p_{j_2} \dots p_{j_r} [V^{4k}]$. If $\pi(k)$ is the number of distinct partitions of k, there are $\pi(k)$ such integers; they are called the PONTRJAGIN numbers of V^{4k}. Consider the ring $\mathfrak{B}$ of 1.1. The module $\mathfrak{B}_k$ has a basis consisting of products of weight k. To each such basis element is associated a corresponding PONTRJAGIN number of V^{4k}, and therefore V^{4k} defines a module homomorphism from $\mathfrak{B}_k$ to the coefficient ring B under which $a \in \mathfrak{B}_k$ maps to $a[V^{4k}] \in B$. If the dimension n of V^n is not divisible by 4 all PONTRJAGIN numbers are defined to be zero.

5.2. Let V^n, W^m be two oriented manifolds, and let $V^n \times W^m$ be the product manifold oriented by the orientations on the ordered pair V^n, W^m. Then

$$_{\mathbf{R}}\theta(V^n \times W^m) = f^*{}_{\mathbf{R}}\theta(V^n) \oplus g^*{}_{\mathbf{R}}\theta(W^m)$$

where $f: V^n \times W^m \to V^n$ and $g: V^n \times W^m \to W^m$ are projection maps and $_{\mathbf{R}}\theta(V^n)$ denotes the tangent $\mathbf{GL}(n, \mathbf{R})$-bundle of V^n (see 4.6). If the PONTRJAGIN classes of V^n, W^m, $V^n \times W^m$ are denoted by p_i, p_i', p_i'' then by 4.5 the following equation holds modulo torsion in the cohomology ring of $V^n \times W^m$:

$$1 + p_1'' + p_2'' + \cdots = f^*(1 + p_1 + p_2 + \cdots)\, g^*(1 + p_1' + p_2' + \cdots)\,. \quad (1)$$

By using an indeterminate z we can write (1) as a "polynomial equation"

$$\sum_{k=0}^{\infty} p_k'' z^k = \sum_{i=0}^{\infty} f^*(p_i)\, z^i \sum_{j=0}^{\infty} g^*(p_j')\, z^j \text{ mod torsion.} \quad (2)$$

In addition we have the equation

$$(f^*(x)\, g^*(y))\, [V^n \times W^m] = x\,[V^n] \cdot y\,[W^m] \quad (3)$$

for all $x \in H^n(V^n, \mathbf{Z}) \otimes B$ and $y \in H^m(W^m, \mathbf{Z}) \otimes B$.

If V^{4k}, W^{4r} are oriented manifolds with dimension divisible by 4, equations (2) and (3) can be used to calculate the PONTRJAGIN numbers of $V^{4k} \times W^{4r}$ in terms of the PONTRJAGIN numbers of V^{4k} and W^{4r}. The result is most easily expressed in terms of the m-sequences of § 1.

Lemma 5.2.1. *Let* $\{K_j(p_1, \ldots, p_j)\}$ *be an* m*-sequence* ($K_j \in \mathfrak{B}_j$ as in 1.2). *Then*

$$K_{k+r}[V^{4k} \times W^{4r}] = K_k[V^{4k}] \cdot K_r[W^{4r}]\,.$$

Proof: Equation (2) and 1.2, (3) and (4), imply (mod torsion)

$$\sum_{j=0}^{k+r} K_j(p_1'', \ldots, p_j'')\, z^j = \sum_{i=0}^{k} f^*(K_i(p_1, \ldots, p_i))\, z^i \cdot \sum_{j=0}^{r} g^*(K_j(p_1', \ldots, p_j'))\, z^j.$$

Equating coefficients of z^{k+r} gives

$$K_{k+r}(p_1'', \ldots, p_{k+r}'') = f^*(K_k(p_1, \ldots, p_k)) \cdot g^*(K_r(p_1', \ldots, p_r'))\,.$$

The result now follows from (3).

Definition: If $\{K_j(p_1, \ldots, p_j)\}$ is an m-sequence let $K(V^{4k}) = K_k[V^{4k}]$. If n is not divisible by 4 let $K(V^n) = 0$. Then $K(V^n)$ is called the *K-genus* of the oriented manifold V^n.

By (2) and (3), the PONTRJAGIN numbers of $V^n \times W^m$ vanish unless both n and m are divisible by 4. Therefore Lemma 5.2.1 can be restated as

Lemma 5.2.2. *The K-genus is multiplicative*:

$$K(V^n \times W^m) = K(V^n) \cdot K(W^m)\,.$$

Consider in particular the m-sequences $\{L_j\}$ and $\{A_j\}$ defined in 1.5 and 1.6. The L-genus and the A-genus of V^n are rational numbers which are denoted by $L(V^n)$ and $A(V^n)$.

Remark: We will show (Theorem 8.2.2) that the L-genus of V^{4k} is equal to the "index" of V^{4k} and hence that $L(V^{4k})$ is an integer. It can also be proved that $A(V^{4k})$ is an integer. These integrality properties are highly non-trivial: look at the denominators which occur in the definitions of L_k and A_k in 1.5 and 1.6! The integrality of L implies that for every manifold V^{4k} certain integral linear combinations (with coprime coefficients) of PONTRJAGIN numbers are divisible by the integer $\mu(L_k)$ defined in 1.5.2. As a consequence, there are conditions which a set of $\pi(k)$ integers must satisfy in order to occur as the set of PONTRJAGIN numbers of a manifold V^{4k}.

§ 6. The ring $\tilde{\Omega} \otimes \mathbf{Q}$

6.1. If V^n, W^n are oriented manifolds of the same dimension, define the sum $V^n + W^n$ to be the disjoint union of V^n and W^n. The sum is oriented in a natural way, because each connected component is oriented, either by the orientation of V^n or by that of W^n. There is also an oriented manifold $-V^n$ defined by reversing the orientation of V^n. For each partition $(j_1, j_2, \ldots, j_r)$ of k we have

$$p_{j_1} p_{j_2} \cdots p_{j_r}[V^{4k} + W^{4k}] = p_{j_1} p_{j_2} \cdots p_{j_r}[V^{4k}] + p_{j_1} p_{j_2} \cdots p_{j_r}[W^{4k}]. \quad (1)$$

Since PONTRJAGIN classes are independent of orientation (4.6),

$$p_{j_1} p_{j_2} \cdots p_{j_r}[-V^{4k}] = -p_{j_1} p_{j_2} \cdots p_{j_r}[V^{4k}]. \quad (2)$$

It follows that the K-genus defined by an m-sequence $\{K_j(p_1, \ldots, p_j)\}$ satisfies

$$K(V^n + W^n) = K(V^n) + K(W^n) \quad (1^*)$$

$$K(-V^n) = -K(V^n). \quad (2^*)$$

6.2. We now introduce an equivalence relation between n-dimensional oriented manifolds:

$V^n \approx W^n$ if and only if each PONTRJAGIN number of V^n is equal to the corresponding PONTRJAGIN number of W^n. (Note that, if $n \not\equiv 0$ modulo 4, there is only one equivalence class, since by definition all PONTRJAGIN numbers vanish.)

By 6.1 the equivalence relation $\approx$ is compatible with the operations $+$, $-$, and the equivalence classes form an additive group $\tilde{\Omega}^n$. If $n \not\equiv 0$ modulo 4 then $\tilde{\Omega}^n = 0$. Let $\tilde{\Omega}$ be the direct sum of all the groups $\tilde{\Omega}^n$ so that each element $a \in \tilde{\Omega}$ is uniquely expressed in the form $a = \sum_{n=0}^{\infty} a_n$ with $a_n \in \tilde{\Omega}^n$ and $a_n = 0$ for n sufficiently large. Then

$$\tilde{\Omega} = \sum_{n=0}^{\infty} \tilde{\Omega}^n = \sum_{k=0}^{\infty} \tilde{\Omega}^{4k}. \quad (3)$$

By 5.2 the equivalence relation $\approx$ is compatible with cartesian product. This defines a product on $\tilde{\Omega}$ for which

$$\tilde{\Omega}^n \tilde{\Omega}^m \subset \tilde{\Omega}^{m+n}. \quad (4)$$

The direct sum decomposition (3) defines a grading on $\tilde{\Omega}$ and we have

Lemma 6.2.1. *$\tilde{\Omega}$ is a graded commutative torsion free ring.*

6.3. Recall that the total PONTRJAGIN class of V^{4k} may be written, using an indeterminate z as in 4.4, in the form

$$1 + p_1 z + p_2 z^2 + \cdots + p_k z^k = \prod_{i=1}^{k} (1 + \beta_i z) \,. \tag{5}$$

We then define the integer $s(V^{4k})$ for an oriented manifold V^{4k} by the formula

$$s(V^{4k}) = (\beta_1^k + \beta_2^k + \cdots + \beta_k^k)\, [V^{4k}] \,.$$

Definition: A sequence $\{V^{4k}\}$ $(k = 0, 1, 2, \ldots)$ of oriented manifolds is a *basis sequence* if $s(V^{4k}) \neq 0$ for all k.

Theorem 6.3.1. *Let $\{V^{4k}\}$ be a basis sequence of oriented manifolds and let B be a ring containing the ring of rational numbers. Then to each sequence a_k of elements of B there corresponds one and only one m-sequence $\{K_j(p_1, \ldots, p_j)\}$ with coefficients in B for which $K(V^{4k}) = a_k$.*

Proof: The m-sequences are in one-one correspondence (see 1.2) with power series $Q(z) = 1 + b_1 z + b_2 z^2 + \cdots$ with coefficients in B. Therefore it is sufficient to show that there is exactly one power series $Q(z)$ such that, for each V^{4k} in the sequence with PONTRJAGIN classes written in the form (5),

$$a_k = K_k[V^{4k}], \text{ where } K_k = \text{coefficient of } z^k \text{ in } \prod_{i=1}^{k} Q(\beta_i z) \,.$$

This equation can be written

$$a_k = s(V^{4k})\, b_k + \text{polynomial in } b_1, b_2, \ldots, b_{k-1} \text{ of weight } k \,. \tag{6_k}$$

The polynomial in (6_k) depends only on V^{4k} and has integer coefficients. The coefficients b_k can now be determined uniquely by induction.

Remark: The proof shows conversely that, *if $\{V^{4k}\}$ is a sequence of oriented manifolds for which the conclusion of* 6.3.1 *holds, then $\{V^{4k}\}$ is a basis sequence.*

Theorem 6.3.2. *The $2k$-dimensional complex projective spaces $\mathbf{P}_{2k}(\mathbf{C})$ form a basis sequence, because $s(\mathbf{P}_{2k}(\mathbf{C})) = 2k + 1$.*

Proof: Let $h \in H^2(\mathbf{P}_{2k}(\mathbf{C}), \mathbf{Z})$ be a generator. By 4.10.2 the PONTRJAGIN class of $\mathbf{P}_{2k}(\mathbf{C})$ is $(1 + h^2)^{2k+1}$. The m-sequence of the power series $1 + z^k$ defines a "genus" (5.2) which for V^{4k} has the value $s(V^{4k})$ and which clearly takes the value $2k + 1$ on $\mathbf{P}_{2k}(\mathbf{C})$.

6.4. In this section we determine the structure of the ring $\tilde{\Omega} \otimes \mathbf{Q}$. Every oriented manifold V^{4k} determines an element (V^{4k}) of $\tilde{\Omega}^{4k} \otimes \mathbf{Q}$. The definition of tensor product implies that every element of $\tilde{\Omega}^{4k} \otimes \mathbf{Q}$ can be written in the form $\frac{1}{m}(V^{4k})$ where m is an integer. The PONTRJAGIN numbers, the K-genus, and the integer $s(V^{4k})$ are all defined in a natural way for elements of $\tilde{\Omega} \otimes \mathbf{Q}$. (In the case of the K-genus it is necessary to assume that the coefficient ring B contains the ring of rational numbers.) The PONTRJAGIN numbers of an element of $\tilde{\Omega}^{4k} \otimes \mathbf{Q}$ are in general non-integral rational numbers. Two elements of $\tilde{\Omega}^{4k} \otimes \mathbf{Q}$ are equal if and only if their corresponding PONTRJAGIN numbers are equal.

Theorem 6.4.1. *Let $\{V^{4k}\}$ be a basis sequence of oriented manifolds. For each partition $(j) = (j_1, j_2, \ldots, j_r)$ of k, let*

$$V_{(j)} = V^{4j_1} \times V^{4j_2} \times \cdots \times V^{4j_r}.$$

Then every element $\alpha \in \tilde{\Omega}^{4k} \otimes \mathbf{Q}$ can be represented uniquely as a sum

$$\alpha = \sum r_{(j)} (V_{(j)}), \quad r_{(j)} \in \mathbf{Q} \tag{7}$$

over all partitions (j) of k. Moreover, to each system $a_{(j)}$ of rational numbers there corresponds an element $\alpha \in \tilde{\Omega}^{4k} \otimes \mathbf{Q}$ whose PONTRJAGIN *numbers satisfy $p_{(j)}[\alpha] = a_{(j)}$.*

Proof: By elementary facts on linear simultaneous equations it is sufficient to prove that a sum $\sum r_{(j)} (V_{(j)})$ over all partitions (j) of k is zero if and only if $r_{(j)} = 0$ for each (j). Suppose that $\sum r_{(j)} (V_{(j)}) = 0$. Let $q_1, q_2, q_3, \ldots$ be a sequence of indeterminates. By 6.3.1 we can find for each integer $t \geqq 0$ an m-sequence which takes the value q_k^t on V^{4k}. This implies that

$$\sum_{(j)} r_{(j)} \, q_{(j)}^t = 0 , \tag{8}$$

where $q_{(j)}$ denotes the product $q_{j_1} q_{j_2} \ldots q_{j_r}$ for $(j) = (j_1, j_2, \ldots, j_r)$. Since the $q_{(j)}$ are pairwise distinct, (8) implies that each $r_{(j)}$ vanishes (VANDERMONDE determinant). Q. E. D.

We also prove the following complement to Theorem 6.4.1.

Theorem 6.4.2. *Let $\{V^{4j}\}$ be an arbitrary sequence of manifolds. Then* I) *the relation $\alpha = \sum r_{(j)} (V_{(j)})$, $r_{(j)} \in \mathbf{Q}$, implies*

$$s(\alpha) = r_k \, s(V^{4k}) \tag{7*}$$

and II) *if, for all k, every element $\alpha \in \tilde{\Omega}^{4k} \otimes \mathbf{Q}$ can be represented as a sum $\alpha = \sum r_{(j)} (V_{(j)})$, $r_{(j)} \in \mathbf{Q}$, then $\{V^{4j}\}$ is a basis sequence.*

Proof of I): Let $\{K_j\}$ be the m-sequence of the power series $1 + z^k$. This m-sequence takes the value $s(\alpha)$ on elements $\alpha \in \tilde{\Omega}^{4k} \otimes \mathbf{Q}$ and the value 0 on elements of $\tilde{\Omega}^{4j} \otimes \mathbf{Q}$ with $1 \leqq j < k$. This implies (7*).

Proof of II): Suppose that, for some k, $s(V^{4k}) = 0$. Then, by I), $s(\alpha) = 0$ for all $\alpha \in \tilde{\Omega}^{4k} \otimes \mathbf{Q}$. But $s(\mathbf{P}_{2k}(\mathbf{C})) = 2k + 1$ by 6.3.2. Contradiction.

An immediate corollary to 6.3.2 and 6.4.1 is

Theorem 6.4.3. *The graded ring $\tilde{\Omega} \otimes \mathbf{Q}$ is isomorphic to the graded ring $\mathbf{Q}[z_1, z_2, \ldots]$ of polynomials in indeterminates z_i with rational coef-*

ficients. The group $\tilde{\Omega}^{4k} \otimes \mathbf{Q}$ is mapped onto the group of products of weight k. Any sequence of elements $\alpha_i \in \tilde{\Omega}^{4i} \otimes \mathbf{Q}$ with $s(\alpha_i) \neq 0$ $(i = 1, 2, \ldots)$ defines by $\alpha_i \to z_i$ an isomorphism of $\tilde{\Omega} \otimes \mathbf{Q}$ on to $\mathbf{Q}[z_1, z_2, \ldots]$, and every isomorphism of $\tilde{\Omega} \otimes \mathbf{Q}$ on to $\mathbf{Q}[z_1, z_2, \ldots]$ can be obtained in this way.

Remark: Theorem 6.4.1 implies in particular that to each system of integers $a_{(j)}$, where (j) runs through all partitions $(j_1, \ldots, j_r)$ of k, there corresponds a positive integer N_k, which depends only on k, such that the system of integers $N_k \cdot a_{(j)}$ occurs as the system of PONTRJAGIN numbers of an oriented manifold V^{4k}. We have already noted in 5.2 that not every system $a_{(j)}$ occurs in this way. This suggests the question: what is the smallest positive integer $\overline{N}_k$ for which every system $\overline{N}_k \cdot a_{(j)}$ with $a_{(j)}$ integral occurs as the system of PONTRJAGIN numbers of a V^{4k}? It follows from work of MILNOR [3] that in fact $\overline{N}_k$ is equal to the denominator $\mu(L_k)$ of the polynomial L_k (see Lemma 1.5.2).

6.5. In this section we consider homomorphisms from the ring $\tilde{\Omega} \otimes \mathbf{Q}$ to the ring $\mathbf{Q}$ of rational numbers. Let $\{K_j(p_1, \ldots, p_j)\}$ be an m-sequence with rational coefficients, and let $K(V^n)$ be the corresponding K-genus of an oriented manifold V^n. The K-genus $K(\alpha)$ is defined for any $\alpha \in \tilde{\Omega} \otimes \mathbf{Q}$ and 5.2.2 and 6.1 (1*), (2*) imply that there is a homomorphism $\tilde{\Omega} \otimes \mathbf{Q} \to \mathbf{Q}$ defined by $\alpha \to K(\alpha)$.

Conversely, any homomorphism $h: \tilde{\Omega} \otimes \mathbf{Q} \to \mathbf{Q}$ arises in this way. Let $h(V^{4k})$ be the values of h on a basis sequence $\{V^{4k}\}$. By 6.3.1 there is a unique m-sequence $\{K_j\}$ with $K(V^{4k}) = h(V^{4k})$. The elements (V^{4k}) generate the ring $\tilde{\Omega} \otimes \mathbf{Q}$, and therefore $K(\alpha) = h(\alpha)$ for every $\alpha \in \tilde{\Omega} \otimes \mathbf{Q}$. This proves

Theorem 6.5.1. *The homomorphisms $\tilde{\Omega} \otimes \mathbf{Q} \to \mathbf{Q}$ are in one-one correspondence with the m-sequences $\{K_j(p_1, \ldots, p_j)\}$ with rational coefficients, and are therefore also in one-one correspondence with formal power series with rational coefficients starting with* 1.

§ 7. The cobordism ring Ω

In § 6 we formed a ring from the set of all oriented manifolds by introducing an equivalence relation $\approx$ compatible with the operations $+$, $-$, $\times$. This equivalence relation is very artificial, and the results of § 6 consist mostly of formal algebra. The only geometrical fact used in § 6 is the existence of a basis sequence of oriented manifolds (Theorem 6.3.2). We now need a deep result from the cobordism theory of THOM which states that the equivalence relation $\approx$ has a direct geometrical significance.

7.1. Recall that the definition of oriented differentiable manifold (2.5) can be extended to include oriented differentiable manifolds with boundary. If X^{n+1} is a compact oriented differentiable manifold with boundary ∂X^{n+1}, then ∂X^{n+1} is a compact manifold with an orientation and differentiable structure induced from that of X^{n+1}.

Definition: An oriented differentiable manifold V^n *bounds* if there exists a compact oriented differentiable manifold X^{n+1} with oriented boundary $\partial X^{n+1} = V^n$. Two manifolds V^n, W^n are *cobordant* if $V^n + (-W^n)$ bounds.

The relation V^n is cobordant to W^n, $V^n \sim W^n$, is an equivalence relation compatible with the operations $+, -, \times$ defined in § 6.1. The equivalence classes of oriented n-dimensional manifolds form an additive group Ω^n whose zero element is the class of manifolds which bound. As in 6.2 we can define the direct sum

$$\Omega = \sum_{n=0}^{\infty} \Omega^n .$$

In this case

$$\Omega^n \, \Omega^m \subset \Omega^{n+m} \text{ and } \alpha \cdot \beta = (-1)^{nm} \beta \cdot \alpha \text{ for } \alpha \in \Omega^n, \ \beta \in \Omega^m \tag{1}$$

and therefore Ω is a graded anti-commutative ring, called the cobordism ring. It is not necessary, for the present application, to know the precise structure of Ω. The original results of THOM are sufficient, and are quoted in the next section.

7.2. We wish to construct an isomorphism $\Omega \otimes \mathbf{Q} \to \tilde{\Omega} \otimes \mathbf{Q}$ between the cobordism ring "modulo torsion" and the ring $\tilde{\Omega} \otimes \mathbf{Q}$ defined in §6. The first step is contained in the following theorem of PONTRJAGIN [2].

Theorem 7.2.1. *If V^n bounds then all the* PONTRJAGIN *numbers of V^n are zero.*

Proof: The PONTRJAGIN numbers of V^n are by definition zero unless $n \equiv 0$ modulo 4. Suppose that V^{4k} is the oriented boundary of X^{4k+1}, and that $j: V^{4k} \to X^{4k+1}$ is the embedding. Let $p_i \in H^{4i}(X^{4k+1}, \mathbf{Z})$ be the PONTRJAGIN classes of the tangent bundle ${}_{\mathbf{R}}\theta(X^{4k+1})$ of X^{4k+1}. Note that this bundle is also defined over points of V^{4k}; in fact, if $\mathbf{1}$ denotes the trivial line bundle,

$$j^* {}_{\mathbf{R}}\theta(X^{4k+1}) = \mathbf{1} \oplus {}_{\mathbf{R}}\theta(V^{4k})$$

where ${}_{\mathbf{R}}\theta(V^{4k})$ is the tangent bundle of V^{4k}. By 4.5 III) the PONTRJAGIN classes of V^{4k} are $j^* p_i$ and every PONTRJAGIN number of V^{4k} is the value of a $4k$-dimensional cocycle of X^{4k+1} on the cycle V^{4k}. But V^{4k} bounds and therefore every PONTRJAGIN number of V^{4k} is zero. Q. E. D.

The theorem of PONTRJAGIN states that the equivalence relation $\sim$ of 7.1 implies the equivalence relation $\approx$ of 6.2. Therefore there is a ring epimorphism $\Omega \to \tilde{\Omega}$ which induces a ring epimorphism

$$\varphi : \Omega \otimes \mathbf{Q} \to \tilde{\Omega} \otimes \mathbf{Q} . \tag{2}$$

The central result of THOM, on which all subsequent work on the cobordism ring is based, is contained in the following theorem.

Theorem 7.2.2. (THOM [2]) *The groups Ω^n are finite for $i \not\equiv 0$* modulo 4. *The group Ω^{4k} is the direct sum of $\pi(k)$* (= number of distinct partitions of k) *groups* $\mathbf{Z}$ *and a finite group.*

We are not able to give the proof of this theorem here, but make the following remarks. THOM's proof divides into two parts

I) Construction of a complex $M(\mathbf{SO}(k))$ and an isomorphism between the group Ω^i and the homotopy group $\pi_{k+i}(M(\mathbf{SO}(k)))$, $i < k$.

II) Calculation of $\pi_{k+i}(M(\mathbf{SO}(k)))$ modulo finite groups by use of the C-theory of SERRE.

The proofs in I) use isotopy and deformation arguments. Let $B(\mathbf{SO}(k))$ be the classifying space of the group $\mathbf{SO}(k)$ (see the bibliographical note to Chapter One). Associated to the universal $\mathbf{SO}(k)$-bundle over $B(\mathbf{SO}(k))$ there is a bundle with fibre $\mathbf{D}^k$, the k-dimensional disc in $\mathbf{R}^k$ defined by $\left\{(x_1, \ldots, x_k);\ \sum_{i=1}^{k} x_i^2 \leqq 1\right\}$, and bundle space $A(\mathbf{SO}(k))$. Let $M(\mathbf{SO}(k))$ be the complex obtained by identifying the boundary of $A(\mathbf{SO}(k))$ to a point. The homomorphism $\Omega^i \to \pi_{i+k}(M(\mathbf{SO}(k)))$ can now be defined. Let V^i be an oriented differentiable manifold. Since $i < k$ there is an embedding of V^i in the $(i+k)$-dimensional sphere $\mathbf{S}^{i+k}$. An isotopy argument shows that two such embeddings have isomorphic normal bundles, and hence that there is a map $f: N \to A(\mathbf{SO}(k))$ of a tubular neighbourhood N of V^i in $\mathbf{S}^{i+k}$ which maps V^i into the zero section of $A(\mathbf{SO}(k))$ and which maps the boundary ∂N of N into the boundary of $A(\mathbf{SO}(k))$. Now consider the composite map

$$\mathbf{S}^{i+k} \to \frac{\mathbf{S}^{i+k}}{\mathbf{S}^{i+k} - N} = \frac{N}{\partial N} \to \frac{A(\mathbf{SO}(k))}{\partial A(\mathbf{SO}(k))} = M(\mathbf{SO}(k)) .$$

This map defines an element of $\pi_{i+k}(M(\mathbf{SO}(k)))$ which actually depends only on the cobordism class of V^i. Deformation arguments are now used to show that the homomorphism $\Omega^i \to \pi_{i+k}(M(\mathbf{SO}(k)))$, $i < k$, is an isomorphism.

The proofs in II) depend on a computation of the cohomology of $M(\mathbf{SO}(k))$ and use properties of EILENBERG-MACLANE complexes and the STEENROD algebra.

Explicit results for $i \leqq 7$ are:

$$\Omega^0 = \mathbf{Z},\ \Omega^1 = \Omega^2 = \Omega^3 = 0,\ \Omega^4 = \mathbf{Z},\ \Omega^5 = \mathbf{Z}_2,\ \Omega^6 = \Omega^7 = 0 .$$

Theorem 7.2.2, together with the formal algebra of § 6, implies immediately

Theorem 7.2.3 (THOM [2]). *The homomorphism $\varphi: \Omega \otimes \mathbf{Q} \to \tilde{\Omega} \otimes \mathbf{Q}$ is an isomorphism, and the structure of the ring $\Omega \otimes \mathbf{Q}$ is therefore deter-*

mined by Theorem 6.4.3. *Two oriented manifolds V^{4k} and W^{4k} have the same* PONTRJAGIN *numbers if and only if some integral multiple of $V^{4k} + (-W^{4k})$ bounds.*

We can also state Theorem 6.5.1 for the cobordism ring. This is important for subsequent applications and can be reformulated as follows:

7.3. *Let ψ be a function which associates a rational number $\psi(V^n)$ to each compact oriented differentiable manifold V^n, which is not identically zero and which has the properties:*

I) $\psi(V^n + W^n) = \psi(V^n) + \psi(W^n)$, $\psi(-V^n) = -\psi(V^n)$

II) $\psi(V^n \times W^n) = \psi(V^n) \cdot \psi(W^n)$

III) *if V^n bounds then* $\psi(V^n) = 0$.

Then $\psi(V^n)$ is zero unless n is divisible by 4, *and there is one and only one m-sequence $\{K_j(p_1, \ldots, p_j)\}$ with rational coefficients such that, for all oriented manifolds V^{4k},*

$$\psi(V^{4k}) = K_k(p_1, \ldots, p_k)\,[V^{4k}]\,,$$

that is, ψ coincides with the K-genus associated to the m-sequence $\{K_j\}$.

By § 1 the m-sequence $\{K_j\}$ corresponds to a uniquely determined power series $Q(z) = 1 + b_1 z + b_2 z^2 + \cdots$. The coefficients b_i of this power series can be calculated inductively using a basis sequence of oriented manifolds. For instance, the sequence of $2k$-dimensional complex projective spaces $\mathbf{P}_{2k}(\mathbf{C})$ can be chosen as a basis sequence (Theorem 6.3.2).

Remark: Property II) is implied by I), III) and the following special case II*) of II):

II*) *There is a basis sequence $\{V^{4k}\}$ such that, for each product of manifolds V^{4j},*

$$\psi(V^{4j_1} \times V^{4j_2} \times \cdots \times V^{4j_r}) = \psi(V^{4j_1})\,\psi(V^{4j_2}) \cdots \psi(V^{4j_r})\,.$$

§ 8. The index of a $4k$-dimensional manifold

8.1. Let $Q(x, y)$ be a real valued symmetric bilinear form on a finite dimensional real vector space. If p^+ is the number of positive eigenvalues of $Q(x, y)$, and p^- is the number of negative eigenvalues, the difference $p^+ - p^-$ is called the *index* of $Q(x, y)$.

8.2. It is well known that there is a symmetric bilinear form associated to every compact oriented $4k$-dimensional manifold: if $x, y \in H^{2k}(M^{4k}, \mathbf{R})$, the cup product xy defines a real number $xy[M^{4k}]$ as in 5.1. The bilinear form $xy[M^{4k}]$ is defined on the real vector space $H^{2k}(M^{4k}, \mathbf{R})$ and is a topological invariant of the oriented manifold M^{4k}. The index of this form is called the *index* of M^{4k} and denoted by $\tau(M^{4k})$. The index of a manifold whose dimension is not divisible by 4 is defined to be zero. We now prove that the function τ satisfies the properties set out in 7.3.

Theorem 8.2.1.

I) $\tau(V^n + W^n) = \tau(V^n) + \tau(W^n)$, $\tau(-V^n) = -\tau(V^n)$

II) $\tau(V^n \times W^m) = \tau(V^n) \cdot \tau(W^m)$

III) *if* V^n *bounds then* $\tau(V^n) = 0$.

Proof: I) follows immediately from the definitions of $V^n + W^n$ and $-V^n$.

II) is known (THOM [2]) but is given there without proof. We therefore prove II) in full. Let $M^{4k} = V^n \times W^m$. Then

$$H^{2k}(M^{4k}, \mathbf{R}) \cong \sum_{s=0}^{2k} H^s(V^n, \mathbf{R}) \otimes H^{2k-s}(W^m, \mathbf{R}) . \tag{1}$$

Elements $x, y \in H^{2k}(M^{4k}, \mathbf{R})$ are said to be orthogonal if $x\,y[M^{4k}] = 0$. Introduce bases $\{v_i^s\}$ for $H^s(V^n, \mathbf{R})$ and $\{w_j^t\}$ for $H^t(W^m, \mathbf{R})$ such that $v_i^s\, v_j^{n-s}[V^n] = \delta_{ij}$ for $s \neq \frac{n}{2}$ and $w_i^t\, w_j^{m-t}[W^m] = \delta_{ij}$ for $t \neq \frac{m}{2}$.

Now consider the group $A = H^{\frac{n}{2}}(V^n, \mathbf{R}) \otimes H^{\frac{m}{2}}(W^m, \mathbf{R})$, taking $A = 0$ if n and m are odd. Then A is orthogonal to the subgroup B of $H^{2k}(M^{4k}, \mathbf{R})$, which consists of all elements of the summation (1) in which no elements of A occur. As a basis for the group B we can take $\{v_i^s \otimes w_j^{2k-s}\}$, $\left(0 \leqq s \leqq n,\ s \neq \frac{n}{2}\right)$. Now

$$(v_i^s \otimes w_j^{2k-s})\,(v_{i'}^{s'} \otimes w_{j'}^{2k-s'})\,[M^{4k}] = \pm 1 \text{ if } s + s' = n,\ i = i', j = j' .$$
$$= 0 \text{ otherwise.}$$

It follows that, with respect to this basis, the restriction of the bilinear form $x\,y[M^{4k}]$ to B is represented by a matrix with blocks $\pm\begin{pmatrix}0 & 1\\ 1 & 0\end{pmatrix}$ down the diagonal and zero elsewhere. Therefore the index of the restriction of $x\,y[M^{4k}]$ to B is 0. Since A and B are orthogonal, $\tau(M^{4k})$ is equal to the index $\tau(A)$ of the restriction of the bilinear form $x\,y[M^{4k}]$ to A. There are now two cases to consider. If n and m are not divisible by 4 then $\tau(A) = 0$. If n and m are divisible by 4 then $\tau(A) = \tau(V^n) \cdot \tau(W^m)$. This completes the proof of II). A more detailed proof can be found in CHERN-HIRZEBRUCH-SERRE [1].

III) is proved by THOM [1]. The proof can be summarised briefly as follows. Suppose that V^{4k} is the oriented boundary of X^{4k+1}, and that $j: V^{4k} \to X^{4k+1}$ is the embedding. THOM considers the diagram of homomorphisms

$$\begin{array}{ccccc} H^{2k}(X^{4k+1}, \mathbf{R}) & \xrightarrow{j^*} & H^{2k}(V^{4k}, \mathbf{R}) & \to & H^{2k+1}(X^{4k+1} \bmod V^{4k}, \mathbf{R}) \\ \downarrow & & \downarrow i & & \downarrow \\ H_{2k+1}(X^{4k+1} \bmod V^{4k}, \mathbf{R}) & \to & H_{2k}(V^{4k}, \mathbf{R}) & \xrightarrow{j_*} & H_{2k}(X^{4k+1}, \mathbf{R}) \end{array}$$

Here the rows are part of exact homology and cohomology sequences, and the vertical arrows are isomorphisms, defined by POINCARÉ duality for V^{4k} and X^{4k+1}, which make the squares commutative.

Let A^{2k} be the image of j^* in $H^{2k}(V^{4k}, \mathbf{R})$ and let K_{2k} be the kernel of j_* in $H_{2k}(V^{4k}, \mathbf{R})$. Then A^{2k} is dual to $H_{2k}(V^{4k}, \mathbf{R})/K_{2k}$ under the duality between $H^{2k}(V^{4k}, \mathbf{R})$ and $H_{2k}(V^{4k}, \mathbf{R})$. On the other hand, the diagram implies that, for $x \in H^{2k}(V^{4k}, \mathbf{R})$,

$$x \in A^{2k} \Leftrightarrow i(x) \in K_{2k}.$$

Therefore, if $b_{2k} = \dim H_{2k}(V^{4k}, \mathbf{R})$ is the $2k$-th BETTI number of V,

$$\dim A^{2k} = \dim K_{2k} = b_{2k} - \dim K_{2k}$$

and

$$\dim A^{2k} = \frac{1}{2} b_{2k}. \tag{2}$$

If $x = j^* y \in A^{2k}$, and v is the fundamental cycle of V^{4k} then $x^2[V^{4k}] = (j^* y^2)[v] = (y^2)[j_* v] = 0$. Therefore the cone $\{x \in H^{2k}(V^{4k}, \mathbf{R}); x^2[V^{4k}] = 0\}$ contains the linear subspace A^{2k} of dimension $\frac{1}{2} b_{2k}$. It follows that the bilinear form $x\,y[V^{4k}]$ has $p^+ = p^-$ (8.1) and hence that $\tau(V^{4k}) = 0$. This proves III) and completes the proof of Theorem 8.2.1.

Theorem **7.2.3** and **7.3** now imply that the index τ can be identified with the K-genus of an m-sequence $\{K_j\}$. For complex projective space $\tau(\mathbf{P}_{2k}(\mathbf{C})) = 1$ for all k. The only m-sequence which takes the value 1 on each $\mathbf{P}_{2k}(\mathbf{C})$ is the sequence $\{L_j(p_1, \ldots, p_j)\}$ (Lemma 1.5.1 and Theorem 4.10.2).

Theorem 8.2.2. *The index $\tau(M^{4k})$ of a compact oriented differentiable manifold M^{4k} can be represented as a linear combination of* PONTRJAGIN *numbers. If $\{L_j\}$ is the m-sequence corresponding to the power series $\frac{\sqrt{z}}{\tanh\sqrt{z}}$ then $\tau(M^{4k}) = L_k(p_1, \ldots, p_k)[M^{4k}]$.* (A list of the first few polynomials L_j is given in 1.5.)

Remark: By the remark at the end of **7.3** it is possible to prove property II) of 8.2.1 by using III) and the fact that the index of any product $\mathbf{P}_{2j_1}(\mathbf{C}) \times \cdots \times \mathbf{P}_{2j_r}(\mathbf{C})$ is 1.

§ 9. The virtual index

9.1. Let M^n be a compact oriented differentiable manifold and let $j: V^{n-k} \to M^n$ be the embedding of an oriented submanifold V^{n-k} of M^n. If ${}_{\mathbf{R}}\theta(V^{n-k})$, ${}_{\mathbf{R}}\theta(M^n)$ are the tangent bundles of V^{n-k}, M^n respectively and $\boldsymbol{\nu}$ is the normal bundle of V^{n-k} in M^n then, by 4.8,

$$j^*{}_{\mathbf{R}}\theta(M^n) = {}_{\mathbf{R}}\theta(V^{n-k}) \oplus \boldsymbol{\nu}.$$

Let $p(V^{n-k})$, $p(M^n)$ be the (total) PONTRJAGIN classes of V^{n-k}, M^n. Then by 4.5 II), III) we have

$$j^* p(M^n) = p(V^{n-k})\, p(\nu) \text{ modulo torsion.} \tag{1}$$

Note that, in a commutative ring of cohomology classes whose odd dimensional components vanish, every element whose 0-dimensional component is 1 has a uniquely determined inverse. This means that, if the PONTRJAGIN classes of M^n and of the normal bundle ν of V^{n-k} in M^n are known, the PONTRJAGIN classes of V^{n-k} can be calculated. For instance, if $k = 1$, since V^{n-1} and M^n are both oriented, ν is trivial and $p_i(V^{n-1}) = i^* p_i(M^n)$ (compare the proof of Theorem 7.2.1).

9.2. For the applications the case $k = 2$ is particularly important. Let $j: V^{n-2} \to M^n$ be as in 9.1 and let $v \in H^2(M^n, \mathbf{Z})$ be the cohomology class dual to the homology class represented by V^{n-2}. In this case, by Theorem 4.8.1,

$$p(\nu) = j^*(1 + v^2)$$

and therefore

$$p(V^{n-2}) = j^*[(1 + v^2)^{-1} p(M^n)] .$$

Since $\{L_j(p_1, \ldots, p_j)\}$ is the m-sequence which corresponds to the power series $\frac{\sqrt{z}}{\tanh \sqrt{z}}$ the definition of m-sequences in 1.2 implies that

$$\sum_{i=0}^{\infty} L_i(p_1(V^{n-2}), \ldots, p_i(V^{n-2})) = j^* \left[\frac{\tanh v}{v} \sum_{i=0}^{\infty} L_i(p_1(M^n), \ldots, p_i(M^n))\right]. \tag{2}$$

We are now in a position to obtain a formula for $\tau(V^{n-2})$. We need the fact (POINCARÉ duality) that if $x \in H^{n-2}(M^n, A) \otimes B$, with A, B additive groups then

$$j^*(x)\,[V^{n-2}] = v\,x\,[M^n] . \tag{3}$$

Theorem 8.2.2, together with (2) and (3), now gives

$$\tau(V^{n-2}) = \varkappa^n \left[\tanh v \sum_{i=0}^{\infty} L_i(p_1(M^n), \ldots, p_i(M^n))\right]. \tag{4}$$

In (4) we use the abbreviation $\varkappa^n$ for the first time. It is used constantly from now on and is defined by the rule:

Let $u^{(n)}$ be the n-dimensional component of an element $u \in \sum_{k=0}^{n} H^k(M^n, A) \otimes B$. Define $\varkappa^n[u] = u^{(n)}[M^n]$. (5)

If $n \not\equiv 2$ modulo 4 formula (4) is trivial, since the left hand side is then zero by definition, while the right hand side is got by evaluating the n-dimensional component $u^{(n)}$ of an expression u which contains no terms of dimension n, so that $\varkappa^n[u] = 0$. In the first few non-trivial cases (4) gives:

$$n = 2, \quad \tau(V^0) = v\,[M^2]$$

$$n = 6, \quad \tau(V^4) = \frac{1}{3}(-v^3 + p_1 v)\,[M^6]$$

$$n = 10, \quad \tau(V^8) = \frac{1}{45}(6v^5 - 5p_1 v^3 + (7p_2 - p_1^2)\,v)\,[M^{10}]\,.$$

9.3. Let M^n be a compact oriented differentiable manifold as in 9.1 and let $v_1, v_2, \ldots, v_r$ be elements of the group $H^2(M^n, \mathbf{Z})$. It will be assumed that v_1 represents a (compact oriented differentiable) submanifold V^{n-2} of M^n, that the restriction of v_2 to V^{n-2} represents a submanifold V^{n-4} of V^{n-2}, ..., and finally that the restriction of v_r to $V^{n-2(r-1)}$ represents a submanifold V^{n-2r} of $V^{n-2(r-1)}$. In this case formula (3) of 9.2 can be generalised: if $x \in H^{n-2r}(M^n, A) \otimes B$ with A, B additive groups and $j: V^{n-2r} \to M^n$ is the embedding then

$$j^*(x)\,[V^{n-2r}] = v_1\, v_2 \ldots v_r\, x\,[M^n]\,. \tag{3'}$$

Successive applications of (2) and (3′) give the following generalisation of (4):

$$\tau(V^{n-2r}) = \varkappa^n \left[\tanh v_1 \tanh v_2 \ldots \tanh v_r \sum_{i=0}^{\infty} L_i(p_1(M^n), \ldots, p_i(M^n))\right]. \tag{4'}$$

According to THOM [2], every 2-dimensional integral cohomology class of a compact oriented differentiable manifold M^n can be represented by a submanifold V^{n-2} of M^n. Successive applications of this theorem show that the above assumptions are justified, so that (4′) holds. As a corollary we see that $\tau(V^{n-2r})$ depends only on the (unordered) set of cohomology classes $v_1, v_2, \ldots, v_r$. We denote the right hand side of (4′) by $\tau(v_1, \ldots, v_r)$, the virtual index of the set $(v_1, \ldots, v_r)$. The theorem of THOM just quoted then implies that every virtual index occurs as the index of a submanifold of M^n and is therefore an integer.

We recall that tanh satisfies the functional equation $\tanh(u+v) = \tanh(u) + \tanh(v) - \tanh(u)\,\tanh(v)\,\tanh(u+v)$ and deduce from (4′):

Theorem 9.3.1. *The virtual index is a function which associates an integer $\tau(v_1, v_2, \ldots, v_r)$ to each (unordered) r-ple $(v_1, v_2, \ldots, v_r)$ of 2-dimensional integral cohomology classes of a compact oriented differentiable manifold M^n. The function τ is zero if $n - 2r \not\equiv 0$ modulo 4, if $2r > n$, or if one of the classes v_i is zero. It satisfies the functional equation*

$$\tau(v_1, \ldots, v_r, u+v) \tag{6}$$
$$= \tau(v_1, \ldots, v_r, u) + \tau(v_1, \ldots, v_r, v) - \tau(v_1, \ldots, v_r, u, v, u+v)\,.$$

In particular, if $n = 4k+2$,

$$\tau(u+v) = \tau(u) + \tau(v) - \tau(u, v, u+v)\,. \tag{6'}$$

9.4. Consider, as an example for Theorem 9.3.1, the product

$$M^{4k+2} = F_1 \times F_2 \times \cdots \times F_{2k+1}$$

of $2k+1$ compact oriented surfaces F_i. Let $x_i \in H^2(M^{4k+2}, \mathbf{Z})$ be the cohomology class which represents the submanifold

$$F_1 \times F_2 \times \cdots \times \hat{F}_i \times \cdots \times F_{2k+1}$$

of M^{4k+2} (where $\hat{F}_i$ means that the factor F_i is omitted). We can calculate $\tau(a_1 x_1 + a_2 x_2 + \cdots + a_{2k+1} x_{2k+1})$, where the a_i are integers, by using (4). Since all the PONTRJAGIN classes of M^{4k+2} except for $p_0 = 1$ are zero,

$$\begin{aligned}
\tau(a_1 x_1 + a_2 x_2 + \cdots + a_{2k+1} x_{2k+1}) \\
&= \varkappa^{4k+2}[\tanh(a_1 x_1 + \cdots + a_{2k+1} x_{2k+1})] \\
&= \frac{\tanh^{(2k+1)}(0)}{(2k+1)!}\, \varkappa^{4k+2}[(a_1 x_1 + \cdots + a_{2k+1} x_{2k+1})^{2k+1}] \\
&= a_1 a_2 \ldots a_{2k+1} \tanh^{(2k+1)}(0)\,.
\end{aligned}$$

This proves that, *if* V^{4k} *is a compact oriented differentiable manifold embedded in the product of* $(2k+1)$ *copies of an oriented 2-sphere* $\mathbf{S}^2$ *which has intersection number* 1 *with each factor, then* $\tau(V^{4k})$ *is the value at* 0 *of the* $(2k+1)$*th derivative of* $\tanh(x)$. By the theorem of THOM quoted in 9.3, such manifolds exist for all k.

Bibliographical note

The results on cobordism used in this chapter are all due to THOM [1], [2]. Actually the differentiability assumptions of THOM are slightly different, but it can be shown that all his results (in particular Theorem 7.2.2) remain true when "differentiable" is taken to mean "C^∞-differentiable". A complete exposition of cobordism theory from this point of view has been given in lectures of MILNOR (Differential Topology, mimeographed notes, Princeton 1958).

THOM also defined the non-oriented cobordism ring $\mathfrak{N} = \sum_{n=0}^{\infty} \mathfrak{N}^n$. Here $\mathfrak{N}^n$ is the group of compact non-oriented differentiable manifolds of dimension n under the equivalence relation: $V^n \sim_2 W^n$ if $V^n + W^n$ bounds a compact non-oriented manifold X^{n+1}. The STIEFEL-WHITNEY classes $w_i \in H^i(V^n, \mathbf{Z}_2)$ define STIEFEL-WHITNEY numbers $w_{j_1} w_{j_2} \ldots w_{j_r}[V^n] \in \mathbf{Z}_2$. THOM proved that $V^n \sim_2 W^n$ if and only if V^n, W^n have the same STIEFEL-WHITNEY numbers, that $\mathfrak{N}$ is a polynomial ring $\mathbf{Z}_2[x_2, x_4, x_5, x_6, x_8, x_9, \ldots]$ over $\mathbf{Z}_2$ with one generator x_i for each $i \neq 2^r - 1$, and that the real projective spaces $\mathbf{P}_{2n}(\mathbf{R})$ give the even dimensional generators x_{2n} (THOM [2]). An explicit construction for the other generators of $\mathfrak{N}$ was given by DOLD [1] (see also MILNOR [7]).

The complete structure of the cobordism ring Ω, and of the graded ring $\widetilde{\Omega}$ defined in 6.2, is now known. MILNOR [3] proved the following more precise version of Theorem 6.4.3: $\widetilde{\Omega}$ is isomorphic to the graded ring $\mathbf{Z}[z_1, z_2, \ldots]$, and an isomorphism $\mathbf{Z}[z_1, z_2, \ldots] \to \widetilde{\Omega}$ is given by associating to z_i a compact oriented differentiable manifold V^{4i} such that

$$s(V^{4i}) = \pm 1 \quad \text{if } 2i+1 \text{ is not a prime power,}$$
$$s(V^{4i}) = \pm q \quad \text{if } 2i+1 \text{ is a power of the prime } q.$$

The cobordism ring Ω^{4k} can be represented as a direct sum

$$\Omega^{4k} = \widetilde{\Omega}^{4k} \oplus T^{4k}$$

where T^j is the group of elements of finite order in Ω^j (and $T^j = \Omega^j$ if $j \not\equiv 0 \bmod 4$). MILNOR [3] proved that T^j has no elements of odd order and gave explicit generators for $\widetilde{\Omega}^{4k}$. Subsequently WALL [1] proved that T_i contains no elements of order 4 and found a complete set of generators for Ω. His results show that $V^n \sim W^n$ if and only if V^n, W^n have the same PONTRJAGIN and STIEFEL-WHITNEY numbers. For a survey of generalisations of the cobordism ring and further developments see ATIYAH [4], CONNER-FLOYD [1], MILNOR [4] and WALL [2].

The index theorem (8.2.2) gives corollaries on the behaviour of the index of an oriented differentiable manifold V. For example, let $f: W \to V$ be a differentiable covering map of degree n. Then $p_i(W) = f^* p_i(V)$ and the index theorem implies that $\tau(W) = n\,\tau(V)$. Does this result remain true if V, W are (non-differentiable) topological manifolds? Let E, B, F be compact *connected* oriented manifolds (not necessarily differentiable). Let $E \to B$ be a fibre bundle with typical fibre F for which the fundamental group $\pi_1(B)$ acts trivially on the cohomology ring $H^*(F, \mathbf{R})$. Then there is a direct topological proof that $\tau(E) = \tau(B)\,\tau(F)$ (CHERN-HIRZEBRUCH-SERRE [1]).

The index theorem implies that the L-genus of an oriented differentiable manifold M depends only on the oriented homotopy type of M. According to KAHN [1] the L-genus is, up to a rational multiple, the only rational linear combination of PONTRJAGIN numbers that is an oriented homotopy type invariant. Far reaching generalizations of the index theorem (applying to differential operators and to finite groups acting on manifolds) have been obtained by ATIYAH and SINGER. These are discussed in the appendix (§ 25).

Chapter Three

The TODD genus

In this chapter M_n will be compact, differentiable of class C^∞ and, in addition, almost complex. The tangent $\mathbf{GL}(n, \mathbf{C})$-bundle of M_n (see 4.6) is denoted by $\theta(M_n)$. We investigate the "genus" associated with the m-sequence $\{T_j(c_1, \ldots, c_j)\}$ of 1.7 as well as the "generalised genus" associated with the m-sequence $\{T_j(y; c_1, \ldots, c_j)\}$ of 1.8.

§ 10. Definition of the TODD genus

10.1. Let X be an admissible space (see 4.2) and let ξ be a continuous $\mathbf{GL}(q, \mathbf{C})$-bundle over X with CHERN classes $c_i \in H^{2i}(X, \mathbf{Z})$. The (total) TODD class of ξ is defined by

$$\operatorname{td}(\xi) = \sum_{j=0}^{\infty} T_j(c_1, \ldots, c_j) \tag{1}$$

where $\{T_j(c_1, \ldots, c_j)\}$ is the m-sequence of 1.7. If ξ' is a continuous $\mathbf{GL}(q', \mathbf{C})$-bundle over X then, by 1.2, the TODD class satisfies

$$\operatorname{td}(\xi \oplus \xi') = \operatorname{td}(\xi)\operatorname{td}(\xi') . \tag{2}$$

If $q = 1$ and $c_1(\xi) = d \in H^2(X, \mathbf{Z})$ then

$$\operatorname{td}(\xi) = \frac{d}{1 - e^{-d}} .$$

Note that $\operatorname{td}(\xi)$ is a series starting with 1 and therefore, since X is finite dimensional, the inverse $(\operatorname{td}(\xi))^{-1}$ exists. The total TODD class can also be defined by means of a formal factorisation: if

$$\sum_{j=0}^{q} c_j x^j = \prod_{i=1}^{q} (1 + \gamma_i x) \quad \text{then} \quad \operatorname{td}(\xi) = \prod_{i=1}^{q} \frac{\gamma_i}{1 - e^{-\gamma_i}} .$$

In a similar way the (total) CHERN character of ξ is defined by

$$\operatorname{ch}(\xi) = \sum_{i=1}^{q} e^{\gamma_i} . \tag{3}$$

By 4.4.3 the CHERN character satisfies

$$\begin{aligned} \operatorname{ch}(\xi \oplus \xi') &= \operatorname{ch}(\xi) + \operatorname{ch}(\xi') , \\ \operatorname{ch}(\xi \otimes \xi') &= \operatorname{ch}(\xi)\operatorname{ch}(\xi') . \end{aligned} \tag{4}$$

If $q = 1$ and $c_1(\xi) = d \in H^2(X, \mathbf{Z})$ then $\operatorname{ch}(\xi) = e^d$.

In general $\mathrm{ch}(\xi) = q + \sum_{k=1}^{\infty} \mathrm{ch}_k(\xi)$ where

$$\mathrm{ch}_k(\xi) = \frac{s_k}{k!} \in H^{2k}(X, \mathbf{Z}) \otimes \mathbf{Q} \quad \text{and} \quad s_k = \sum_{i=1}^{q} \gamma_i^k \quad (k \geqq 1) .$$

The symmetric functions s_k and c_i are related by NEWTON formulae [compare 1.4 (10)]

$$s_k - c_1 s_{k-1} + \cdots + (-1)^k c_k k = 0 \quad (k \geqq 1) .$$

The CHERN character is related to the TODD class td by

Theorem 10.1.1. *Let ξ be a continuous $\mathbf{GL}(q, \mathbf{C})$-bundle over an admissible space X. Then*

$$\sum_{r=0}^{q} (-1)^r \mathrm{ch}\, \lambda^r \xi^* = (\mathrm{td}(\xi))^{-1} c_q(\xi) .$$

Proof: If $\sum_{j=0}^{q} c_j(\xi)\, x^j = \prod_{i=1}^{q} (1 + \gamma_i x)$ then by 4.4.3

$$\mathrm{ch}\, \lambda^r \xi^* = \sum e^{-(\gamma_{i_1} + \cdots + \gamma_{i_r})}$$

where the sum is over all combinations $i_1, \ldots, i_r$ with $1 \leqq i_1 < \cdots < i_r \leqq q$. Therefore

$$\begin{aligned}\sum_{r=0}^{q} (-1)^r \mathrm{ch}\, \lambda^r \xi^* &= \prod_{i=1}^{q} (1 - e^{-\gamma_i}) \\ &= (\gamma_1 \cdots \gamma_q) \prod_{i=1}^{q} \frac{1 - e^{-\gamma_i}}{\gamma_i} \\ &= (\mathrm{td}(\xi))^{-1} c_q(\xi) .\end{aligned}$$

10.2. Let M_n be an almost complex manifold (4.6). The almost complex structure defines a particular orientation of M_n. If $u \in H^*(M_n)$ and $u^{(2n)}$ is the $2n$-dimensional component of u we write $\varkappa_n(u) = u^{(2n)}[M_n]$. Let $c_i \in H^{2i}(M_n, \mathbf{Z})$ be the CHERN classes of $\theta(M_n)$. Every product $c_{j_1} c_{j_2} \ldots c_{j_r}$ of weight $n = j_1 + j_2 + \cdots + j_r$ defines an integer $c_{j_1} c_{j_2} \ldots c_{j_r}[M_n]$. If $\pi(n)$ is the number of distinct partitions of n there are $\pi(n)$ such integers; they are called the CHERN numbers of M_n. For example (Theorem 4.10.1) the CHERN number $c_n[M_n]$ is precisely the EULER-POINCARÉ characteristic of M_n. Consider the ring $\mathfrak{B} = B[c_1, c_2, \ldots]$ of 1.1 (see also 1.3). As in 5.1, each element $b \in \mathfrak{B}_n$ determines an element $b[M_n] \in B$.

The cartesian product $V_n \times W_m$ of two almost complex manifolds is almost complex in a natural way: the tangent $\mathbf{GL}(n + m, \mathbf{C})$-bundle of the product is the WHITNEY sum $f^*(\theta(V_n)) \oplus g^*(\theta(W_m))$ where $f: V_n \times W_m \to V_n$ and $g: V_n \times W_m \to W_m$ are projection maps. As in 5.2 we have

Lemma 10.2.1. *Let* $\{K_j(c_1, \ldots, c_j)\}$ *be an m-sequence* ($K_j \in \mathfrak{B}_j$, as in 1.2, 1.3). *Then*

$$K_{n+m}[V_n \times W_m] = K_n[V_n] \cdot K_m[W_m] .$$

$K_n[M_n]$ is called the K-genus of M_n. Now consider the m-sequences

$$\{T_j(c_1, \ldots, c_j)\}, \quad \{T_j(y; c_1, \ldots, c_j)\}$$

defined in 1.7, 1.8 and associated to the power series

$$Q(x) = \frac{x}{1-\exp(-x)}, \quad Q(y; x) = \frac{x(y+1)}{1-\exp(-x(y+1))} - xy .$$

The rational number $T_n[M_n]$ is called the TODD genus (or T-genus) of M_n and written $T(M_n)$. Thus

$$T(M_n) = \varkappa_n[\mathrm{td}(\theta(M_n))] .$$

By 1.8, $T_n(y; c_1, \ldots, c_n)$ $[M_n]$ is a polynomial of degree n in y with rational coefficients. It can therefore be written in the form

$$T_y(M_n) = \sum_{p=0}^{n} T^p(M_n)\, y^p .$$

The polynomial $T_y(M_n)$ is called the generalised TODD genus (or T_y-genus) of M_n. By definition, $T_0(M_n) = T^0(M_n) = T(M_n)$.

Lemma 10.2.1 implies that $T_y(V_n \times W_m) = T_y(V_n)\, T_y(W_m)$ and in particular that $T(V_n \times W_m) = T(V_n)\, T(W_m)$. By 1.8 (13) the rational numbers $T^p(M_n)$ satisfy the "duality formula"

$$T^p(M_n) = (-1)^n\, T^{n-p}(M_n) .$$

By 1.8 (16), together with Theorem 8.2.2,

$$T_{-1}(M_n) = \sum_{p=0}^{n} (-1)^p\, T^p(M_n) = c_n[M_n] , \tag{5}$$

$$T_1(M_n) = \sum_{p=0}^{n} T^p(M_n) = \tau(M_n) . \tag{6}$$

Thus $T_{-1}(M_n)$ is the EULER-POINCARÉ characteristic of M_n while $T_1(M_n)$ is the index of M_n [notice that the above "duality formula" shows that $T_1(M_n) = 0$ for n odd].

10.3. The (total) CHERN class of the complex projective space $\mathbf{P}_n(\mathbf{C})$ is $(1+h_n)^{n+1}$ by Theorem 4.10.2. Lemma 1.7.1 and Lemma 1.8.1 therefore imply

Theorem 10.3.1. *The T-genus is the only genus associated to an m-sequence with rational coefficients which takes the value* 1 *on every complex projective space* $\mathbf{P}_n(\mathbf{C})$. *The* T_y*-genus is the only one associated to an m-sequence with coefficients in* $\mathbf{Q}[y]$ *which takes the value* $1 - y + y^2 - \cdots + (-1)^n y^n$ *on* $\mathbf{P}_n(\mathbf{C})$.

§ 11. The virtual generalised TODD genus

11.1. Let V_{n-k} be a (compact) almost complex submanifold of the almost complex manifold M_n and $j: V_{n-k} \to M_n$ the embedding. By 4.9 there is a normal $\mathbf{GL}(k, \mathbf{C})$-bundle ν over V_{n-k} such that $j^* \theta(M_n) = \theta(V_{n-k}) \oplus \nu$. By 4.4.3, II)

$$j^* c(M_n) = c(V_{n-k})\, c(\nu)\,.$$

Consider the special case $k = 1$. Theorem 4.8.1 gives $c(\nu) = 1 + j^* v$ where $v \in H^2(M_n; \mathbf{Z})$ is the cohomology class determined by the fundamental class of the oriented submanifold V_{n-1}. Therefore

$$\begin{aligned} &1 + c_1(V_{n-1}) + c_2(V_{n-1}) + \cdots \\ &\qquad = j^* [(1 + c_1(M_n) + c_2(M_n) + \cdots)\,(1 + v)^{-1}]\,. \end{aligned} \tag{1}$$

It is now possible to calculate the T_y-genus of V_{n-1}. This genus is associated with the power series $Q(y; x) = \dfrac{x}{R(y; x)}$ where

$$R(y; x) = \frac{e^{x(y+1)} - 1}{e^{x(y+1)} + y} \tag{2}$$

$$R(1; x) = \tanh x,\ R(-1; x) = x(1 + x)^{-1},\ R(0; x) = 1 - e^{-x}\,.$$

Then (1) implies

$$\sum_{i=0}^{\infty} T_i(y; c_1(V_{n-1}), \ldots, c_i(V_{n-1})) = j^* \left(\frac{R(y; v)}{v} \sum_{i=0}^{\infty} T_i(y; c_1(M_n), \ldots) \right) \tag{3}$$

and hence, as in 9.2,

$$T_y(V_{n-1}) = \varkappa_n \left[R(y; v) \sum_{j=0}^{\infty} T_j(y; c_1(M_n), \ldots, c_j(M_n)) \right]. \tag{4}$$

In the case $y = 1$ formula (4) is [in view of 1.8 (16)] exactly 9.2 (4). In the case $y = -1$ it gives a formula for the EULER-POINCARÉ characteristic $E(V_{n-1}) = c_{n-1}(V_{n-1})\,[V_{n-1}]$:

$$(-1)^{n-1} E(V_{n-1}) = \sum_{i=0}^{n-1} (-1)^i v^{n-i} c_i(M_n)\,[M_n]. \tag{5}$$

Formula (5) can naturally be obtained directly from (1). In the case $y = 0$ we have

$$T(V_{n-1}) = \varkappa_n[(1 - e^{-v})\,\mathrm{td}(\theta(M_n))]. \tag{6}$$

11.2. We now come to the definition of the virtual T_y-genus. For $v_1, \ldots, v_r \in H^2(M_n, \mathbf{Z})$ let

$$T_y(v_1, \ldots, v_r)_M = \varkappa_n \left[R(y; v_1) \ldots R(y; v_r) \sum_{j=0}^{\infty} T_j(y; c_1(M_n), \ldots) \right]. \tag{7}$$

Here the subscript M denotes the manifold in which the virtual genus is defined; in subsequent paragraphs this subscript will be omitted if it is clear from the context which manifold is meant.

By 1.8, $T_y(v_1, \ldots, v_r)_M$ is a polynomial of degree $n - r$ in y with rational coefficients. $T_y(v_1, \ldots, v_n)_M = 0$ for $r > n$ because $R(y; x)$ is divisible by x. For $r = n$, $T_y(v_1, \ldots, v_n)_M = v_1 v_2 \ldots v_n [M_n]$. We call $T_y(v_1, \ldots, v_r)_M$ the (virtual) T_y-genus of the r-ple $(v_1, \ldots, v_r)$. It is independent of the ordering $v_1, \ldots, v_r$. An unordered r-ple of elements of $H^2(M_n, \mathbf{Z})$ is also called a virtual almost complex $(n - r)$-dimensional submanifold of M_n. We write

$$T_y(v_1, \ldots, v_r)_M = \sum_{p=0}^{n-r} T^p(v_1, \ldots, v_r)_M \, y^p . \tag{8a}$$

The rational number

$$T(v_1, \ldots, v_r)_M = T_0(v_1, \ldots, v_r)_M = T^0(v_1, \ldots, v_r)_M \tag{8b}$$

is called the virtual TODD genus of the virtual submanifold $(v_1, \ldots, v_r)$. The duality formula

$$T^p(v_1, \ldots, v_r)_M = (-1)^{n-r} \, T^{n-r-p}(v_1, \ldots, v_r)_M \tag{9}$$

holds, and Formula 11.1 (3) implies immediately

Theorem 11.2.1. *Let V_{n-1} be an almost complex submanifold of M_n, $v \in H^2(M_n, \mathbf{Z})$ the cohomology class determined by V_{n-1} and $j : V_{n-1} \to M_n$ the embedding. Let $v_2, \ldots, v_r \in H^2(M_n, \mathbf{Z})$. Then $T_y(j^* v_2, \ldots, j^* v_r)_V = T_y(v, v_2, \ldots, v_r)_M$. In particular*

$$T_y(V_{n-1}) = T_y(v)_M .$$

11.3. The functional equation of the index [9.3 (6)] is a special case of a functional equation satisfied by the virtual T_y-genus. If a, y are indeterminates and $R(x) = \dfrac{e^{ax} - 1}{e^{ax} + y}$ then

$$R(u+v) = R(u) + R(v) + (y-1) \, R(u) \, R(v) - y R(u) \, R(v) \, R(u+v) . \tag{10}$$

The substitution $a = 1 + y$ then yields a functional equation for $R(y; x)$. For $y = 1$ this is the functional equation of $\tanh x$, for $y = 0$ the functional equation of $1 - e^{-x}$, and for $y = -1$ the functional equation of $x(1 + x)^{-1}$. As a corollary we have

Theorem 11.3.1. *The virtual T_y-genus satisfies the functional equation*

$$\begin{aligned} T_y(v_1, \ldots, v_r, u + v) &= T_y(v_1, \ldots, v_r, u) + T_y(v_1, \ldots, v_r, v) + \\ &+ (y-1) \, T_y(v_1, \ldots, v_r, u, v) - y \, T_y(v_1, \ldots, v_r, u, v, u + v) \end{aligned}$$

where $v_1, \ldots, v_r, u, v$ are elements of $H^2(M_n, \mathbf{Z})$. In the special case $r = 0$ the equation becomes

$$T_y(u + v) = T_y(u) + T_y(v) - (y-1) \, T_y(u, v) - y \, T_y(u, v, u + v).$$

For $y = 1$ *this implies that the virtual index satisfies*

$$\tau(u+v) = \tau(u) + \tau(v) - \tau(u, v, u+v),$$

for $y = 0$ *the virtual* TODD *genus satisfies*

$$T(u+v) = T(u) + T(v) - T(u, v),$$

and for $y = -1$ *the virtual* EULER-POINCARÉ *characteristic satisfies*

$$T_{-1}(u+v) = T_{-1}(u) + T_{-1}(v) - 2T_{-1}(u, v) + T_{-1}(u, v, u+v).$$

§ 12. The T-characteristic of a $\mathbf{GL}(q, \mathbf{C})$-bundle

12.1. Let ξ be a continuous $\mathbf{GL}(q, \mathbf{C})$-bundle over M_n. Any differentiable, or complex analytic, $\mathbf{GL}(q, \mathbf{C})$-bundle over M_n can also be regarded as a continuous $\mathbf{GL}(q, \mathbf{C})$-bundle so that the results of this paragraph apply, in particular, to these cases. Let

$$c(M_n) = \sum_{i=0}^{n} c_i, \quad c(\xi) = \sum_{i=0}^{q} d_i \tag{1}$$

where $c_i, d_i \in H^{2i}(M_n, \mathbf{Z})$ and $c_0 = d_0 = 1$.

The TODD class of $\theta(M_n)$ and the CHERN character of ξ (see 10.1) are used to define the rational number

$$T(M_n, \xi) = \varkappa_n\left[\operatorname{ch}(\xi)\operatorname{td}(\theta(M_n))\right]. \tag{2}$$

$T(M_n, \xi)$ is called the T-characteristic of the $\mathbf{GL}(q, \mathbf{C})$-bundle ξ over M_n. In the special case where ξ is a $\mathbf{C}^*$-bundle with CHERN class $1 + d$, $d \in H^2(M_n, \mathbf{Z})$, equation (2) becomes

$$T(M_n, \xi) = \varkappa_n\left[e^d \operatorname{td}(\theta(M_n))\right]. \tag{3}$$

Now the $\mathbf{C}^*$-bundles over M_n are in one-one correspondence with elements d of $H^2(M_n, \mathbf{Z})$ by 3.8 and Theorem 4.3.1. Therefore we may write $T(M_n, d)$ for $T(M_n, \xi)$ in (3). The definitions imply

$$T(M, d) = T(M) - T(-d)_M. \tag{4}$$

If ξ is a $\mathbf{GL}(q, \mathbf{C})$-bundle and ξ' is a $\mathbf{GL}(q', \mathbf{C})$-bundle over M_n then the first equation of 10.1 (4) gives

$$T(M_n, \xi \oplus \xi') = T(M_n, \xi) + T(M_n, \xi'). \tag{5}$$

If ξ is a $\mathbf{GL}(q, \mathbf{C})$-bundle over V_n and η is a $\mathbf{GL}(r, \mathbf{C})$-bundle over W_m then 10.1 (2) and 10.1 (4) give

$$T(V_n \times W_m, f^*(\xi) \otimes g^*(\eta)) = T(V_n, \xi)\, T(W_m, \eta), \tag{6}$$

where f and g are projection maps as in 5.2.

12.2. In order to extend the results of 12.1 to apply to a T_y-characteristic $T_y(M_n, \xi)$ it is necessary to consider the dual tangent $\mathbf{GL}(n, \mathbf{C})$-bundle $\theta^* = \theta(M_n)^*$ of M_n. Let $\lambda^p(\theta^*)$ be the p-th exterior product (3.6) of θ^*. Consider the formal factorisations [see 12.1 (1)]

$$\sum_{j=0}^{n} c_j x^j = \prod_{i=1}^{n} (1 + \gamma_i x) \quad \text{and} \quad \sum_{j=0}^{q} d_j x^j = \prod_{i=1}^{q} (1 + \delta_i x) .$$

Then the CHERN classes of θ^* are the elementary symmetric functions in the $-\gamma_i$ and, by 4.4.3, the CHERN classes of $\lambda^p(\theta^*)$ are the elementary symmetric functions in the formal roots $-(\gamma_{i_1} + \gamma_{i_2} + \cdots + \gamma_{i_p})$. Then 1.8 (15) implies that

$$T(M_n, \lambda^p(\theta^*)) = T^p(M_n) . \tag{7}$$

Now consider the tensor product $\lambda^p(\theta^*) \otimes \xi$. We denote the rational number $T(M_n, \lambda^p(\theta^*) \otimes \xi)$ also by $T^p(M_n, \xi)$ and define

$$T_y(M_n, \xi) = \sum_{p=0}^{n} T^p(M_n, \xi)\, y^p . \tag{8}$$

Equations (2) and 10.1 (4) imply that

$$T^p(M_n, \xi) = \varkappa_n\big[\operatorname{ch}(\xi)\operatorname{ch}(\lambda^p(\theta^*))\operatorname{td}(\theta(M_n))\big] . \tag{9}$$

A trivial generalisation of the argument used to prove 1.8 (15) gives

$$T_y(M_n, \xi) = \varkappa_n\left[\left(\sum_{i=1}^{q} e^{(1+y)\delta_i}\right)\left(\sum_{j=0}^{n} T_j(y; c_1, \ldots, c_j)\right)\right] . \tag{10}$$

Notice that, if $y = -1$, the number $T_{-1}(M_n, \xi)$ does not depend on the CHERN classes of ξ but only on the rank q of ξ. In this case $T_{-1}(M_n, \xi) = q\, E(M_n)$ where $E(M_n)$ is the EULER-POINCARÉ characteristic of M_n.

Substituting $\frac{1}{y}$ for y in (10) and multiplying both sides of the equation by $(-y)^n$, we can rewrite the right hand side of (10) as

$$\varkappa_n\left[\left(\sum_{i=1}^{q} e^{-(1+y)\delta_i}\right)\left(\sum_{j=0}^{n} T_j(y; c_1, \ldots, c_j)\right)\right]$$

and obtain, applying Theorem 4.4.3, the duality formula

$$y^n T_{\frac{1}{y}}(M_n, \xi) = (-1)^n T_y(M_n, \xi^*) .$$

Therefore

$$T^p(M_n, \xi) = (-1)^n T^{n-p}(M_n, \xi^*) \tag{11}$$

and, in particular, when $p = 0$

$$T(M_n, \xi) = (-1)^n T(M_n, \lambda^n(\theta^*) \otimes \xi^*) . \tag{12}$$

In fact it is possible to deduce (11) from (12): replace ξ by $\xi \otimes \lambda^p(\theta^*)$ in (12) and recall that, by Theorem 3.6.1, $\lambda^n(\theta^*) \otimes (\xi \otimes \lambda^p(\theta^*))^* = \xi^* \otimes \lambda^n(\theta^*) \otimes \lambda^p(\theta) = \xi^* \otimes \lambda^{n-p}(\theta^*)$.

The bundle $\lambda^n(\theta^*)$ is a $\mathbf{C}^*$-bundle, and is called the canonical $\mathbf{C}^*$-bundle over M_n. By Theorem 4.4.3 it has (total) CHERN class $1 - c_1(M_n)$.

Let $\mathrm{ch}_{(y)}(\xi) \in H^*(M_n, \mathbf{Z}) \otimes \mathbf{Q}[y]$ be defined by the equation

$$\mathrm{ch}_{(y)}(\xi) = e^{(1+y)\delta_1} + \cdots + e^{(1+y)\delta_q}$$

where the CHERN classes of ξ are the elementary symmetric functions in $\delta_1, \ldots, \delta_q$. Then, as in 10.1 (4),

$$\mathrm{ch}_{(y)}(\xi \oplus \xi') = \mathrm{ch}_{(y)}(\xi) + \mathrm{ch}_{(y)}(\xi') \quad \text{and} \quad \mathrm{ch}_{(y)}(\xi \otimes \xi') = \mathrm{ch}_{(y)}(\xi)\,\mathrm{ch}_{(y)}(\xi').$$

These equations imply that

$$T_y(M_n, \xi \oplus \xi') = T_y(M_n, \xi) + T_y(M_n, \xi') \tag{13}$$

and

$$T_y(V_n \times W_m, f^*(\xi) \otimes g^*(\eta)) = T_y(V_n, \xi)\, T_y(W_m, \eta)\,. \tag{14}$$

In equations (13) and (14) we use again without comment the notations of equations (5) and (6) of 12.1.

12.3. Let ξ be a $\mathbf{GL}(q, \mathbf{C})$-bundle over M_n. If $v_1, \ldots, v_r$ are elements of $H^2(M_n, \mathbf{Z})$ then the virtual T_y-characteristic of ξ with respect to the "virtual submanifold" $(v_1, \ldots, v_r)$ can be defined by the following generalisation of 11.2 (7):

$$T_y(v_1, \ldots, v_r|, \xi)_M = \varkappa_n \left[\mathrm{ch}_{(y)}(\xi) \prod_{i=1}^{r} R(y; v_i) \sum_{j=0}^{\infty} T_j(y; c_1(M_n), \ldots)\right]. \tag{15}$$

As in 11.1, the subscript M will be dropped if it is clear which manifold is meant. If ξ is the trivial $\mathbf{GL}(q, \mathbf{C})$-bundle then $T_y(v_1, \ldots, v_r|, \xi) = q\, T_y(v_1, \ldots, v_r)$. Naturally we write, when $y = 0$,

$$T_0(v_1, \ldots, v_r|, \xi) = T(v_1, \ldots, v_r|, \xi)$$

and call $T(v_1, \ldots, v_r|, \xi)$ the virtual T-characteristic. As a generalisation of Theorem 11.2.1 we have

THEOREM 12.3.1. *Let V_{n-1} be an almost complex submanifold of M_n, $v \in H^2(M_n, \mathbf{Z})$ the cohomology class determined by V_{n-1} and $j: V_{n-1} \to M_n$ the embedding. Let $v_2, \ldots, v_r \in H^2(M_n, \mathbf{Z})$ and let ξ be a $\mathbf{GL}(q, \mathbf{C})$-bundle over M_n. Then*

$$T_y(j^* v_2, \ldots, j^* v_r|, j^* \xi)_V = T_y(v, v_2, \ldots, v_r|, \xi)_M\,.$$

In particular

$$T_y(V_{n-1}, j^* \xi) = T_y(v|, \xi)_M\,.$$

The functional equation of Theorem 11.3.1 can also be extended to apply to the virtual T-characteristic.

Theorem 12.3.2. *Let ξ be a $\mathbf{GL}(q, \mathbf{C})$-bundle over M_n. The virtual T_y-characteristic of ξ satisfies the functional equation*

$$\begin{aligned} T_y(v_1, \ldots, v_r, u+v|, \xi)_M &= T_y(v_1, \ldots, v_r, u|, \xi)_M + \\ &+ T_y(v_1, \ldots, v_r, v|, \xi)_M + (y-1)\, T_y(v_1, \ldots, v_r, u, v|, \xi)_M - \\ &- y\, T_y(v_1, \ldots, v_r, u, v, u+v|, \xi)_M \end{aligned}$$

where $v_1, \ldots, v_r, u, v$ are elements of $H^2(M_n, \mathbf{Z})$.

The proof uses the functional equation 11.3 (10) and the remark that the expression inside the square brackets [ ] of (15), $\varkappa_n$ of which is the T_y-characteristic, always contains the factor $\mathrm{ch}_{(y)}(\xi)$.

$T_y(v_1, \ldots, v_r|, \xi)_M$ is a polynomial of degree $n-r$ in y with rational coefficients. It is identically zero for $r > n$. If $r = n$ then

$$T_y(v_1, \ldots, v_n|, \xi)_M = q \cdot (v_1 \ldots v_n [M_n]) .$$

The duality formula, in the case of "virtual submanifolds", becomes

$$y^{n-r}\, T_{\frac{1}{y}}(v_1, \ldots, v_r|, \xi)_M = (-1)^{n-r}\, T_y(v_1, \ldots, v_r|, \xi^*)_M . \tag{16}$$

Theorem 12.3.3. *Let η be a $\mathbf{C}^*$-bundle over M_n with total* Chern *class $1+v$, $v \in H^2(M^n, \mathbf{Z})$ and let ξ be a $\mathbf{GL}(q, \mathbf{C})$-bundle over M_n. Let $v_1, \ldots, v_r$ be elements of $H^2(M^n, \mathbf{Z})$. Then*

$$\begin{aligned} T_y(v_1, \ldots, v_r|, \xi)_M &= T_y(v_1, \ldots, v_r, v|, \xi)_M + T_y(v_1, \ldots, v_r|, \xi \otimes \eta^{-1})_M + \\ &+ y\, T_y(v_1, \ldots, v_r, v|, \xi \otimes \eta^{-1})_M . \end{aligned}$$

Proof: In formula (15) for the T_y-characteristic the expression in square brackets [ ] contains the factor $\sum_{j=0}^{\infty} T_j(y; c_1(M_n), \ldots)$. This factor is the same for each of the four T_y-characteristics which occur in the equation to be proved. Similarly, each of the four terms contains the factor $\prod_{i=1}^{r} R(y; v_i)$ and, since $\mathrm{ch}_{(y)}(\xi \otimes \eta^{-1}) = \mathrm{ch}_{(y)}(\xi)\, \mathrm{ch}_{(y)}(\eta)^{-1}$, the factor $\mathrm{ch}_{(y)}(\xi)$. It is therefore sufficient to prove the equation

$$1 = R(y; v) + \mathrm{ch}_{(y)}(\eta^{-1}) + y R(y; v)\, \mathrm{ch}_{(y)}(\eta^{-1}) .$$

But $\mathrm{ch}_{(y)}(\eta^{-1}) = e^{-(1+y)v}$ and therefore this equation follows from 11.1(2).

Consider the special case of 12.3.3 in which $r = 0$ and $j : V_{n-1} \to M_n$ is the embedding of an almost complex submanifold of M_n. Let $v \in H^2(M_n, \mathbf{Z})$ be the cohomology class determined by V_{n-1} and let η be the $\mathbf{C}^*$-bundle with Chern class $1+v$. Then by Theorem 12.3.1,

$$T_y(M_n, \xi) = T_y(V_{n-1}, j^* \xi) + T_y(M_n, \xi \otimes \eta^{-1}) + y\, T_y(V_{n-1}, j^*(\xi \otimes \eta^{-1})) \tag{17}$$

and, equating coefficients,

$$T^p(M_n, \xi) \\ = T^p(V_{n-1}, j^* \xi) + T^p(M_n, \xi \otimes \eta^{-1}) + T^{p-1}(V_{n-1}, j^*(\xi \otimes \eta^{-1})) . \quad (18)$$

In (18) it is understood that $T^p(M_n, \xi) = 0$ for $p < 0$ and $p > n$, and that $T^p(V_{n-1}, j^* \xi) = 0$ for $p < 0$ and $p > n - 1$. Formula (4) of **12.1** is a special case of (18).

§ 13. Split manifolds and splitting methods

13.1. The discussion in this section is valid for continuous, differentiable or complex analytic bundles (cf. 3.1, 3.2). It is to be understood in these cases that X is respectively a topological space, differentiable manifold or complex manifold.

Let ξ be a $\mathbf{GL}(q, \mathbf{C})$-bundle over X. We consider a principal bundle L over X associated to ξ with $\mathbf{GL}(q, \mathbf{C})$ as fibre and construct the fibre bundle

$$E = L/\Delta(q, \mathbf{C}) \quad \text{[cf. 3.4. b) and 4.1. a)]}$$

with the flag manifold $\mathbf{F}(q) = \mathbf{GL}(q, \mathbf{C})/\Delta(q, \mathbf{C})$ as fibre:

$$\varphi : E \to X, \quad \text{fibre } \mathbf{F}(q) . \quad (1)$$

The tangent principal bundle of the complex manifold $\mathbf{F}(q)$ will be denoted by $\mathbf{T}(q)$:

$$\mathbf{T}(q) \to \mathbf{F}(q), \quad \text{fibre } \mathbf{GL}(m, \mathbf{C}) . \quad (2)$$

Here $m = q(q-1)/2$ is the complex dimension of $\mathbf{F}(q)$.

The group $\mathbf{GL}(q, \mathbf{C})$ operates by left translation on $\mathbf{F}(q)$ and hence also in a natural way on $\mathbf{T}(q)$. The method of 3.2. d) therefore allows us to construct, from the action of $\mathbf{GL}(q, \mathbf{C})$ on the cartesian product $L \times \mathbf{T}(q)$, a fibre bundle $\mathfrak{E}(q)$ associated to ξ:

$$\mathfrak{E}(q) \to X, \quad \text{fibre } \mathbf{T}(q) . \quad (3)$$

$\mathfrak{E}(q)$ is a principal bundle over E with $\mathbf{GL}(m, \mathbf{C})$ as fibre. The result is the following commutative diagram in which each arrow is the projection map of a fibre bundle.

$$\begin{array}{ccc} \mathfrak{E}(q) & \xrightarrow{\text{fibre } \mathbf{GL}(m,\mathbf{C})} & E \\ & {\scriptstyle \text{fibre } \mathbf{T}(q)} \searrow \quad \swarrow {\scriptstyle \varphi \text{, fibre } \mathbf{F}(q)} & \\ & X & \end{array} \quad (4)$$

Over each point of X the situation is as in (2).

The $\mathbf{GL}(m, \mathbf{C})$-bundle over E which is associated to the principal bundle $\mathfrak{E}(q)$ will be denoted by ξ^Δ and called the "bundle along the fibres $\mathbf{F}(q)$ of E".

The bundle $\varphi^* \xi$ over E admits the group $\Delta(q, \mathbf{C})$ as structure group in a natural way (Theorem 3.4.4). By 4.1. c) this defines an ordered sequence $\xi_1, \xi_2, \ldots, \xi_q$ of q diagonal $\mathbf{C}^*$-bundles over E.

Theorem 13.1.1. In terms of the above notations: *The* $\mathbf{GL}(m, \mathbf{C})$-*bundle* ξ^Δ *over* E *admits* $\Delta(m, \mathbf{C})$ *as structure group in such a way that the* m *diagonal* $\mathbf{C}^*$-*bundles are the bundles* $\xi_i \otimes \xi_j^{-1}$ $(i > j)$ *in the following order*: $\xi_i \otimes \xi_j^{-1}$ *is before* $\xi_{i'} \otimes \xi_{j'}^{-1}$ *if either* $j > j'$ *or* $(j = j'$ *and* $i < i')$.

The proof will be by induction on q. The theorem is trivial for $q = 1$.

a) Construct the fibre bundle $\overline{X} = L/\mathbf{GL}(1, q-1; \mathbf{C})$. The fibre of $\overline{X}$ is the complex projective space $\mathbf{P}_{q-1}(\mathbf{C})$, since by 4.1. a),

$$\mathfrak{G}(1, q-1; \mathbf{C}) = \mathbf{P}_{q-1}(\mathbf{C}) = \mathbf{GL}(q, \mathbf{C})/\mathbf{GL}(1, q-1; \mathbf{C}) . \tag{5}$$

A matrix of $\mathbf{GL}(1, q-1; \mathbf{C})$ has the form [see 4.1. a)]

$$A = \left(\begin{array}{c|c} a & a_{12}, \ldots, a_{1q} \\ \hline 0 & A'' \end{array}\right) .$$

The homomorphism $h: \mathbf{GL}(1, q-1; \mathbf{C}) \to \mathbf{GL}(q-1, \mathbf{C})$ which associates to $A \in \mathbf{GL}(1, q-1; \mathbf{C})$ the matrix $A'' \in \mathbf{GL}(q-1, \mathbf{C})$ maps $\Delta(q, \mathbf{C})$ on to $\Delta(q-1, \mathbf{C})$. Therefore

$$\mathbf{GL}(1, q-1; \mathbf{C})/\Delta(q, \mathbf{C}) = \mathbf{GL}(q-1, \mathbf{C})/\Delta(q-1, \mathbf{C}) = \mathbf{F}(q-1) . \tag{6}$$

b) Clearly E is a fibre bundle over $\overline{X}$ with

$$\mathbf{F}(q-1) = \mathbf{GL}(1, q-1; \mathbf{C})/\Delta(q, \mathbf{C})$$

as fibre and $\mathbf{GL}(1, q-1; \mathbf{C})$ as structure group [cf. 3.2. c)]. Since the kernel of the homomorphism $h: \mathbf{GL}(1, q-1; \mathbf{C}) \to \mathbf{GL}(q-1, \mathbf{C})$ operates trivially on $\mathbf{F}(q-1)$ it follows [cf. (6)] that E admits the group $\mathbf{GL}(q-1, \mathbf{C})$ as structure group in a natural way.

If X is a point then $E = \mathbf{F}(q)$, $\overline{X} = \mathbf{P}_{q-1}(\mathbf{C})$. In this case the conclusion is that $\mathbf{F}(q)$ is a fibre bundle over $\mathbf{P}_{q-1}(\mathbf{C})$ with $\mathbf{F}(q-1)$ as fibre and $\mathbf{GL}(q-1, \mathbf{C})$ as structure group:

$$\pi: \mathbf{F}(q) \to \mathbf{P}_{q-1}(\mathbf{C}), \quad \text{fibre } \mathbf{F}(q-1) . \tag{7}$$

There is a commutative diagram

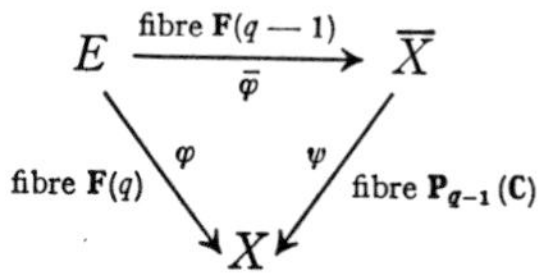

$$\begin{array}{ccc} E & \xrightarrow[\bar{\varphi}]{\text{fibre } \mathbf{F}(q-1)} & \overline{X} \\ {\scriptstyle \text{fibre } \mathbf{F}(q)}\ \searrow^{\varphi} & & \swarrow^{\psi}\ {\scriptstyle \text{fibre } \mathbf{P}_{q-1}(\mathbf{C})} \\ & X & \end{array} \tag{8}$$

Over each point of X the situation is as in (7).

c) The structure group of $\psi^* \xi$ can be reduced to $\mathbf{GL}(1, q-1, \mathbf{C})$ in a natural way; let η be the resulting $\mathbf{C}^*$-sub-bundle over $\overline{X}$ and let $\bar{\xi}$ be the $\mathbf{GL}(q-1, \mathbf{C})$-quotient-bundle over $\overline{X}$. The fibre bundles E and $\overline{X}$

over X are both associated to ξ, while the fibre bundle E over $\overline{X}$ is associated to $\bar{\xi}$. The bundle $\bar{\varphi}^* \bar{\xi}$ over E admits $\Delta(q-1, \mathbf{C})$ as structure group in a natural way. The corresponding diagonal $\mathbf{C}^*$-bundles are given by the sequence $\xi_2, \xi_3, \ldots, \xi_q$. Moreover $\bar{\varphi}^* \eta = \xi_1$.

d) Now consider the principal tangent bundle $\mathbf{T}$ of the complex manifold $\mathbf{P}_{q-1}(\mathbf{C})$:

$$\mathbf{T} \to \mathbf{P}_{q-1}(\mathbf{C}), \quad \text{fibre } \mathbf{GL}(q-1, \mathbf{C}) . \tag{9}$$

The group $\mathbf{GL}(q, \mathbf{C})$ operates on $\mathbf{P}_{q-1}(\mathbf{C})$ and hence also in a natural way on $\mathbf{T}$. It can be shown that $\mathbf{GL}(q, \mathbf{C})$ operates transitively on $\mathbf{T}$; that is, given any two points y_1, y_2 of $\mathbf{T}$ there is an element of $\mathbf{GL}(q, \mathbf{C})$ which sends y_1 to y_2. Therefore $\mathbf{T}$ can be represented as a quotient space $\mathbf{GL}(q, \mathbf{C})/H$ of $\mathbf{GL}(q, \mathbf{C})$, where H is the subgroup which leaves fixed a given point y_0 of $\mathbf{T}$. If an element of $\mathbf{GL}(q, \mathbf{C})$ leaves y_0 fixed then it must also leave fixed the whole fibre of (9) through y_0. We represent $\mathbf{P}_{q-1}(\mathbf{C})$ as a quotient space by (5) and choose as y_0 a point which lies in the fibre of (9) over the point of $\mathbf{P}_{q-1}(\mathbf{C})$ corresponding to the coset $\mathbf{GL}(1, q-1; \mathbf{C})$. The required group H is then a subgroup of $\mathbf{GL}(1, q-1; \mathbf{C})$ and it is now easy to show that H is actually the subgroup of matrices of the form

$$\left(\begin{array}{c|c} a & a_{12}, \ldots, a_{1q} \\ \hline 0 & aI \end{array}\right), \quad I = \text{identity matrix.}$$

H is a normal subgroup of $\mathbf{GL}(1, q-1; \mathbf{C})$, and is the kernel of the homomorphism $\mathbf{GL}(1, q-1; \mathbf{C}) \to \mathbf{GL}(q-1, \mathbf{C})$ which, in the notation of a), maps A to $a^{-1} A''$. Now dividing the "numerator and denominator" of (5) by H we obtain

$$(\mathbf{GL}(q, \mathbf{C})/H)/(\mathbf{GL}(1, q-1; \mathbf{C})/H) = \mathbf{P}_{q-1}(\mathbf{C}) .$$

It follows that (9) is identical to the fibre bundle given by

$$\mathbf{T} = \mathbf{GL}(q, \mathbf{C})/H \to (\mathbf{GL}(q, \mathbf{C})/H)/(\mathbf{GL}(1, q-1; \mathbf{C})/H) . \tag{9*}$$

e) From the principal bundle L over X we can construct the space L/H. There is a commutative diagram

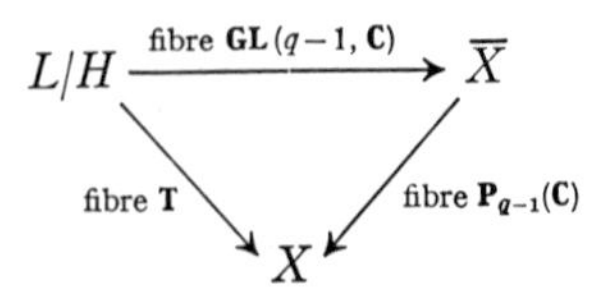

Over each point of X the situation is as in (9).

L/H is a principal bundle over $\overline{X}$. By c) and d) it is associated to the bundle $\eta^{-1} \otimes \bar{\xi}$ over $\overline{X}$. We call $\eta^{-1} \otimes \bar{\xi}$ the "bundle along the fibres $\mathbf{P}_{q-1}(\mathbf{C})$ of $\overline{X}$" [see (8)].

f) We now carry out construction (1)–(4) for the $\mathbf{GL}(q-1, \mathbf{C})$-bundle $\bar{\xi}$ over $\bar{X}$, marking everything which arises from $\bar{\xi}$ by adding a bar. Thus let $\bar{m} = (q-1)(q-2)/2$. Then $m = q(q-1)/2 = \bar{m} + (q-1)$. It is easy to show that the structure group of the $\mathbf{GL}(m, \mathbf{C})$-bundle ξ^{Δ} [bundle along the fibres $\mathbf{F}(q)$ of E] can be reduced to the group $\mathbf{GL}(\bar{m}, q-1; \mathbf{C})$ so that $\bar{\xi}^{\Delta}$ [bundle along the fibres $\mathbf{F}(q-1)$ of E] is the corresponding subbundle and $\bar{\varphi}^*(\eta^{-1} \otimes \bar{\xi})$ the corresponding quotient bundle. Here $\eta^{-1} \otimes \bar{\xi}$ is the bundle along the fibres $\mathbf{P}_{q-1}(\mathbf{C})$ of $\bar{X}$.

We assume that the theorem is proved for $q-1$. The diagonal $\mathbf{C}^*$-bundles of $\bar{\varphi}^* \bar{\xi}$ are $\xi_2, \ldots, \xi_q$ in that order. Therefore $\bar{\xi}^{\Delta}$ admits the group $\boldsymbol{\Delta}(\bar{m}, \mathbf{C})$ as structure group in such a way that the diagonal $\mathbf{C}^*$-bundles $\xi_i \otimes \xi_j^{-1}$ $(i > j \geqq 2)$ are in the order given by the statement of the theorem. But

$$\bar{\varphi}^*(\eta^{-1} \otimes \bar{\xi}) = \xi_1^{-1} \otimes \bar{\varphi}^* \bar{\xi} .$$

Therefore $\bar{\varphi}^*(\eta^{-1} \otimes \bar{\xi})$ admits the group $\boldsymbol{\Delta}(q-1, \mathbf{C})$ as structure group with the diagonal $\mathbf{C}^*$-bundles

$$\xi_2 \otimes \xi_1^{-1}, \ldots, \xi_q \otimes \xi_1^{-1} .$$

This completes the proof for q. Q. E. D.

13.2. Theorem 13.1.1 holds in the complex analytic case. Since this fact will be particularly important in the sequel, we restate it as a separate theorem.

Theorem 13.2.1. *Let X be a complex manifold, ξ a complex analytic $\mathbf{GL}(q, \mathbf{C})$-bundle over X, and L a complex analytic principal bundle over X associated to ξ. Consider the fibre bundle $E = L/\boldsymbol{\Delta}(q, \mathbf{C})$ with the flag manifold $\mathbf{F}(q) = \mathbf{GL}(q, \mathbf{C})/\boldsymbol{\Delta}(q, \mathbf{C})$ as fibre*:

$$\varphi : E \to X, \quad \text{fibre } \mathbf{F}(q) . \tag{1*}$$

E is a complex manifold and φ is a holomorphic map of E on to X. The structure group of the complex analytic bundle $\varphi^* \xi$ can be complex analytically reduced to the group $\boldsymbol{\Delta}(q, \mathbf{C})$ in a natural way. Let $\xi_1, \xi_2, \ldots, \xi_q$ (in that order) be the q diagonal complex analytic $\mathbf{C}^*$-bundles. The bundle ξ^{Δ} along the fibres of (1) is a complex analytic $\mathbf{GL}(m, \mathbf{C})$-bundle $[m = q(q-1)/2]$, whose structure group can be complex analytically reduced to $\boldsymbol{\Delta}(m, \mathbf{C})$; in this case the m diagonal complex analytic $\mathbf{C}^*$-bundles are the bundles $\xi_i \otimes \xi_j^{-1}$ $(i > j)$ in the order specified in* Theorem 13.1.1.

Remark: The proof of Theorem 13.1.1 given in the previous section is direct, but left a number of details to the reader. It has been pointed out by A. Borel that the fact that the structure group of the bundle ξ^{Δ} can be reduced to $\boldsymbol{\Delta}(m, \mathbf{C})$ follows immediately from a theorem of Lie. The statement of the theorem (see for instance C. Chevalley: Théorie des groupes de Lie, Tome III. Paris: Hermann 1955, especially p. 100 and p. 104) is:

Let H be a solvable connected complex LIE *group and* $\varrho: H \to \mathbf{GL}(m, \mathbf{C})$ *a holomorphic homomorphism. Then there is an element* $a \in \mathbf{GL}(m, \mathbf{C})$ *such that* $a\, \varrho(H)\, a^{-1} \subset \boldsymbol{\Delta}(m, \mathbf{C})$.

The statement about the structure group of ξ^{Δ} is deduced as follows. Let $e_0 \in \mathbf{F}(q) = \mathbf{GL}(q, \mathbf{C})/\boldsymbol{\Delta}(q, \mathbf{C})$ be the point corresponding to the coset $\boldsymbol{\Delta}(q, \mathbf{C})$. The group $\mathbf{GL}(q, \mathbf{C})$ operates on $\mathbf{F}(q)$, and $\boldsymbol{\Delta}(q, \mathbf{C})$ is the isotropy group of e_0 [*i. e.* the subgroup consisting of all elements of $\mathbf{GL}(q, \mathbf{C})$ which leave e_0 fixed]. $\boldsymbol{\Delta}(q, \mathbf{C})$ operates on the contravariant tangent space $\mathbf{C}_m(e_0)$ of $e_0 \in \mathbf{F}(q)$ and this operation defines a holomorphic homomorphism $\boldsymbol{\Delta}(q, \mathbf{C}) \to \mathbf{GL}(m, \mathbf{C})$, $m = \frac{1}{2} q(q-1)$. Since $\boldsymbol{\Delta}(q, \mathbf{C})$ is solvable the theorem of LIE implies that there is in $\mathbf{C}_m(e_0)$ a flag of linear subspaces $L_0 \subset L_1 \subset \cdots \subset L_m = \mathbf{C}_m(e_0)$ such that each L_i is mapped into itself by every element of the group $\boldsymbol{\Delta}(q, \mathbf{C})$, *i. e.* the flag is invariant under the operation of the group. Now $\mathbf{GL}(q, \mathbf{C})$ operates transitively on $\mathbf{F}(q)$ so that the flag can be transplanted to any point of $\mathbf{F}(q)$. This transplanting is unambiguous because the flag remains invariant under the operation of the isotropy group. The conclusion is that $\mathbf{F}(q)$ admits a tangential complex analytic field of flags which is left invariant, *i. e.* goes over into itself under the operations of $\mathbf{GL}(q, \mathbf{C})$. The required statement about ξ^{Δ} now follows. For generalisations of Theorem 13.1.1 and for its connection with the theory of roots of LIE groups we refer to BOREL-HIRZEBRUCH [1].

13.3. Let X be a (differentiable) almost complex manifold of complex dimension n and ξ a differentiable $\mathbf{GL}(q, \mathbf{C})$ bundle over X. The construction of 13.1 yields a differentiable manifold E which is a fibre bundle over X with fibre $\mathbf{F}(q)$ and differentiable projection map $\varphi: E \to X$. It is clear that E admits an almost complex structure, whose tangent $\mathbf{GL}(n+m, \mathbf{C})$ bundle $\theta(E)$, $m = q(q-1)/2$, has the "bundle along the fibres" ξ^{Δ} as subbundle and the bundle $\varphi^* \theta(X)$ as the corresponding quotient bundle. Let ξ_i $(i = 1, \ldots, q)$ be the diagonal $\mathbf{C}^*$-bundles of $\varphi^* \xi$ over E, and let $c(\xi_i) = 1 + \gamma_i$, $\gamma_i \in H^2(E, \mathbf{Z})$. Theorem 13.1.1 implies that the total CHERN class of E is given by

$$c(E) = \varphi^* c(X) \prod_{q \geqq i > j \geqq 1} (1 + \gamma_i - \gamma_j)\,. \tag{10}$$

If in particular ξ is chosen as the tangent bundle $\theta(X)$ of X we denote the almost complex manifold E by X^{Δ}. In this case $\varphi^* \xi = \varphi^* \theta(X)$ admits the group $\boldsymbol{\Delta}(n, \mathbf{C})$ as structure group, the corresponding n diagonal $\mathbf{C}^*$-bundles are $\xi_1, \ldots, \xi_n$, and $\theta(E)$ admits the group $\boldsymbol{\Delta}(n(n+1)/2, \mathbf{C})$ as structure group with the $n(n+1)/2$ diagonal bundles $\xi_i \otimes \xi_j^{-1}$, $\xi_1, \ldots, \xi_n$ $(n \geqq i > j \geqq 1)$. The total CHERN class of X^{Δ} is therefore given by

$$c(X^{\Delta}) = \prod_{i=1}^{n} (1 + \gamma_i) \prod_{n \geq i > j \geq 1} (1 + \gamma_i - \gamma_j) . \tag{11}$$

13.4. The discussion of the previous section can be carried over to the complex analytic case. Let X be a complex manifold of complex dimension n with complex analytic tangent bundle $\theta(X)$ and let ξ be a complex analytic $\mathbf{GL}(q, \mathbf{C})$-bundle over X. Then E is in a natural way a complex manifold of dimension $n + m$, $m = q(q-1)/2$, and $\varphi: E \to X$ is a holomorphic map. E is a complex analytic fibre bundle over X with fibre $\mathbf{F}(q)$ and projection map φ. The complex analytic tangent $\mathbf{GL}(n+m, \mathbf{C})$-bundle $\theta(E)$ admits $\mathbf{GL}(m, n; \mathbf{C})$ as structure group in a natural way, since E admits a complex analytic field of complex m-dimensional plane elements (the field tangent to the fibres of E). The complex analytic subbundle is the $\mathbf{GL}(m, \mathbf{C})$-bundle ξ^{Δ}, and the corresponding complex analytic quotient bundle is the $\mathbf{GL}(n, \mathbf{C})$-bundle $\varphi^* \theta(X)$. The CHERN class of the complex manifold E is given by (10). In the special case in which $\xi = \theta(X)$ we again write $E = X^{\Delta}$. In this case both the subbundle ξ^{Δ} and the quotient bundle $\varphi^* \xi = \varphi^* \theta(X)$ admit the corresponding group of triangular matrices as complex analytic structure group. This shows that the structure group of the complex analytic bundle $\theta(X^{\Delta})$ can be reduced complex analytically to the group $\Delta(n(n+1)/2, \mathbf{C})$, and that the corresponding $n(n+1)/2$ diagonal $\mathbf{C}^*$-bundles are $\xi_i \otimes \xi_j^{-1}$, $\xi_1, \ldots, \xi_n$ $(n \geq i > j \geq 1)$. If $c(\xi_i) = 1 + \gamma_i$ then the total CHERN class of X^{Δ} is given by (11).

13.5. a) An almost complex manifold X of complex dimension n is called a *split manifold* if the (differentiable) tangent $\mathbf{GL}(n, \mathbf{C})$-bundle $\theta(X)$ admits the group $\Delta(n, \mathbf{C})$ of triangular matrices as structure group. This defines n diagonal bundles $\xi_1, \ldots, \xi_n \in H^1(X, \mathbf{C}^*_{\mathfrak{d}})$ and in fact $\theta(X)$ is the WHITNEY sum of the bundles ξ_i. If $c(\xi_i) = 1 + a_i$, $a_i \in H^2(X, \mathbf{Z})$ then

$$c(X) = \prod_{i=1}^{n} (1 + a_i) . \tag{12}$$

b) A complex manifold X of complex dimension n is called a *complex analytic split manifold* if the complex analytic $\mathbf{GL}(n, \mathbf{C})$-bundle $\theta(X)$ admits the group $\Delta(n, \mathbf{C})$ of triangular matrices as complex analytic structure group, *i. e.* $\theta(X)$ is an element in the image of the map

$$H^1(X, \Delta(n, \mathbf{C})_\omega) \to H^1(X, \mathbf{GL}(n, \mathbf{C})_\omega) .$$

This defines n diagonal bundles $\xi_1, \ldots, \xi_n \in H^1(X, \mathbf{C}^*_\omega)$. In general $\theta(X)$ is not the complex analytic WHITNEY sum of $\xi_1, \ldots, \xi_n$. Nevertheless, if all bundles are regarded as continuous (or differentiable) bundles, then $\theta(X)$ is the WHITNEY sum $\xi_1 \oplus \cdots \oplus \xi_n$. The CHERN class of X is therefore given by (12).

The process described in sections 13.3 and 13.4 therefore associates to each almost complex manifold X an (almost complex) split manifold X^Δ, and to each complex manifold X a complex analytic split manifold X^Δ. This fact will be of decisive significance in the sequel. It will appear that certain theorems hold for X whenever they hold for X^Δ, and therefore need to be proved only in the case that X is a split manifold.

13.6. Let X be a compact almost complex split manifold of complex dimension n. We use the notations of 13.5 a) and derive a formula which will imply that the Todd genus $T(X)$ can be expressed in terms of virtual indices:

$$(1+y)^n\, T(X) = \sum_{l=0}^{n} y^l \sum_{1 \leq i_1 < \cdots < i_l \leq n} T_y(a_{i_1}, \ldots, a_{i_l})_X\,. \tag{13}$$

For the proof we recall the definition of the virtual T_y-genus in 11.2 (7), apply (12), and obtain for the right hand side of (13)

$$\begin{aligned} &\varkappa_n \left[\prod_{i=1}^{n} (1 + yR(y; a_i)) \prod_{i=1}^{n} Q(y; a_i)\right] \\ &= \varkappa_n \left[\prod_{i=1}^{n} (Q(y; a_i) + a_i\, y)\right] \\ &= \varkappa_n \left[\prod_{i=1}^{n} \frac{(1+y)\, a_i}{1 - \exp(-(1+y)\, a_i)}\right] \\ &= (1+y)^n\, \varkappa_n \left[\prod_{i=1}^{n} \frac{a_i}{1 - \exp(-a_i)}\right] = (1+y)^n\, T(X)\,. \end{aligned}$$

For $y = 1$ the virtual T_y-genus becomes the virtual index:

$$2^n\, T(X) = \sum_{l=0}^{n} \sum_{1 \leq i_1 < \cdots < i_l \leq n} \tau(a_{i_1}, \ldots, a_{i_l})_X\,. \tag{13*}$$

Since the virtual index is an integer (Theorem 9.3.1) a corollary is

Theorem 13.6.1. *The* Todd *genus of a compact almost complex split manifold multiplied by* 2^n *is an integer.*

Formula (13) can be generalised to apply to the virtual T-genus. If $b_1, \ldots, b_r \in H^2(X, \mathbf{Z})$, $(r \leq n)$, then

$$\begin{aligned} &(1+y)^{n-r}\, T(b_1, b_2, \ldots, b_r)_X \\ &= \varkappa_n \left[\prod_{i=1}^{r} R(y; b_i)\,(1 + yR(y; b_i))^{-1} \prod_{j=1}^{n} (1 + yR(y; a_j)) \prod_{k=1}^{n} Q(y; a_k)\right]. \end{aligned} \tag{14}$$

This formula, and the definition of the virtual T_y-genus, imply that $(1+y)^{n-r}\, T(b_1, \ldots, b_r)_X$ can be expressed as a sum of terms each of which is a virtual T_y-genus multiplied by a polynomial in y with integral coefficients. If $y = 1$ then $2^{n-r}\, T(b_1, \ldots, b_r)_X$ is expressed as a sum of terms each of which is an integral multiple of a virtual index. This proves:

Theorem 13.6.2. *Let X be a compact almost complex split manifold, and let $b_1, \ldots, b_r$ be elements of $H^2(X, \mathbf{Z})$. The virtual* TODD *genus $T(b_1, \ldots, b_r)_X$ multiplied by 2^{n-r} is an integer.*

§ 14. Multiplicative properties of the TODD genus

14.1. Some algebraic remarks: Let $\mathbf{K}$ be a field of characteristic 0, and let $c_1, \ldots, c_n$ be indeterminates. We consider the field $\mathbf{K}(c_1, \ldots, c_n)$ and an indeterminate x, and adjoin elements $\gamma_1, \ldots, \gamma_n$ to $\mathbf{K}(c_1, \ldots, c_n)$ such that

$$1 + c_1 x + \cdots + c_n x^n = (1 + \gamma_1 x) \ldots (1 + \gamma_n x) .$$

The field $\mathbf{K}(c_1, \ldots, c_n)(\gamma_1, \ldots, \gamma_n)$ is then an algebraic extension of $\mathbf{K}(c_1, \ldots, c_n)$ of degree $n!$. The $n!$ elements $\gamma_1^{a_1} \gamma_2^{a_2} \cdots \gamma_{n-1}^{a_{n-1}}$ $(0 \leqq a_i \leqq \leqq n - i)$ form an additive basis for the extension field. It is easy to prove the following lemma:

Lemma 14.1.1. *Every formal power series P in $\gamma_1, \ldots, \gamma_n$ with coefficients in $\mathbf{K}$ can be expressed uniquely in the form:*

$$P = \sum_{0 \leqq a_i \leqq n-i} \varrho_{a_1 a_2 \ldots a_{n-1}} \gamma_1^{a_1} \gamma_2^{a_2} \cdots \gamma_{n-1}^{a_{n-1}} \tag{1}$$

where the $\varrho_{a_1 a_2 \ldots a_{n-1}}$ are formal power series in $c_1, \ldots, c_n$ with coefficients in $\mathbf{K}$. If P has integer coefficients then each $\varrho_{a_1 a_2 \ldots a_{n-1}}$ has integer coefficients.

We define the "indicator" $\varrho(P)$ by

$$\varrho(P) = (-1)^{n(n-1)/2} \varrho_{n-1, n-2, \ldots, 1} \cdot$$

If $s: (\gamma_1, \gamma_2, \ldots, \gamma_n) \to (\gamma_{j_1}, \gamma_{j_2}, \ldots, \gamma_{j_n})$ is a permutation there is an expression corresponding to (1)

$$P = \sum_{0 \leqq a_i \leqq n-i} \varrho^{(s)}_{a_1 a_2 \ldots a_{n-1}} \gamma_{j_1}^{a_1} \gamma_{j_2}^{a_2} \cdots \gamma_{j_{n-1}}^{a_{n-1}} \tag{1, s}$$

and the s-indicator of P is defined by

$$\varrho^{(s)}(P) = (-1)^{n(n-1)/2} \varrho^{(s)}_{n-1, n-2, \ldots, 1} \cdot$$

The $n!$ elements $\gamma_{j_1}^{a_1} \gamma_{j_2}^{a_2} \cdots \gamma_{j_{n-1}}^{a_{n-1}}$ $(0 \leqq a_i \leqq n - i)$ form another basis for the extension field, and it is clear that all of these elements have indicator 0 with the exception of $\gamma_{j_1}^{n-1} \gamma_{j_2}^{n-2} \ldots \gamma_{j_{n-1}}$ which has indicator ± 1. Therefore

$$\varrho^{(s)}(P) = \varrho(s(P)) = \pm \varrho(P) \tag{2}$$

where $s(P)$ denotes the power series got by applying the permutation s to P.

Lemma 14.1.2. *If P remains invariant under the interchange of γ_i and γ_j for some $i \neq j$ then the indicator $\varrho(P)$ of P is zero.*

Proof: It is sufficient to prove the lemma for P a polynomial. Suppose that P remains invariant under the interchange of γ_i and γ_j, $(i \neq j)$. By (2) we can assume, without loss of generality that $i = n-1$, $j = n$. Now GALOIS theory implies that P is an element of the field extension of $\mathbf{K}(c_1, c_2, \ldots, c_n)$ generated by $\gamma_1, \ldots, \gamma_{n-2}$. Therefore the indicator $\varrho(P)$ is zero.

Corollary: *Let s be the permutation $(\gamma_1, \gamma_2, \ldots, \gamma_n) \to (\gamma_{j_1}, \gamma_{j_2}, \ldots, \gamma_{j_n})$. Then*

$$\varrho(\gamma_1^{n-1}\gamma_2^{n-2}\ldots\gamma_{n-1}) = \operatorname{sign}(s)\cdot\varrho(s(\gamma_1^{n-1}\gamma_2^{n-2}\ldots\gamma_{n-1}))\,. \tag{3}$$

Proof: It is sufficient to prove (3) for the case where s is an interchange (i, j). In this case

$$\gamma_1^{n-1}\gamma_2^{n-2}\ldots\gamma_{n-1} + \gamma_{j_1}^{n-1}\gamma_{j_2}^{n-2}\ldots\gamma_{j_{n-1}}$$

remains invariant under s, and the result follows from Lemma 14.1.2. It is now easy to give a formula for $\varrho(P)$. By (2) and (3)

$$\varrho(P) = \operatorname{sign}(s)\cdot\varrho^{(s)}(P) = \operatorname{sign}(s)\cdot\varrho(s(P))\,. \tag{2*}$$

This implies

$$n!\,\varrho(P) = \varrho\Big(\sum_s \operatorname{sign}(s)\cdot s(P)\Big) \tag{4}$$

where the summation is over all $n!$ permutations s. The expression $\sum_s \operatorname{sign}(s)\cdot s(P)$ is clearly alternating. The quotient

$$\mathfrak{q}(P) = \Big(\sum_s \operatorname{sign}(s)\cdot s(P)\Big)\Big/\prod_{i>j}(\gamma_i - \gamma_j)$$

is therefore symmetric and hence a power series in $c_1, \ldots, c_n$. This gives

$$n!\,\varrho(P) = \varrho\Big(\prod_{i>j}(\gamma_i - \gamma_j)\Big)\cdot\mathfrak{q}(P)\,. \tag{4*}$$

If $P = \gamma_1^{n-1}\gamma_2^{n-2}\ldots\gamma_{n-1}$ then $\varrho(P) = (-1)^{n(n-1)/2}$ and

$$\sum_s \operatorname{sign}(s)\cdot s(P) = (-1)^{n(n-1)/2}\prod_{i>j}(\gamma_i - \gamma_j)\,.$$

Now (4) implies that $\varrho\Big(\prod_{i>j}(\gamma_i - \gamma_j)\Big) = n!$ and (4*) then gives the required formula for an arbitrary power series P:

$$\varrho(P) = \Big(\sum_s \operatorname{sign}(s)\cdot s(P)\Big)\Big/\prod_{i>j}(\gamma_i - \gamma_j)\,. \tag{5}$$

Lemma 14.1.3. *Let* $P = \prod_{i>j}\dfrac{\gamma_j - \gamma_i}{\exp(\gamma_j - \gamma_i) - 1}$. *Then* $\varrho(P) = 1$.

Proof: Let $2a = \sum_{i>j}(\gamma_i - \gamma_j)$. Then by 1.7

$$P = e^a \prod_{i>j}\frac{\gamma_i - \gamma_j}{2\sinh((\gamma_i - \gamma_j)/2)}$$

and by (5)

$$\varrho(P) = \left(\sum_s \operatorname{sign}(s)\, e^{s(a)}\right) \Big/ \prod_{i>j} 2\sinh((\gamma_i - \gamma_j)/2)\,.$$

Let $x_i = \exp(-\gamma_i/2)$. Then

$$e^a (x_1 \ldots x_n)^{n-1} = (x_1^2)^{n-1} (x_2^2)^{n-2} \ldots x_{n-1}^2$$

and

$$2x_i\, x_j \sinh((\gamma_i - \gamma_j)/2) = x_j^2 - x_i^2\,.$$

The result now follows (VANDERMONDE determinants).

14.2. We now return to the flag manifold $\mathbf{F}(n) = \mathbf{GL}(n, \mathbf{C})/\Delta(n, \mathbf{C})$. Theorem 13.2.1, with X a point, shows that $\mathbf{F}(n)$ is a complex analytic split manifold. The (total) CHERN class of $\mathbf{F}(n)$ is

$$c(\mathbf{F}(n)) = \prod_{i>j} (1 + \gamma_i - \gamma_j)\,. \tag{6}$$

The elements $\gamma_i \in H^2(\mathbf{F}(n), \mathbf{Z})$ satisfy $c(\xi_i) = 1 + \gamma_i$ (see 13.2) and

$$\prod_{i=1}^{n} (1 + \gamma_i) = 1\,. \tag{7}$$

According to BOREL [2] the cohomology ring $H^*(\mathbf{F}(n), \mathbf{Z})$ is generated by the γ_i with (7) as the only relation:

$$H^*(\mathbf{F}(n), \mathbf{Z}) = \mathbf{Z}[\gamma_1, \ldots, \gamma_n]/I^+(c_1, \ldots, c_n)$$

where $\gamma_1, \ldots, \gamma_n$ are regarded as indeterminates and where I^+ is the ideal generated by the elementary symmetric functions $c_1, \ldots, c_n$ in the γ_i. Applying the results of 14.1 we see that the $n!$ elements $\gamma_1^{a_1} \gamma_2^{a_2} \ldots \gamma_{n-1}^{a_{n-1}}$, $0 \leqq a_i \leqq n - i$, form an additive basis for the cohomology ring $H^*(\mathbf{F}(n), \mathbf{Z})$. A polynomial P in the γ_i with integer coefficients defines an element of the cohomology ring. To express this element in terms of the given basis use the expression (1) obtained in the previous section: the coefficients ϱ in (1) are equal to their constant terms modulo the ideal I^+.

The EULER-POINCARÉ characteristic of $\mathbf{F}(n)$ is $n!$ since $H^*(\mathbf{F}(n), \mathbf{Z})$ contains only elements of even degree. Therefore (6) and Theorem 4.10.1 imply that

$$n! = \prod_{i>j} (\gamma_i - \gamma_j)\, [\mathbf{F}(n)]\,. \tag{8}$$

In 14.1 it was shown that $\varrho\left(\prod_{i>j} (\gamma_i - \gamma_j)\right) = n!$. Therefore $(-1)^{n(n-1)/2}\, \gamma_1^{n-1} \gamma_2^{n-2} \ldots \gamma_{n-1}$ is the generator of $H^m(\mathbf{F}(n), \mathbf{Z})$, $m = n(n-1)/2$, determined by the natural orientation of $\mathbf{F}(n)$.

Finally Lemma 14.1.3 and (6) give the TODD genus of $\mathbf{F}(n)$:

$$T(\mathbf{F}(n)) = 1\,.$$

14.3. We now return to the situation discussed in 13.3.

Theorem 14.3.1. *Let ξ be a differentiable $\mathbf{GL}(q, \mathbf{C})$-bundle over a compact n-dimensional almost complex manifold X and let L be a principal bundle associated to ξ. The fibre bundle $E = L/\Delta(q, \mathbf{C})$ has the flag manifold $\mathbf{F}(q)$ as fibre and can be regarded in a natural way as an almost complex manifold of dimension $n + \frac{1}{2} q(q-1)$. Let ζ be a $\mathbf{GL}(l, \mathbf{C})$-bundle over X and let $\varphi: E \to X$ be the projection. Then the T-characteristic of ζ satisfies*

$$T(E, \varphi^* \zeta) = T(X, \zeta)\, T(\mathbf{F}(q)) = T(X, \zeta)\,. \tag{9}$$

Let $b_1, \ldots, b_r \in H^2(X, \mathbf{Z})$. Then the virtual T-characteristic satisfies

$$T(\varphi^* b_1, \ldots, \varphi^* b_r|, \varphi^* \zeta)_E = T(b_1, \ldots, b_r|, \zeta)_X\,. \tag{10}$$

Proof: Since (9) is a special case it is sufficient to prove (10). Let $c(\varphi^* \xi) = 1 + \varphi^* c_1 + \cdots + \varphi^* c_q = \prod_{i=1}^{q} (1 + \gamma_i)$ and let $m = q(q-1)/2$. Then the definition of the virtual T-characteristic, 12.3 (15), together with 13.3 (10) gives

$$\begin{aligned} &T(\varphi^* b_1, \ldots, \varphi^* b_r|, \varphi^* \zeta)_E \\ &\quad = \varkappa_{n+m}\left[\varphi^*\left(\operatorname{ch}\zeta \cdot \prod_{i=1}^{r}(1 - e^{-b_i}) \cdot \operatorname{td}(X)\right) \prod_{i>j} \frac{\gamma_j - \gamma_i}{\exp(\gamma_j - \gamma_i) - 1}\right]. \end{aligned}$$

We denote the first factor $\varphi^*(\ )$ of the expression in $[\ ]$ by $\varphi^* A$ and the second factor $\prod_{i>j}$ by P. Now apply the algebraic remarks of 14.1 with n replaced by q and the indeterminates $c_1, \ldots, c_n$ replaced by $\varphi^* c_1, \ldots, \varphi^* c_q$. Then P is of the form 14.1 (1). The coefficients $\varrho_{a_1 a_2 \ldots a_{q-1}}$ are polynomials in $\varphi^* c_1, \ldots, \varphi^* c_q$. We have to take the terms of complex dimension $n + m$ in $\varphi^*(A) \cdot P$ and note that any term of the form $\varphi^* x$ with $x \in H^*(X, \mathbf{Z}) \otimes \mathbf{Q}$ is zero if it has complex dimension $> n$. Therefore

$$\varkappa_{n+m}[\varphi^*(A) \cdot P]_E = \varkappa_{n+m}[(-1)^m \varphi^*(A) \cdot \varrho(P)\, \gamma_1^{q-1} \gamma_2^{q-2} \ldots \gamma_{q-1}]\,.$$

Now by Lemma 14.1.3, $\varrho(P) = 1$, and by 14.2 the restriction of $(-1)^m \gamma_1^{q-1} \gamma_2^{q-2} \ldots \gamma_{q-1}$ to a fibre $\mathbf{F}(q)$ is the natural generator of $H^m(\mathbf{F}(q), \mathbf{Z})$. Therefore $\varkappa_{n+m}[\varphi^*(A) \cdot P]_E = \varkappa_n[A]_X$ and the proof of (10) is complete.

Formulae (9) and (10) imply respectively that the Todd genus of X is equal to that of E, and that the virtual Todd genus of $(b_1, \ldots, b_r)$ in X is equal to that of $(\varphi^* b_1, \ldots, \varphi^* b_r)$ in E. If X is an arbitrary compact almost complex manifold we can choose E to be the split manifold X^Δ (see 13.3). Therefore Theorems 13.6.1 and 13.6.2 imply

Theorem 14.3.2. *The* TODD *genus of a compact almost complex manifold X multiplied by 2^n is an integer. More generally the virtual* TODD *genus of $(b_1, \ldots, b_r)$, $b_i \in H^2(X, \mathbf{Z})$, multiplied by 2^{n-r} is an integer.*

14.4. Formula (10) of Theorem 14.3.1 can be generalized:

$$T_y(\varphi^* b_1, \ldots, \varphi^* b_r |, \varphi^* \zeta)_E = T_y(b_1, \ldots, b_r |, \zeta) \cdot T_y(\mathbf{F}(q)) . \quad (10^*)$$

In the proof of Theorem 14.3.1 it is sufficient to generalise Lemma 14.1.3 as follows:

Let y be an indeterminate over the field of rationals, and replace the groundfield $\mathbf{K}$ of 14.1 by the ring of polynomials in y over the rationals. If we let

$$P = \prod_{i>j} Q(y; \gamma_i - \gamma_j) ,$$

where $Q(y; x) = \dfrac{x(y+1)}{1-\exp(-x(y+1))} - x y$, then

$$\varrho(P) = \frac{1-(-1)^n y^n}{1+y} \cdot \frac{1-(-1)^{n-1} y^{n-1}}{1+y} \cdots \frac{1-y^2}{1+y} . \quad (11)$$

Therefore $T_y(\mathbf{F}(n))$ is precisely the formula given in (11), that is

$$T_y(\mathbf{F}(n)) = T_y(\mathbf{P}_{n-1}(\mathbf{C})) \cdot T_y(\mathbf{P}_{n-2}(\mathbf{C})) \ldots T_y(\mathbf{P}_1(\mathbf{C})) .$$

More generally let $\{K_j(c_1, \ldots, c_j)\}$ be an m-sequence with characteristic power series $B(x) = K(1+x)$. Then the proof of Theorem 14.3.1 can be used to prove the equation

$$K(E) = K(X) \cdot K(\mathbf{F}(q))$$

provided that $\varrho\Big(\prod_{q \geq i > j \geq 1} B(\gamma_i - \gamma_j)\Big)$ is an element of the groundfield, and so independent of $c_1, \ldots, c_q$. It is then clear that this element is equal to $K(\mathbf{F}(q))$. (We are using the notations of 14.1, with n replaced by q.)

The T_y-genus, as a genus in the sense of 10.2, has the property

$$T_y(V \times W) = T_y(V)\, T_y(W) .$$

By (10*), if E is a fibre bundle over X with the flag manifold $\mathbf{F}(q)$ as fibre then the T_y-genus behaves multiplicatively

$$T_y(E) = T_y(X)\, T_y(\mathbf{F}(q)) .$$

This raises the question: for what fibre bundles E over X with a given fibre F is it true that $T_y(E) = T_y(X)\, T_y(F)$? It is naturally assumed that E, X, F are compact almost complex manifolds and that the fibration is "compatible with the almost complex structures". We give a special case in which the T_y-genus does behave multiplicatively:

Let ξ be a differentiable $\mathbf{GL}(q, \mathbf{C})$*-bundle over a compact almost complex manifold X, and let L be a principal bundle associated to ξ.* $E' = L/\mathbf{GL}(1, q-1; \mathbf{C})$ *is a fibre bundle over X with fibre* $\mathbf{P}_{q-1}(\mathbf{C})$ *associated to L, and has a natural almost complex structure. Then* $T_y(E') = T_y(X)\, T_y(\mathbf{P}_{q-1}(\mathbf{C}))$.

Proof: $E = L/\boldsymbol{\Delta}(q, \mathbf{C})$ is a fibre bundle over E' with $\mathbf{F}(q-1)$ as fibre [see 13.1 (8)]. Since T_y behaves multiplicatively for fibre bundles with flag manifolds as fibre

$$T_y(E) = T_y(X)\, T_y(\mathbf{F}(q)) \quad \text{and} \quad T_y(E) = T_y(E')\, T_y(\mathbf{F}(q-1))\,.$$

But $T_y(\mathbf{F}(q)) = T_y(\mathbf{F}(q-1))\, T_y(\mathbf{P}_{q-1}(\mathbf{C}))$ and therefore [since $T_y(\mathbf{F}(q-1))$ starts with 1]

$$T_y(E') = T_y(X)\, T_y(\mathbf{P}_{q-1}(\mathbf{C}))\,. \tag{12}$$

For further results on multiplicative properties of the T_y-genus we refer to BOREL-HIRZEBRUCH [1].

Bibliographical note

The analogue for almost complex manifolds of the cobordism ring is due to MILNOR [3]. It can be defined using the concept of weak complex structure (BOREL-HIRZEBRUCH [1], Part III). A weak complex structure of a real vector bundle ξ consists of a trivial bundle α and a complex structure for $\xi \oplus \alpha$, *i.e.* a complex vector bundle η and a specific isomorphism $\varrho(\eta) = \xi \oplus \alpha$ (see 4.5). A compact differentiable manifold X is weakly almost complex if its tangent bundle ${}_{\mathbf{R}}\theta$ has been endowed with a weak complex structure. In this case $\varrho(\eta) = {}_{\mathbf{R}}\theta \oplus \alpha$ and $c(\eta)$ is called the total CHERN class of the weakly almost complex manifold X. The weak complex structure induces an orientation on X and the integers $c_{i_1} c_{i_2} \ldots c_{i_r}[X]$, $2(i_1 + i_2 + \cdots + i_r) = \dim X$ are called the CHERN numbers of X.

The definition of weakly almost complex extends to manifolds with boundary and can be used to define an equivalence relation $V \sim W$ between weakly almost complex manifolds. The equivalence classes form the complex cobordism ring Γ. For a treatment which generalises immediately to other structures on manifolds see MILNOR [4]. Results of NOVIKOV and MILNOR [3] imply that $V \sim W$ if and only if V, W have the same CHERN numbers. In particular the TODD genus $T(V)$ is an invariant of the complex cobordism class of V.

MILNOR [3] proves that Γ is isomorphic to $\mathbf{Z}[y_1, y_2, \ldots]$. An isomorphism $\mathbf{Z}[y_1, y_2, \ldots] \to \Gamma$ is given by associating to y_n a compact almost complex manifold Y_n satisfying the following conditions:

Y_n has tangent $\mathbf{GL}(n, \mathbf{C})$-bundle θ and, in the notation of 10.1,

$s(Y_n) = s_n(\theta)\,[Y_n] = \pm 1$ if $n+1$ is not a prime power,

$s(Y_n) = s_n(\theta)\,[Y_n] = \pm q$ if $n+1$ is a power of the prime q.

In fact the manifolds Y_n can always be chosen to lie in a particular set generated by taking inverses, sums and products of manifolds of the following type (HIRZEBRUCH [6]): complex projective spaces $\mathbf{P}_r(\mathbf{C})$ for which $s(\mathbf{P}_r(\mathbf{C})) = r+1$, and hypersurfaces $\mathbf{H}_{(r,t)}$ of degree (1,1) in $\mathbf{P}_r(\mathbf{C}) \times \mathbf{P}_t(\mathbf{C})$, $r > 1$, $t > 1$, for which $s(\mathbf{H}_{(r,t)}) = -\binom{r+t}{r}$. Thus it is possible to choose generators Y_n of Γ which are linear combinations of algebraic manifolds. The manifolds Y_{2k} provide generators for the torsion-free part $\tilde{\Omega}$ of the cobordism ring Ω (see the bibliographical note to Chapter Two).

It is a corollary of Theorem 20.2.2 that the TODD genus is an integer for every algebraic manifold, and hence for every linear combination of algebraic manifolds. So the above results imply that $T(X)$ is an integer for every compact almost complex manifold X. The second part of Theorem 14.3.2 holds similarly without reference to 2^{n-r}, and can be generalised to include the T_y-characteristic of a continuous $\mathbf{GL}(q, \mathbf{C})$-bundle ξ over X: the virtual T_y-characteristic $T_y(b_1, \ldots, b_r | ,\xi)$, $b_i \in H^2(X, \mathbf{Z})$, is a polynomial in y with integer coefficients. For further integrality theorems, which can be deduced from the integrality of the TODD genus, see Parts II and III of BOREL-HIRZEBRUCH [1]. For another approach to the integrality theorems for arbitrary differentiable manifolds see ATIYAH-HIRZEBRUCH [1, 2] and the appendix (§ 26).

Chapter Four

The RIEMANN-ROCH theorem for algebraic manifolds

In this chapter V is a complex n-dimensional manifold. The proof of the RIEMANN-ROCH theorem depends on results on compact complex manifolds which are due to CARTAN, DOLBEAULT, KODAIRA, SERRE and SPENCER. These results are summarised in § 15. At two points in the proof it becomes necessary to make additional assumptions on V: first that V is a KÄHLER manifold (15.6—15.9) and then that V is algebraic.

§ 15. Cohomology of compact complex manifolds

15.1. Let W be a complex analytic vector bundle over V, and let $\Omega(W)$ be the sheaf of germs of local holomorphic sections of W (see 3.5). The cohomology groups of V with coefficients in $\Omega(W)$ will be denoted more shortly by $H^i(V, W)$. The groups $H^i(V, W)$ are complex vector spaces. It will be shown that they are zero if i is greater than the complex dimension of V and that they are finite dimensional over $\mathbf{C}$ if V is compact. If W, W' are isomorphic vector bundles then $\Omega(W)$, $\Omega(W')$ are isomorphic sheaves and it follows that the cohomology groups $H^i(V, W)$, $H^i(V, W')$ are isomorphic. (For this reason isomorphic vector bundles will often be identified.)

The trivial line bundle is denoted by $\mathbf{1}$. The sheaf $\Omega(\mathbf{1})$ is just the sheaf $\mathbf{C}_\omega$ (see 2.5 and 3.1) of germs of local holomorphic functions on V; it will also be denoted by Ω.

$H^0(V, W)$ is the complex vector space of all global (*i. e.* defined on the whole of V) holomorphic sections of W. In particular, $H^0(V, \mathbf{1})$ is the vector space of all holomorphic functions defined on the whole of V. The dimension of $H^0(V, \mathbf{1})$ is equal to the number of connected components of V if V is compact.

15.2. Consider the sheaf $\mathbf{C}_\omega^*$ of germs of local holomorphic never zero functions on the complex manifold V (see 2.5). The complex analytic $\mathbf{C}^*$-bundles over V form the abelian group $H^1(V, \mathbf{C}_\omega^*)$ in which the addition is given by tensor product of bundles (see 3.7).

A divisor D of V is traditionally defined by a system $\{f_i\}$ of meromorphic "place functions" on V:

Let $\mathfrak{U} = \{U_i\}_{i \in I}$ be an open covering of V. For each $i \in I$ let f_i be a meromorphic (not identically zero) function defined on U_i such that on $U_i \cap U_j$ the function f_i/f_j has neither zeros nor poles.

It is then necessary to state under what circumstances two such systems of meromorphic functions define the same divisor. This can of course be done in the usual way. Alternatively, the divisors of V can be defined by means of sheaves:

Let $\mathfrak{G}$ be the sheaf of germs of local meromorphic (not identically zero) functions. The sheaf multiplication in $\mathfrak{G}$ is the usual multiplication of germs. $\mathbf{C}_\omega^*$ is a subsheaf of $\mathfrak{G}$ and so there is a sheaf $\mathfrak{D} = \mathfrak{G}/\mathbf{C}_\omega^*$ defined by the exact sequence

$$0 \to \mathbf{C}_\omega^* \to \mathfrak{G} \to \mathfrak{D} \to 0 . \tag{1}$$

The *divisors* are the elements of the abelian group $H^0(V, \mathfrak{D})$. We write this group additively: if $\mathfrak{U} = \{U_i\}_{i \in I}$ is an open covering and D, D' are divisors defined by meromorphic functions f_i, f'_i on U_i then $D + D'$ is the divisor defined by the meromorphic functions $f_i f'_i$ on U_i. The exact cohomology sequence of (1) gives

$$H^0(V, \mathfrak{G}) \xrightarrow{h} H^0(V, \mathfrak{D}) \xrightarrow{\delta^0_*} H^1(V, \mathbf{C}_\omega^*) . \tag{2}$$

$H^0(V, \mathfrak{G})$ is the multiplicative group of meromorphic functions on V which are not identically zero on any connected component of V. A meromorphic function $f \in H^0(V, \mathfrak{G})$ defines a divisor $(f) = h f$ which is called the divisor of the meromorphic function f. Two divisors are said to be linearly equivalent if their difference is the divisor of a meromorphic function $f \in H^0(V, \mathfrak{G})$. It follows that *the divisor classes* (with respect to linear equivalence) *are represented by elements of the abelian group* $H^0(V, \mathfrak{D})/h H^0(V, \mathfrak{G})$. By the exactness of (2), *this group is isomorphic to a subgroup of* $H^1(V, \mathbf{C}_\omega^*)$.

If D is a divisor we denote by $[D]$ the complex analytic $\mathbf{C}^*$-bundle $(\delta_0^* D)^{-1}$. The complex analytic line bundle determined, up to isomorphism, by $[D]$ is denoted by $\{D\}$. If D is represented, with respect to some open covering $\mathfrak{U} = \{U_i\}$, by meromorphic place functions f_i then $[D]$ is given by the cocycle

$$f_{ij} = f_i/f_j \qquad (f_{ij} : U_i \cap U_j \to \mathbf{C}^*) . \tag{3}$$

A divisor D is said to be holomorphic if it can be represented by place functions f_i which are all holomorphic. Clearly this property depends only on the divisor D.

Remark: In the literature holomorphic divisors are called "non-negative" or, if at least one place function has zeros, "positive". We avoid this terminology because the word "positive" is given a different meaning in 18.1.

A holomorphic divisor D is said to be *non-singular* if, with respect to some open covering $\mathfrak{U} = \{U_i\}$, it is represented by place functions f_i with the property:

Either $f_i \equiv 1$ *or* U_i *admits a system of local complex coordinates for which* f_i *is one of the coordinates.*

Let D be a non-singular divisor and $\dim V = n$. The set of all points $x \in V$ such that $f_i(x) = 0$ for at least one i with $x \in U_i$, and hence for all i with $x \in U_i$, is a complex manifold of dimension $n - 1$. We denote this complex submanifold by the same symbol D, in agreement with the terminology used in 4.9.

Now let D be an arbitrary divisor of V defined by place functions f_i. Consider the set $L(D)$ of all meromorphic functions g on V for which the functions $g f_i$ on U_i are holomorphic. Note that we do not require that $g \in H^0(V, \mathfrak{G})$. The set $L(D)$ depends only on the divisor D. Addition of meromorphic functions defines the structure of a complex vector space on $L(D)$. We can now state the

RIEMANN-ROCH problem: *Determine the dimension of* $L(D)$.

Theorem 15.2.1. *Let D be a divisor of a complex manifold V. The complex vector spaces $L(D)$ and $H^0(V, \{D\})$ are isomorphic.*

Proof: $H^0(V, \{D\})$ is the vector space of global holomorphic sections of the line bundle $\{D\}$. Let D be represented, with respect to an open covering $\mathfrak{U} = \{U_i\}$, by place functions f_i. Then by (3) and 3.2 a) the line bundle $\{D\}$ is got from $\bigcup(U_i \times \mathbf{C})$ by identifying $u \times k \in U_j \times \mathbf{C}$ with $u \times \frac{f_i(u)}{f_j(u)} k \in U_i \times \mathbf{C}$ for $u \in U_i \cap U_j$. A section s of $\{D\}$ is given by holomorphic functions s_i on U_i such that $s_i = \frac{f_i}{f_j} s_j$ on $U_i \cap U_j$. Associate to s the global meromorphic function

$$h(s) = \frac{s_i}{f_i} = \frac{s_j}{f_j} \in L(D) .$$

Then the map $h: H^0(V, \{D\}) \to L(D)$ is an isomorphism.

Remark: Let $|D|$ be the complex projective space associated to the vector space $H^0(V, \{D\})$. It is obtained by identifying $c\,a$ and a for $a \in H^0(V, \{D\})$, $a \neq 0$, $c \in \mathbf{C}$, $c \neq 0$. Then $\dim |D| + 1 = \dim H^0(V, \{D\})$. The proof shows that, *if V is compact and connected*, the points of $|D|$ are in one-one correspondence with the holomorphic divisors contained in the divisor class of D.

Theorem 15.2.1 suggests a generalisation of the RIEMANN-ROCH problem. Let W be a complex analytic vector bundle over V and let $H^0(V, W)$ be the vector space of holomorphic sections of W introduced in 15.1.

Generalised RIEMANN-ROCH problem: *Determine the dimension of the vector space* $H^0(V, W)$.

15.3. a). Let $\dim V = n$ and let $\lambda^p \boldsymbol{T}$ be the complex analytic vector bundle of covariant tangent p-vectors (see 4.7). Then $\boldsymbol{T} = \lambda^1 \boldsymbol{T}$ is the vector bundle of covariant tangent vectors and $\lambda^0 \boldsymbol{T}$ is the trivial line bundle. $\lambda^n \boldsymbol{T}$ is also a line bundle; it is called the canonical line bundle of V and denoted by K.

If V admits a meromorphic n-form with divisor E then K is associated with the complex analytic $\mathbf{C}^*$-bundle $[E]$, so that $K = \{E\}$.

If W is a complex analytic vector bundle over V we shall also denote the cohomology groups $H^q(V, W \otimes \lambda^p \boldsymbol{T})$ by $H^{p,q}(V, W)$. Thus

$$H^{0,q}(V, W) = H^q(V, W).$$

15.3. b). Given a complex vector bundle W over V, the conjugate vector bundle $\overline{W}$ over V can be defined by the following construction. Let

$$g_{ij}: U_i \cap U_j \to \mathbf{GL}(q, \mathbf{C})$$

be coordinate transformations which define W by identifications on the disjoint union $\bigcup (U_i \times \mathbf{C}_q)$. Then $\overline{W}$ is defined by coordinate transformations

$$\overline{g_{ij}}: U_i \cap U_j \to \mathbf{GL}(q, \mathbf{C}).$$

Here $\overline{g_{ij}}(x) \in \mathbf{GL}(q, \mathbf{C})$ denotes the matrix obtained from $g_{ij}(x)$ by the conjugation of every coefficient.

If W is complex analytic then $\overline{W}$ is no longer complex analytic, but is regarded as a differentiable vector bundle. As differentiable vector bundles W, $\overline{W}$ are "anti-isomorphic". That is, there is a differentiable homeomorphism $\varkappa: W \to \overline{W}$ which maps fibres W_x into fibres $\overline{W}_x$ such that

$$\varkappa(a + a') = \varkappa(a) + \varkappa(a'), \ \varkappa(ca) = \bar{c}\varkappa(a) \ \text{ for } a, a' \in W_x, \ c \in \mathbf{C}.$$

In terms of the local product structure $U_i \times \mathbf{C}_q$, the anti-isomorphism $\varkappa: W \to \overline{W}$ can be represented by conjugation. Clearly, if W, W' are isomorphic vector bundles then so are the vector bundles $\overline{W}$, $\overline{W}'$.

15.3. c). Let W be a differentiable vector bundle over X given, for some open covering $\mathfrak{U} = \{U_i\}$, by differentiable coordinate transformations

$$f_{ij}: U_i \cap U_j \to \mathbf{GL}(q, \mathbf{C}).$$

Then the structure groups can be reduced to $\mathbf{U}(q)$. That is [see 4.1. b)], there are differentiable maps

$$h_i: U_i \to \mathbf{GL}(q, \mathbf{C})$$

such that

$$h_i(x)\, f_{ij}(x)\, h_j^{-1}(x) \in \mathbf{U}(q) \quad \text{for} \quad x \in U_i \cap U_j.$$

Let $\dagger$ denote transposition of matrices and define

$$g_i = \bar{h}_i^\dagger\, h_i: U_i \to \mathbf{GL}(q, \mathbf{C}).$$

The dual vector bundle W^* can be defined by the coordinate transformations

$$(f_{ij}^{-1})^\dagger: U_i \cap U_j \to \mathbf{GL}(q, \mathbf{C}).$$

In terms of the local product structure $U_i \times \mathbf{C}_q$ we can define an anti-isomorphism

$$\psi : W \to W^*$$

by

$$\psi(u \times t) = u \times \overline{g_i(u) \cdot t}, \quad u \in U_i, \quad t \in \mathbf{C}_q .$$

We call ψ the "hermitian" anti-isomorphism defined by the above reduction of the structure groups. ψ defines a hermitian metric on each fibre W_x of W. The corresponding (positive definite) hermitian form is given by $\psi(a) \cdot a$. Here $a \in W_x$ and $\psi(a) \cdot a$ is the value of the linear form $\psi(a)$ on a. Similarly there is a hermitian anti-isomorphism $\psi^{-1} : W^* \to W$.

15.4. In this section we sketch results on the cohomology groups $H^{p,q}(V, W)$ due to DOLBEAULT [1, 2], KODAIRA [3] and SERRE [3].

Let $\mathfrak{A}^{p,q}$ be the sheaf of germs of local differentiable differential forms of type (p, q) on the complex manifold V. Then $\mathfrak{A}^{p,q}$ is precisely (see 4.7) the sheaf of germs of local differentiable sections of the (differentiable) vector bundle $\lambda^p \boldsymbol{T} \otimes \lambda^q \overline{\boldsymbol{T}}$. Note that, by 15.3 b), $\lambda^q \overline{\boldsymbol{T}} = \overline{\lambda^q \boldsymbol{T}}$.

The operator d on differential forms can be written as a sum

$$d = \partial + \bar{\partial}$$

where ∂ = differentiation with respect to the z-variables,
$\bar{\partial}$ = differentiation with respect to the $\bar{z}$-variables,
and $\partial\partial = \bar{\partial}\bar{\partial} = \partial\bar{\partial} + \bar{\partial}\partial = 0$.

The operator $\bar{\partial}$ transforms forms of type (p, q) into forms of type $(p, q+1)$ and therefore induces sheaf homomorphisms

$$\bar{\partial} : \mathfrak{A}^{p,q} \to \mathfrak{A}^{p,q+1} .$$

The kernel of $\bar{\partial} : \mathfrak{A}^{p,0} \to \mathfrak{A}^{p,1}$ is the sheaf $\Omega(\lambda^p \boldsymbol{T})$ of germs of local holomorphic p-forms, since for a form of type $(p, 0)$ the statements "$\bar{\partial}$ vanishes" and "holomorphic" are immediately equivalent. The embedding of $\Omega(\lambda^p \boldsymbol{T})$ in $\mathfrak{A}^{p,0}$, together with the homomorphisms $\bar{\partial}$, gives the following sequence of sheaves over V:

$$0 \to \Omega(\lambda^p \boldsymbol{T}) \to \mathfrak{A}^{p,0} \to \mathfrak{A}^{p,1} \to \cdots \to \mathfrak{A}^{p,q} \to \cdots . \qquad (4)$$

It has just been shown that the beginning of this sequence is exact. A "POINCARÉ lemma" first proved by GROTHENDIECK shows that the whole sequence (4) is exact (see CARTAN [4], DOLBEAULT [1]).

Now let W be a *complex analytic* vector bundle over V with fibre $\mathbf{C}_r$. We consider the differentiable vector bundle $W \otimes \lambda^p \boldsymbol{T} \otimes \overline{\lambda^q \boldsymbol{T}}$ and denote the sheaf of germs of local differentiable sections of this vector bundle by $\mathfrak{A}^{p,q}(W)$. Thus $\mathfrak{A}^{p,q}(1) = \mathfrak{A}^{p,q}$. Sections of $\mathfrak{A}^{p,q}(W)$, that is differentiable sections of the vector bundle $W \otimes \lambda^p \boldsymbol{T} \otimes \overline{\lambda^q \boldsymbol{T}}$ are called differentiable differential forms (or simply: forms) of type (p, q) with coefficients in W. We let

$$\begin{aligned} A^{p,q}(W) &= \Gamma(V, \mathfrak{A}^{p,q}(W)) = \mathbf{C}\text{-module of global forms of type } (p,q) \text{ with coefficients in } W \\ A^{p,q} &= A^{p,q}(\mathbf{1}) \qquad = \mathbf{C}\text{-module of ordinary global forms of type } (p,q). \end{aligned} \tag{5}$$

Let W be given, over some open set U_i, by a local product structure $U_i \times \mathbf{C}_r$. A local form of type (p, q) with coefficients in W can be represented by an r-ple of ordinary local forms of type (p, q). The operation $\bar{\partial}$ acts on this r-ple. The identifications between $U_i \times \mathbf{C}_r$ and $U_j \times \mathbf{C}_r$ are given by holomorphic functions

$$U_i \cap U_j \to \mathbf{GL}(r, \mathbf{C}) .$$

But $\bar{\partial}$ is zero on holomorphic functions, and therefore the action of $\bar{\partial}$ is independent of the choice of local product structure. Therefore $\bar{\partial}$ induces a sheaf homomorphism

$$\bar{\partial} : \mathfrak{A}^{p,q}(W) \to \mathfrak{A}^{p,q+1}(W) .$$

The exactness of (4) now implies that the following sequence is exact:

$$0 \to \Omega(W \otimes \lambda^p \boldsymbol{T}) \to \mathfrak{A}^{p,0}(W) \to \mathfrak{A}^{p,1}(W) \to \cdots \to \mathfrak{A}^{p,q}(W) \to \cdots. \tag{6}$$

The sheaf of germs of local differentiable sections of a vector bundle over V is fine (see 3.5). Therefore (6) is a fine resolution of the sheaf $\Omega(W \otimes \lambda^p \boldsymbol{T})$, and Theorem 2.12.1 implies

Theorem 15.4.1 (Dolbeault-Serre). *The complex vector space* $H^{p,q}(V, W) = H^q(V, W \otimes \lambda^p \boldsymbol{T})$ *is isomorphic to the q-th cohomology module of the* $\bar{\partial}$*-resolution* (6). *That is,*

$$H^{p,q}(V, W) \cong Z^{p,q}(W)/\bar{\partial}(A^{p,q-1}(W)) \tag{7}$$

where $Z^{p,q}(W)$ *is the module of all those global forms of type* (p, q) *with coefficients in* W *which vanish under* $\bar{\partial}$.

An immediate corollary is the fact that $H^{p,q}(V, W)$ is zero if p or q is greater than the complex dimension of V.

For the remainder of this paragraph it will be assumed that V *is compact.* Let $n = \dim V$. Consider the vector bundle W^* dual to W. A product

$$A^{p,q}(W) \times A^{r,s}(W^*) \to A^{p+r,\, q+s}(\mathbf{1})$$

can be defined in a natural way. The product of $\alpha \in A^{p,q}(W)$ and $\beta \in A^{r,s}(W^*)$ is denoted by $\alpha \wedge \beta$. For $W = \mathbf{1}$ it is the usual exterior product of forms. The product satisfies

$$\begin{aligned} \bar{\partial}(\alpha \wedge \beta) &= \bar{\partial}\alpha \wedge \beta + (-1)^{p+q}\, \alpha \wedge \bar{\partial}\beta \\ \alpha \wedge \beta &= (-1)^{(p+q)\,(r+s)}\, \beta \wedge \alpha . \end{aligned} \tag{8}$$

If $r = n - p$ and $s = n - q$ then $\alpha \wedge \beta \in A^{n,n}(1)$ and the integral

$$\iota(\alpha, \beta) = \int_V \alpha \wedge \beta$$

is well defined. If $\alpha = \bar{\partial}\gamma$, $\gamma \in A^{p,q-1}(W)$, and $\bar{\partial}\beta = 0$ then by (8) and STOKES' Theorem

$$\iota(\alpha, \beta) = \int_V \bar{\partial}(\gamma \wedge \beta) = \int_V d(\gamma \wedge \beta) = 0 .$$

Similarly $\iota(\alpha, \beta) = 0$ if $\beta = \bar{\partial}\gamma$, $\gamma \in A^{n-p,n-q-1}(W^*)$ and $\bar{\partial}\alpha = 0$. Therefore by (7) the bilinear form ι induces a pairing of $H^{p,q}(V, W)$ and $H^{n-p,n-q}(V, W^*)$ with values in $\mathbf{C}$. Thus if $a \in H^{p,q}(V, W)$, $b \in H^{n-p,n-q}(V, W^*)$ the complex number $\iota(a, b)$ is defined, depends only on (a, b), and is linear in a and b.

KODAIRA has extended the theory of harmonic forms to apply to forms with coefficients in W. Introduce a fixed hermitian metric on V. This induces [15.3 c)] an isomorphism $\overline{\boldsymbol{T}} \cong \boldsymbol{T}^*$.
Using this isomorphism and Theorem 3.6.1 we obtain isomorphisms

$$\begin{aligned} \lambda^p \boldsymbol{T} \otimes \overline{\lambda^q \boldsymbol{T}} &\cong \lambda^p \boldsymbol{T} \otimes \lambda^n \boldsymbol{T}^* \otimes \lambda^n \boldsymbol{T} \otimes \lambda^q \boldsymbol{T}^* \\ &\cong \lambda^{n-p} \boldsymbol{T}^* \otimes \lambda^{n-q} \boldsymbol{T} \\ &\cong \lambda^{n-q} \boldsymbol{T} \otimes \lambda^{n-p} \overline{\boldsymbol{T}} . \end{aligned}$$

The result is a duality operator

$$* : \lambda^p \boldsymbol{T} \otimes \overline{\lambda^q \boldsymbol{T}} \to \lambda^{n-q} \boldsymbol{T} \otimes \overline{\lambda^{n-p} \boldsymbol{T}} .$$

The isomorphism $**$ from $\lambda^p \boldsymbol{T} \otimes \overline{\lambda^q \boldsymbol{T}}$ on to itself is multiplication by $(-1)^{p+q}$.

Now let the structure groups of the vector bundle W be reduced to the unitary group. By 15.3 c) there are hermitian anti-isomorphisms

$$\psi : W \to W^*, \quad \psi^{-1} : W^* \to W .$$

Let $\varkappa$ (conjugation) be the anti-isomorphism from $\lambda^r \boldsymbol{T} \otimes \overline{\lambda^s \boldsymbol{T}}$ to $\lambda^s \boldsymbol{T} \otimes \overline{\lambda^r \boldsymbol{T}}$. We define

$$\# = \psi \otimes (\varkappa *), \quad \widetilde{\#} = \psi^{-1} \otimes (\varkappa *)$$

and obtain anti-isomorphisms

$$\begin{aligned} \# &: W \otimes \lambda^p \boldsymbol{T} \otimes \overline{\lambda^q \boldsymbol{T}} \to W^* \otimes \lambda^{n-p} \boldsymbol{T} \otimes \overline{\lambda^{n-q} \boldsymbol{T}} \\ \widetilde{\#} &: W^* \otimes \lambda^r \boldsymbol{T} \otimes \overline{\lambda^s \boldsymbol{T}} \to W \otimes \lambda^{n-r} \boldsymbol{T} \otimes \overline{\lambda^{n-s} \boldsymbol{T}} . \end{aligned}$$

For $r = n - p$, $s = n - q$ the isomorphism $\widetilde{\#}\,\#$ is multiplication by $(-1)^{p+q}$.

$\#$ and $\widetilde{\#}$ induce anti-isomorphisms of the corresponding sheaves

$$\begin{aligned} \# &: \mathfrak{A}^{p,q}(W) \to \mathfrak{A}^{n-p,n-q}(W^*) \\ \widetilde{\#} &: \mathfrak{A}^{r,s}(W^*) \to \mathfrak{A}^{n-r,n-s}(W) . \end{aligned}$$

Since W and W^* are complex analytic there are sheaf homomorphisms

$$\bar{\partial}: \mathfrak{A}^{p,q}(W) \to \mathfrak{A}^{p,q+1}(W), \quad \bar{\partial}: \mathfrak{A}^{r,s}(W^*) \to \mathfrak{A}^{r,s+1}(W^*).$$

We define the homomorphism

$$\vartheta: \mathfrak{A}^{p,q}(W) \to \mathfrak{A}^{p,q-1}(W)$$

by

$$\vartheta = -\,\widetilde{\#}\,\bar{\partial}\,\#\,.$$

If $\alpha, \beta \in A^{p,q}(W)$ are global forms of type (p, q) with coefficients in W, the scalar product

$$(\alpha, \beta) = \iota(\alpha, \#\,\beta) = \int_V \alpha \wedge \#\,\beta$$

can be introduced. Then $(\alpha, \alpha) \geqq 0$, and $(\alpha, \alpha) = 0$ if and only if $\alpha = 0$.

With respect to this scalar product, ϑ and $\bar{\partial}$ are adjoint operations:

$$(\alpha, \vartheta\,\beta) = (\bar{\partial}\,\alpha, \beta) \quad \text{for} \quad \alpha \in A^{p,q}(W), \quad \beta \in A^{p,q-1}(W)\,. \tag{9}$$

Proof: $(\alpha, \vartheta\,\beta) = -\int_V \alpha \wedge \#\,\widetilde{\#}\,\bar{\partial}\,\#\,\beta = (-1)^{p+q+1} \int_V \alpha \wedge \bar{\partial}\,\#\,\beta.$

Therefore

$$\begin{aligned}(\bar{\partial}\,\alpha, \beta) - (\alpha, \vartheta\,\beta) &= \int_V (\bar{\partial}\,\alpha \wedge \#\,\beta + (-1)^{p+q}\,\alpha \wedge \bar{\partial}\,\#\,\beta) \\ &= \int_V \bar{\partial}(\alpha \wedge \#\,\beta) \\ &= \int_V d(\alpha \wedge \#\,\beta) \\ &= 0 \text{ by STOKES' theorem.}\end{aligned}$$

We now define the complex LAPLACE-BELTRAMI operator $\square: A^{p,q}(W) \to A^{p,q}(W)$ by $\square = \vartheta\,\bar{\partial} + \bar{\partial}\,\vartheta$. The subspace of elements $\alpha \in A^{p,q}(W)$ for which $\square\,\alpha = 0$ will be denoted by $B^{p,q}(W)$. This is the subspace of "complex harmonic" forms. As in the usual case, (9) implies: $\square\,\alpha = 0$ if and only if $\vartheta\,\alpha = \bar{\partial}\,\alpha = 0$.

The methods of the theory of harmonic integrals now show that with respect to the scalar product $A^{p,q}(W)$ can be represented as the direct sum of three mutually orthogonal components:

$$A^{p,q}(W) = \bar{\partial}\,A^{p,q-1}(W) \oplus \vartheta\,A^{p,q+1}(W) \oplus B^{p,q}(V, W)\,.$$

Hence $Z^{p,q}(W) = \bar{\partial}\,A^{p,q-1}(W) \oplus B^{p,q}(V, W)$ and therefore, by Theorem 15.4.1,

$$H^{p,q}(V, W) \cong Z^{p,q}(W)/\bar{\partial}\,A^{p,q-1}(W) \cong B^{p,q}(V, W)\,.$$

From the fact that $\square$ is an elliptic partial differential operator over the compact manifold V, KODAIRA deduces that $B^{p,q}(V, W)$ is finite dimensional, and hence that $H^{p,q}(V, W)$ is finite dimensional [see also SPENCER [2]; a general definition of elliptic differential operator is given in the appendix (25.1) together with references to proofs of finite dimensionality (25.2)]. The operators ϑ, $\bar{\partial}$, $\square$ are defined equally for the sheaves $\mathfrak{A}^{p,q}(W^*)$ and the operator induces an anti-isomorphism from $B^{p,q}(V, W)$ to $B^{n-p,n-q}(V, W^*)$.

We collect the results, of whose proofs we have given a bare outline, in two theorems.

Theorem 15.4.2 (KODAIRA [3]). *Let W be a complex analytic vector bundle over a compact complex manifold V. Then $H^{p,q}(V, W)$ is a finite dimensional vector space which* (after the introduction of a hermitian metric on V and a unitary structure for W; see 15.3 c)) *is isomorphic to the vector space of "complex harmonic" forms of type (p, q) with coefficients in W. In particular $H^p(V, W) = H^{0,p}(V, W)$ is finite dimensional. If $p > n$ or $q > n$ then $H^{p,q}(V, W) = 0$.*

Theorem 15.4.3 (SERRE [3]). Let V, W be as in the previous theorem. *The bilinear form ι is a dual pairing of the vector spaces $H^{p,q}(V, W)$ and $H^{n-p,n-q}(V, W^*)$. In particular if $K = \lambda^n \boldsymbol{T}$ is the canonical line bundle then $H^p(V, W)$ and $H^{n-q}(V, K \otimes W^*)$ are dual vector spaces.*

We write $\dim H^{p,q}(V, W) = h^{p,q}(V, W)$ and $\dim H^{p,q}(V, 1) = h^{p,q}(V)$ [= the "number "of complex harmonic forms of type (p, q) on V].

Remarks: Counter-examples show that it is not true in general that $h^{p,q}(V) = h^{q,p}(V)$. It will be shown in 15.6 that this is however true when V is a KÄHLER manifold. This fact will be used in the proof of Theorem 15.8.2. There is a generalisation of Theorem 15.4.2, due to CARTAN-SERRE [1], which is mentioned in the appendix (23.1).

15.5. Let W be a complex analytic vector bundle over a compact complex manifold V_n. Since the groups $H^i(V, W)$ are finite dimensional, and zero for $i > n$, the EULER-POINCARÉ characteristic

$$\chi(V, W) = \sum_{i=0}^{\infty} (-1)^i \dim H^i(V, W) = \sum_{i=0}^{n} (-1)^i \dim H^i(V, W).$$

is defined (2.10). Define $\chi^p(V, W)$ by

$$\chi^p(V, W) = \chi(V, W \otimes \lambda^p \boldsymbol{T}) = \sum_{q=0}^{n} (-1)^q h^{p,q}(V, W). \tag{10}$$

Then

$$\chi^0(V, W) = \chi(V, W) \text{ and } \chi^p(V, W) = 0 \text{ for } p < 0 \text{ and } p > n. \tag{11}$$

For $W = \mathbf{1}$ we naturally write

$$\chi^p(V, 1) = \chi^p(V) = \sum_{q=0}^{n} (-1)^q h^{p,q}(V)$$

By using an indeterminate y we can define

$$\chi_y(V, W) = \sum_{p=0}^{n} \chi^p(V, W) y^p, \quad \chi_y(V) = \sum_{p=0}^{n} \chi^p(V) y^p. \tag{12}$$

We call $\chi_y(V, W)$ the χ_y-characteristic of the vector bundle W and

$\chi_y(V)$ the χ_y-genus of V. By definition

$$\chi_0(V, W) = \chi^0(V, W) = \chi(V, W) \quad \text{and} \quad \chi_0(V) = \chi^0(V) = \chi(V) .$$

$\chi(V) = \sum_{q=0}^{n} (-1)^q h^{0,q}(V)$ *is called the arithmetic genus of* V. (13)

The SERRE duality theorem (15.4.3) implies that

$$\begin{aligned} \chi^p(V, W) &= (-1)^n \chi^{n-p}(V, W^*) \\ \chi(V, W) &= (-1)^n \chi(V, K \otimes W^*) . \end{aligned} \tag{14}$$

We emphasise that *the arithmetic genus* $\chi(V)$ *of a compact complex manifold* V *is defined as the* EULER-POINCARÉ *characteristic of the cohomology with coefficients in the sheaf of germs of local holomorphic functions on* V.

15.6. Let V_n be a compact complex manifold. A hermitian metric on V has the form

$$d s^2 = 2 \sum g_{\alpha\beta}(z, \bar{z}) \, (d z^\alpha \cdot d \bar{z}^\beta), \quad \overline{g_{\alpha\beta}} = g_{\beta\alpha} \tag{15}$$

with respect to local coordinates z^α $(\alpha = 1, \ldots, n)$. To each hermitian metric $d s^2$ is associated an exterior differential form

$$\omega = i \sum g_{\alpha\beta}(z, \bar{z}) \, d z^\alpha \wedge d \bar{z}^\beta \tag{16}$$

which can be written as a real differential form by using real coordinates x^α $(\alpha = 1, \ldots, 2n)$ for which $z^\alpha = x^{2\alpha-1} + i x^{2\alpha}$. The hermitian metric $d s^2$ is called a KÄHLER metric if $d\omega = 0$ (KÄHLER [2]). The form ω then represents an element of the cohomology group $H^2(V, \mathbf{R})$ which is called the fundamental class of the KÄHLER metric (here we are of course using the DE RHAM isomorphism).

In the present work we adopt the following terminology: by a manifold with a KÄHLER metric we mean a compact complex manifold with a particular choice of KÄHLER metric; by a KÄHLER manifold we mean a compact complex manifold which admits at least one KÄHLER metric. We summarise briefly the properties of KÄHLER manifolds needed for the present work. A fuller account can be found in WEIL [2].

15.7. Let V be a manifold with a KÄHLER metric. Then the $h^{p,q}(V)$ can be calculated with the help of the KÄHLER metric by choosing $W = 1$ in 15.4. The following discussion is concerned with this case.

For a KÄHLER metric the complex LAPLACE-BELTRAMI operator $\square$ is equal to $\frac{\triangle}{2}$, where $\triangle$ is the real LAPLACE operator $d\delta + \delta d$, $(\delta = - * d *)$. The operator $\square$ therefore commutes with conjugation, and $\alpha \to \bar{\alpha}$ defines an anti-isomorphism from $B^{p,q}$ [harmonic forms of type (p, q)] on to $B^{q,p}$ [harmonic forms of type (q, p)]. Therefore a (compact) KÄHLER manifold V has

$$h^{p,q}(V) = h^{q,p}(V), \quad h^{p,q}(V) = \dim B^{p,q} . \tag{17}$$

The theory of DE RHAM and HODGE gives a natural isomorphism

$$H^r(V, \mathbf{C}) \cong \sum_{p+q=r} B^{p,q}. \tag{18}$$

Therefore the r-th BETTI number $b_r(V)$ satisfies

$$b_r(V) = \sum_{p+q=r} h^{p,q}(V). \tag{18*}$$

Under the isomorphism (18) the subspace $B^{p,q}$ of $H^{p+q}(V, \mathbf{C})$ is represented, in the sense of DE RHAM, by the subspace of forms α of type (p, q) with $d\alpha = 0$. Elements of this subspace, which clearly does not depend on the particular choice of KÄHLER metric, are said to be of type (p, q).

An element of $H^{p+q}(V, \mathbf{Z})$ or $H^{p+q}(V, \mathbf{R})$ is said to be of type (p, q) if when regarded as an element of $H^{p+q}(V, \mathbf{C})$ it is of type (p, q).

Formulae (17), (18*) are in general false for arbitrary compact complex manifolds. For a KÄHLER manifold V, (17) gives $h^{0,q} = h^{q,0}$. For an arbitrary compact complex manifold $h^{q,0}$ is by definition $\dim H^0(V, \lambda^q \boldsymbol{T})$, that is the dimension of the complex vector space of holomorphic q-forms on V. These are also called the forms of the first kind of degree q. Let $g_q = h^{q,0}$. Then we have proved

Theorem 15.7.1. *The arithmetic genus $\chi(V_n)$ of a compact* KÄHLER *manifold V_n is equal to $\sum_{i=0}^{n} (-1)^i g_i$, where g_i is the number of forms of the first kind of degree i on V_n linearly independent over* $\mathbf{C}$.

15.8. We have associated (in 15.5) a polynomial $\chi_y(V)$ to each compact complex manifold V. For $y = 0$ the value of this polynomial is the arithmetic genus of V. The next two theorems give an interpretation of the value of $\chi_y(V)$ for $y = -1$ and for $y = 1$.

Theorem 15.8.1. *If V_n is a compact complex manifold then*

$$\chi_{-1}(V_n) = \sum_{p=0}^{n} (-1)^p \chi^p(V_n) = \sum_{p,q} (-1)^{p+q} h^{p,q}(V_n)$$

is equal to the (ordinary) EULER-POINCARÉ *characteristic $E(V_n)$.*

Proof (due to SERRE [3], p. 26): Let $\Omega^p = \Omega(\lambda^p \boldsymbol{T})$ be the sheaf of germs of local holomorphic p-forms. The operator d defines an exact sequence

$$0 \to \mathbf{C} \to \Omega^0 \to \Omega^1 \to \cdots \to \Omega^n \to 0.$$

$E(V_n)$ is the EULER-POINCARÉ characteristic of cohomology with coefficients in the constant sheaf $\mathbf{C}$. The result now follows from Theorem 2.10.3.

Remark: If V_n is a KÄHLER manifold then Theorem 15.8.1 is an immediate consequence of (18*).

Theorem 15.8.2 (see HODGE [4]). *If V_n is a (compact)* KÄHLER *manifold then*

$$\chi_1(V_n) = \sum_{p=0}^{n} \chi^p(V_n) = \sum_{p,q} (-1)^q h^{p,q}(V_n)$$

is equal to the index $\tau(V_n)$ defined in 8.2.

Proof: If n is odd then by SERRE duality (Theorem 15.4.3)

$$\chi^p(V_n) = (-1)^n \chi^{n-p}(V_n) = -\chi^{n-p}(V_n)$$

and therefore $\sum_{p=0}^{n} \chi^p(V_n) = 0$. On the other hand $\tau(V_n) = 0$ by definition. Thus for n odd the theorem is true for arbitrary compact complex manifolds.

Now suppose n is even. We shall use a number of facts on manifolds with a KÄHLER metric. For these we refer to ECKMANN-GUGGENHEIMER [1, 2], GUGGENHEIMER [1], HODGE [1] and WEIL [2]. If $z_j = x_{2j-1} + i x_{2j}$ are local complex coordinates then ECKMANN-GUGGENHEIMER and HODGE use the orientation for V_n given by $dx_1 \wedge dx_3 \wedge \cdots \wedge dx_{2n-1} \wedge dx_2 \wedge dx_4 \wedge \cdots \wedge dx_{2n}$. We use the orientation given by the natural order $dx_1 \wedge dx_2 \wedge \cdots \wedge dx_{2n}$. The two orientations differ by a sign $(-1)^{\frac{n(n-1)}{2}}$. To simplify the subsequent formulae we assume that $n = 2m$.

Let $B^{p,q}$ be the complex vector space of harmonic forms of type (p, q). The fundamental form ω defined in 15.6 is a particular harmonic form of type (1, 1) whose product with any other harmonic form is again harmonic.

Define a homomorphism

$$L: B^{p,q} \to B^{p+1,q+1}$$

by associating to each form $\alpha \in B^{p,q}$ the form $L\alpha = \omega\,\alpha \in B^{p+1,q+1}$. Then, since ω is real, $\overline{L\alpha} = L\bar{\alpha}$. By 15.4 there is an anti-isomorphism

$$\#: B^{p,q} \to B^{n-p,n-q}$$

for which $\#\,\alpha = \overline{*\,\alpha} = *\,\bar{\alpha}$. We consider the homomorphism

$$\Lambda: B^{p,q} \to B^{p-1,q-1}$$

defined by $\Lambda = (-1)^{p+q}\,\#\,L\,\#$. Then $\Lambda = (-1)^{p+q} * L *$ and $\overline{\Lambda\alpha} = \Lambda\bar{\alpha}$.

The kernel of Λ is denoted by $B_0^{p,q}$ and called the subspace of effective harmonic forms of type (p, q).

(a) $\Lambda L^k: B_0^{p-k,q-k} \to B^{p-1,q-1}$ $(p + q \leqq n,\ k \geqq 1)$ is (up to a non-zero scalar factor) equal to L^{k-1}.

(b) $L^k: B_0^{p-k,q-k} \to B^{p,q}$ $(p + q \leqq n)$ is a monomorphism.

For $p + q \leqq n$ there is a direct sum decomposition

(c) $B^{p,q} = B_0^{p,q} \oplus L B_0^{p-1,q-1} \oplus \cdots \oplus L^r B_0^{p-r,q-r}$ $(r = \min(p, q))$.

We define $B_k^{p,q} = L^k B_0^{p-k,q-k}$. The elements of $B_k^{p,q}$ are called harmonic forms of type (p, q) and class k. The following formula is then decisive for the proof:

(d) $\# \varphi = (-1)^{q+k} \bar{\varphi}$ for $\varphi \in B_k^{p,q}$ and $p + q = n$. Note that $\bar{\varphi}$ is an element of $B_k^{q,p}$.

The cohomology group $H^n(V_n, \mathbf{C})$ is a complex vector space [see 15.7 (18)]

(e) $H^n(V_n, \mathbf{C}) = \sum\limits_{\substack{p+q=n \\ k \leqq \min(p,q)}} B_k^{p,q}$.

We recall that the scalar product

$$(\alpha, \beta) = \int_{V_n} \alpha \wedge \# \beta$$

is defined for harmonic forms α, β of the same total degree.

(f) *The summands in the direct sum decomposition* (e) *are mutually orthogonal with respect to the scalar product.*

Proof: The scalar product can be non-zero only if $\alpha \wedge \# \beta$ is of type (n, n). Therefore $B_k^{p,q}$, $B_{k'}^{p',q'}$ are orthogonal for $(p, q) \neq (p', q')$. If $\alpha \in B_k^{p,q}$ and $\beta \in B_{k'}^{p,q}$ for $k > k'$ and $p + q = n$ then

$$(\alpha, \beta) = (L^k \alpha_0, L^{k'} \beta_0) \text{ with } \alpha_0, \beta_0 \text{ effective } (\Lambda \alpha_0 = \Lambda \beta_0 = 0)\,.$$

Since L and Λ are adjoint operators, $(L\alpha, \varphi) = (\alpha, \Lambda\varphi)$, and therefore, by (a), $(\alpha, \beta) = (\alpha_0, \Lambda^k L^{k'} \beta_0) = 0$.

The cohomology groups $H^n(V_n, \mathbf{R})$ can be identified with the real vector space of real harmonic forms. There is a direct sum decomposition

(g) $H^n(V_n, \mathbf{R}) = \sum E_k^{p,q} \quad (p + q = n,\ k \leqq p \leqq q)$

where $E_k^{p,q}$ is the real vector space of real harmonic forms α which can be written in the form $\alpha = \varphi + \bar{\varphi}$ with $\varphi \in B_k^{p,q}$ (and hence $\bar{\varphi} \in B_k^{q,p}$). Clearly $\tau(V_n)$ is the index (see 8.1) of the quadratic form

$$Q(\alpha, \beta) = \int_{V_n} \alpha \wedge \beta \quad (\alpha, \beta \in H^n(V_n, \mathbf{R}))\,.$$

By (d) and (f) the real vector space summands in the sum (g) are mutually orthogonal with respect to this quadratic form. Now (d) implies that the quadratic form $(-1)^{q+k} Q(\alpha, \beta)$ is positive definite when restricted to $E_k^{p,q}$.

Therefore

$$\tau(V_n) = \sum (-1)^{q+k} \dim_{\mathbf{R}} E_k^{p,q}$$

(the sum is over $p + q = n$, $k \leqq p \leqq q$).

Clearly $\dim_{\mathbf{R}} E_k^{p,q} = 2 \dim_{\mathbf{C}} B_k^{p,q}$ for $p < q$. If $n = 2m$ then $\dim_{\mathbf{R}} E_k^{m,m} = \dim_{\mathbf{C}} B_k^{m,m}$.

Therefore

(h) $\tau(V_n) = \sum (-1)^{q+k} \dim_{\mathbf{C}} B_k^{p,q} \quad (p + q = n,\ k \leqq \min(p, q))$.

Now let $h^{p,q} = \dim_{\mathbf{C}} B^{p,q}$ as before. It follows from (b) and (c) that

(i) $h^{p-k,q-k} - h^{p-k-1,q-k-1} = \dim_{\mathbf{C}} B_k^{p,q}$ for $p+q \leqq n$.

Since $h^{r,s} = h^{s,r} = h^{n-r,n-s}$ we have

(j) $h^{p-k-1,q-k-1} = h^{p+k+1,q+k+1}$ for $p+q=n$.

Finally (h), (i) and (j) imply

$$\begin{aligned}\tau(V_n) &= \sum_{\substack{k\geqq 0\\ p+q=n}} (-1)^{q-k} h^{p-k,q-k} + \sum_{\substack{k\geqq 0\\ p+q=n}} (-1)^{q+k+1} h^{p+k+1,q+k+1}\\ &= \sum_{p+q\leqq n} (-1)^q h^{p,q} + \sum_{p+q>n} (-1)^q h^{p,q}\\ &= \sum_{p,q} (-1)^q h^{p,q}. \qquad \text{Q. E. D.}\end{aligned}$$

Theorem 15.8.2 is used in 19.5 to give an essential step in the proof of the RIEMANN-ROCH theorem.

Problem: Find a direct proof of Theorem 15.8.2 which is valid for an arbitrary compact complex manifold V_n. A somewhat indirect proof is sketched in the appendix (25.4).

15.9. Let V be a KÄHLER manifold (15.6). The exact sequence $0 \to \mathbf{Z} \to \mathbf{C}_\omega \to \mathbf{C}_\omega^* \to 0$ defines an exact cohomology sequence [see 2.5 (11) and Theorem 2.10.1; by definition $\mathbf{C}_\omega = \Omega$]:

$$H^1(V, \mathbf{C}_\omega^*) \xrightarrow{\delta_*^1} H^2(V, \mathbf{Z}) \to H^2(V, \Omega). \tag{19}$$

Now $H^2(V, \Omega) = H^2(V, 1) \cong B^{0,2}(V)$. Therefore (KODAIRA-SPENCER [2]): an element $a \in H^2(V, Z)$ is mapped on to the zero element of $H^2(V, \Omega)$ if and only if a is of type (1, 1).

By Theorem 4.3.1, if $\xi \in H^1(V, \mathbf{C}_\omega^*)$ is a complex analytic $\mathbf{C}^*$-bundle then $\delta_*^1 \xi = c_1(\xi)$. If F is a complex line bundle over V and ξ is the associated $\mathbf{C}^*$-bundle then $c_1(\xi)$ is called the *cohomology class* of F. The exactness of (19) then implies

Theorem 15.9.1 (LEFSCHETZ-HODGE, KODAIRA-SPENCER [2]). *Let V be a compact* KÄHLER *manifold. An element $a \in H^2(V, \mathbf{Z})$ is the cohomology class of a complex analytic line bundle over V if and only if a is of type* (1, 1).

Remark: This theorem has also been proved in the non-KÄHLER case by DOLBEAULT ([2], Théorème 2.3).

15.10. Let V be a KÄHLER manifold with $h^{p,q} = 0$ for $p \neq q$. Then $\chi_y(V)$ is essentially equal to the POINCARÉ polynomial $P(t; V) = \sum b_r t^r$ of V (b_r = r-th BETTI number of V). More precisely,

$$\chi^p(V) = \sum_q (-1)^q h^{p,q} = (-1)^p h^{p,p} = (-1)^p b_{2p}.$$

The odd BETTI numbers of V are zero and so

$$\chi_{-t^2}(V) = P(t; V) = \sum b_r t^r. \tag{20}$$

KÄHLER manifolds with this property include the complex projective spaces and the flag manifolds $\mathbf{F}(n)$. For $\mathbf{F}(n)$ this can be seen as follows: the cohomology ring $H^*(\mathbf{F}(n), \mathbf{Z})$ is generated by elements $\gamma_i \in H^2(\mathbf{F}(n), \mathbf{Z})$. By 14.2 there are complex analytic $\mathbf{C}^*$-bundles ξ_i over $\mathbf{F}(n)$ with $c_1(\xi_i) = \gamma_i$. By the "only if" of Theorem 15.9.1 the γ_i are of type $(1, 1)$ and therefore any cohomology class of $\mathbf{F}(n)$ is of type (p, p). Notice in particular that for the complex projective spaces and for the flag manifolds the polynomials χ_y and T_y (see 14.4) agree, since both are essentially equal to the POINCARÉ polynomial.

15.11. If V_n and V'_m are KÄHLER manifolds then

$$h^{p,q}(V_n \times V'_m) = \sum_{\substack{r+u=p \\ s+v=q}} h^{r,s}(V_n)\, h^{u,v}(V'_m)\,. \tag{21}$$

Let y, z be indeterminates and associate to each KÄHLER manifold V the polynomial $\Pi_{y,z}(V) = \sum_{p,q} h^{p,q} y^p z^q$. Then (21) is equivalent to

$$\Pi_{y,z}(V_n \times V'_m) = \Pi_{y,z}(V_n) \cdot \Pi_{y,z}(V'_m)\,. \tag{22}$$

Let $z = -1$ in (22). Then $\Pi_{y,-1} = \chi_y$ and

$$\chi_y(V_n \times V'_m) = \chi_y(V_n) \cdot \chi_y(V'_m)\,, \tag{23}$$

another property common to χ_y and T_y.

§ 16. Further properties of the χ_y-characteristic

In this paragraph V is always a complex manifold.

16.1. Consider an exact sequence

$$0 \to W' \xrightarrow{h'} W \xrightarrow{h} W'' \to 0 \tag{1}$$

of complex analytic vector bundles over V [see 4.1 d)]. The sequence of sheaves

$$0 \to \Omega(W') \xrightarrow{h'} \Omega(W) \xrightarrow{h} \Omega(W'') \to 0\,, \tag{2}$$

obtained from (1) by taking sheaves of germs of local holomorphic sections, is also exact.

Proof: Every germ $s' \in \Omega(W')$ of a local holomorphic section of W' is mapped to a germ $h'(s') \in \Omega(W)$, every germ $s \in \Omega(W)$ to a germ $h(s) \in \Omega(W'')$. The sequence $0 \to \Omega(W') \to \Omega(W) \to \Omega(W'')$ is clearly exact, so it remains to prove that every germ $s'' \in \Omega(W'')$ can be written in the form $s'' = h(s)$, $s \in \Omega(W)$. This is a consequence of Remark 2 of 4.1 d).

Theorem 16.1.1. *Let*

$$0 \to W' \to W \to W'' \to 0 \tag{1}$$

be an exact sequence of complex analytic vector bundles over a compact complex manifold V. Then

$$\chi(V, W) = \chi(V, W') + \chi(V, W'') . \tag{3}$$

More generally

$$\chi^p(V, W) = \chi^p(V, W') + \chi^p(V, W'') , \tag{3*}$$

so that

$$\chi_y(V, W) = \chi_y(V, W') + \chi_y(V, W'') .$$

Proof: The sheaves which occur in (2) are of type (F) by Theorem 15.4.2, and therefore (3) follows from Theorem 2.10.2. To obtain (3*) it is sufficient to replace (1) by the sequence

$$0 \to W' \otimes \lambda^p \boldsymbol{T} \to W \otimes \lambda^p \boldsymbol{T} \to W'' \otimes \lambda^p \boldsymbol{T} \to 0 \tag{1*}$$

which is exact by Theorem 4.1.2. (3*) follows by applying (3) to (1*).

Theorem 16.1.2. *Let W be a complex analytic vector bundle (fibre $\mathbf{C}_q$) over a compact complex manifold V, and suppose that the structure group of W can be complex analytically reduced to the triangular group $\Delta(q, \mathbf{C})$. Let $A_1, A_2, \ldots, A_q$ be the corresponding diagonal line bundles* [see 4.1 e)]. *Let W' be another complex analytic vector bundle over V. Then*

$$\chi(V, W' \otimes W) \\ = \chi(V, W' \otimes A_1) + \chi(V, W' \otimes A_2) + \cdots + \chi(V, W' \otimes A_q) .$$

Proof by induction on q: The theorem is trivial for $q = 1$. Suppose it is proved for $q - 1$. There is an exact sequence

$$0 \to A_1 \to W \to W/A_1 \to 0$$

in which the vector bundle W/A_1 admits $\Delta(q-1, \mathbf{C})$ as structure group with $A_2, \ldots, A_q$ as diagonal line bundles. The induction hypothesis implies that

$$\chi(V, W' \otimes W/A_1) = \chi(V, W' \otimes A_2) + \cdots + \chi(V, W' \otimes A_q) .$$

Now (3), applied to the exact sequence

$$0 \to W' \otimes A_1 \to W' \otimes W \to W' \otimes W/A_1 \to 0 ,$$

implies that $\chi(V, W' \otimes W) = \chi(V, W' \otimes A_1) + \chi(V, W' \otimes W/A_1)$. This completes the proof.

16.2. Let W be a vector bundle over the complex manifold V and S a non-singular divisor of V (see 15.2). Let $\mathfrak{U} = \{U_i\}$ be an open covering of V for which S is given by holomorphic functions s_i on U_i. The $\mathbf{C}^*$-bundle $[S]$ is then represented by the cocycle $\{s_{ij}\} = \{s_i/s_j\}$. Let $\{S\}$ be the associated line bundle constructed from the cocycle $\{s_{ij}\}$ by identifications on $\bigcup (U_i \times \mathbf{C})$ [see 3.2. a) and 15.2]. The maps $s_i : U_i \to \mathbf{C}$

define a global section s of $\{S\}$ which is zero at points of S and non-zero elsewhere. Let $(W \otimes \{S\})_S$ be the restriction to S of the vector bundle $W \otimes \{S\}$, and let $\Omega((W \otimes \{S\})_S)$ be the sheaf over S of germs of locally holomorphic sections of $(W \otimes \{S\})_S$. The extension by zero of this sheaf from S to the whole of V will be denoted by $\hat{\Omega}((W \otimes \{S\})_S)$ as in Theorem 2.4.3.

Theorem 16.2.1. *Let V be a complex manifold and let S be a non-singular divisor of V. Let W be a complex analytic vector bundle over V. There is an exact sequence*

$$0 \to \Omega(W) \to \Omega(W \otimes \{S\}) \to \hat{\Omega}((W \otimes \{S\})_S) \to 0 \tag{4}$$

of complex analytic sheaves on V.

Proof: Associate to each local section s' of W the local section $s' \otimes s$ of $W \otimes \{S\}$. Since s is a global section of $\{S\}$, which is not identically zero on any open set of V, this defines a monomorphism $h': \Omega(W) \to$ $\to \Omega(W \otimes \{S\})$. Over the complement of S in V the section s is never zero and therefore h' is onto. Hence the quotient sheaf $\Omega(W \otimes \{S\})/\Omega(W)$ is zero over the complement of S, and is uniquely defined by its restriction to S. It is therefore sufficient to prove the exactness of the sequence

$$0 \to \Omega(W)\,|S \xrightarrow{h'} \Omega(W \otimes \{S\})|\,S \xrightarrow{h} \Omega((W \otimes \{S\})_S) \to 0 \tag{5}$$

where $\ldots\,|S$ denotes restriction of the sheaf $\ldots$ to S, and where h is the homomorphism which restricts a section of $W \otimes \{S\}$ over an open set U of V to the corresponding section of $(W \otimes \{S\})_S$ over the open set $U \cap S$ of S.

To prove the exactness of (5) associate to each point $x \in S$ a neighbourhood U_x in V over which W and $\{S\}$ are represented as product bundles $U_x \times \mathbf{C}_q$ and $U_x \times \mathbf{C}$. U_x can be chosen so small that $U_x \subset U_i$ for some set U_i of the covering. The section s is given by the holomorphic function $s_x = s_i|U_x$. Now $W \otimes \{S\}$ is represented by the product bundle $U_x \times (\mathbf{C}_q \otimes \mathbf{C})$. Consider the map $\mathbf{C}_q \otimes \mathbf{C} \to \mathbf{C}_q$ defined by $(z_1, \ldots, z_q) \otimes z \to (z_1 z, \ldots, z_q z)$. This defines a product structure $U_x \times \mathbf{C}_q$ for $W \otimes \{S\}$. With respect to these product structures local holomorphic sections of W and $W \otimes \{S\}$ are represented by q-ples $(g_1, \ldots, g_q)$ and $(f_1, \ldots, f_q)$ of local holomorphic functions. The homomorphism h' is then defined by

$$(f_1, \ldots, f_q) = h'(g_1, \ldots, g_q) = (s_x g_1, \ldots, s_x g_q)\,.$$

The homomorphism h is the restriction of $(f_1, \ldots, f_q)$ to S and is onto, since the germ of a local holomorphic function on S is always the restriction of the germ of some local holomorphic function on V. This restriction is zero if and only if the local holomorphic functions $f_1, \ldots, f_q$ are each divisible by s_x, that is, if and only if $(f_1, \ldots, f_q)$ lies in the image of h'. This proves the exactness of (5).

If V is compact then the non-singular divisor S is itself a compact complex manifold, and W_S is a complex analytic vector bundle over S. In the sequel we write simply $\chi(S, W)$ for $\chi(S, W_S)$, and similarly for $\chi^p(S, W)$ and $\chi_y(S, W)$. With these notations, if we replace W in (4) by $W \otimes \{S\}^{-1}$ and apply Theorems 2.6.3 and 2.10.2, we obtain (see KODAIRA-SPENCER [3])

Theorem 16.2.2. *Let V be a compact complex manifold, S a non-singular divisor of V and W a complex analytic vector bundle over V. Then*

$$\chi(V, W) = \chi(V, W \otimes \{S\}^{-1}) + \chi(S, W) . \tag{6}$$

In particular, when W is the trivial line bundle

$$\chi(V) = \chi(V, \{S\}^{-1}) + \chi(S) . \tag{6*}$$

16.3. Let V, S be as in Theorem 16.2.2. For the rest of this paragraph it will be assumed that V is compact. Denote the complex analytic contravariant tangent bundles of V, S by $\mathfrak{T}(V)$, $\mathfrak{T}(S)$. Then the vector bundles $\lambda^p(\mathfrak{T}(V))$, $\lambda^p(\mathfrak{T}(S))$ are the complex analytic vector bundles of contravariant p-vectors on V, S. The corresponding bundles of covariant p-vectors are denoted by $\lambda^p(\boldsymbol{T}(V))$, $\lambda^p(\boldsymbol{T}(S))$ as in 4.7. There is an exact sequence (see 4.9)

$$0 \to \mathfrak{T}(S) \to \mathfrak{T}(V)_S \to \{S\}_S \to 0 . \tag{7}$$

By Theorem 4.1.3* there is a corresponding exact sequence for bundles of contravariant p-vectors

$$0 \to \lambda^p(\mathfrak{T}(S)) \to \lambda^p(\mathfrak{T}(V)_S) \to \lambda^{p-1}(\mathfrak{T}(S)) \otimes \{S\}_S \to 0 , \tag{8}$$

and, by dualising, for bundles of covariant p-vectors

$$0 \to \lambda^{p-1}(\boldsymbol{T}(S)) \otimes \{S\}_S^{-1} \to \lambda^p(\boldsymbol{T}(V)_S) \to \lambda^p(\boldsymbol{T}(S)) \to 0 . \tag{8'}$$

Let W be a complex analytic vector bundle over V, and consider the sequence obtained from (8') by tensoring each term by W_S, the restriction of W to S. By Theorem 16.1.1 the exact sequence obtained gives a formula

$$\chi(S, W \otimes \lambda^p(\boldsymbol{T}(V))) = \chi^{p-1}(S, W \otimes \{S\}^{-1}) + \chi^p(S, W) . \tag{9}$$

Now replace W in formula (6) by $W \otimes \lambda^p(\boldsymbol{T}(V))$. A comparison with (9) gives the important "four term formula" [KODAIRA-SPENCER [3], Formula (14)]

$$\chi^p(V, W) = \chi^p(V, W \otimes \{S\}^{-1}) + \chi^p(S, W) + \chi^{p-1}(S, W \otimes \{S\}^{-1}) . \tag{10$_p$}$$

This formula holds for all $p \geqq 0$ provided that if $p = 0$ the last term is interpreted as 0 [in this case we get (6)]. The term $\chi^p(S, W)$ is zero for

$p = n = \dim V$, and for $p > n$ all four terms are zero. If y is an indeterminate, (10_p) holds with each term multiplied by y^p, and summing over all $p \geqq 0$ gives

$$\chi_y(V, W) = \chi_y(V, W \otimes \{S\}^{-1}) + \chi_y(S, W) + y\, \chi_y(S, W \otimes \{S\}^{-1}) \,. \quad (10^*)$$

16.4. By repeated application of equation (10_p), the integer $\chi^p(S, W)$ $(p \geqq 0)$ can be expressed as a linear combination of integers of the form $\chi^q(V, A)$ where each A is a certain complex analytic vector bundle over V. For example $(10_0) = (6)$ gives

$$\chi^0(S, W) = \chi^0(V, W) - \chi^0(V, W \otimes \{S\}^{-1}) \,. \qquad (11_0)$$

The last term $\chi^0(S, W \otimes \{S\}^{-1})$ in formula (10_1) for $\chi^1(S, W)$ can be calculated by replacing W in (11_0) by $W \otimes \{S\}^{-1}$. Thus

$$\begin{aligned}\chi^1(S, W) = \chi^1(V, W) - \chi^1(V, W \otimes \{S\}^{-1}) - \\ - \chi^0(V, W \otimes \{S\}^{-1}) + \chi^0(V, W \otimes \{S\}^{-2}) \,. \end{aligned} \qquad (11_1)$$

A continuation of this method gives the formula

$$\chi^p(S,W) = \sum_{i=0}^{p} (-1)^i [\chi^{p-i}(V, W \otimes \{S\}^{-i}) - \chi^{p-i}(V, W \otimes \{S\}^{-(i+1)})] \,. \quad (11_p)$$

This formula holds for all $p \geqq 0$. The left hand side of (11_p) is zero for $p \geqq n$ because S has complex dimension $n-1$, but it is not immediate that the terms on the right hand side cancel for $p \geqq n$. In other words, given a vector bundle W and a non-singular divisor S, certain relations hold between the integers $\chi^k(V, W \otimes \{S\}^r)$. Do these relations still hold if $\{S\}$ is replaced by an arbitrary line bundle F over V? We shall see that the answer is yes if V is an algebraic manifold.

16.5. Let $\mathbf{Z}\{y\}$ be the integral domain of all formal power series $a_0 + a_1 y + a_2 y^2 + \cdots$ with integers a_i as coefficients. The polynomial ring $\mathbf{Z}[y]$ is a subring of $\mathbf{Z}\{y\}$.

It is not possible to deduce from (11_p) an expression of $\chi_y(S, W)$ as a linear combination of a finite number of polynomials of type $\chi_y(V, A)$. Nevertheless in the domain $\mathbf{Z}\{y\}$ of formal power series it is true that

$$\chi_y(S, W) = \sum_{i=0}^{\infty} (-y)^i [\chi_y(V, W \otimes \{S\}^{-i}) - \chi_y(V, W \otimes \{S\}^{-(i+1)})]. \quad (11^*)$$

The right hand side of (11^*) is a formal power series which in fact terminates. The coefficient of y^p in this power series is given by (11_p) and is zero for $p \geqq n$.

§ 17. The virtual χ_y-characteristic

17.1. The definition of the virtual χ_y-genus and the virtual χ_y-characteristic, as well as the associated calculations, are simplified by introducing the following formalism.

Let E be an extension ring of the ring $\mathbf{Z}$ of rational integers, and let the integer 1 be the identity element of E. We consider the rings $\mathbf{Z}\{y\}$ and $E\{y\}$ of formal power series with coefficients in $\mathbf{Z}$ and E respectively. $\mathbf{Z}\{y\}$ is then a subring of $E\{y\}$. We call a map

$$h: E\{y\} \to \mathbf{Z}\{y\}$$

an allowable additive homomorphism (or *d-homomorphism*) if:

I) $h(u+v) = h(u) + h(v)$ for $u, v \in E\{y\}$,

II) $h(u\,v) = u\,h(v)$ for $u \in \mathbf{Z}\{y\}$, $v \in E\{y\}$.

In other words: $E\{y\}$ and $\mathbf{Z}\{y\}$ are regarded as $\mathbf{Z}\{y\}$-modules. A d-homomorphism is a homomorphism from the $\mathbf{Z}\{y\}$-module $E\{y\}$ to the $\mathbf{Z}\{y\}$-module $\mathbf{Z}\{y\}$. Condition II) implies that $h(u) = u\,h(1)$ for $u \in \mathbf{Z}\{y\}$.

Lemma 17.1.1. *Let h_0 be an additive homomorphism from E to $\mathbf{Z}\{y\}$. Then there is one and only one d-homomorphism h from $E\{y\}$ to $\mathbf{Z}\{y\}$ which agrees with h_0 on E.*

Proof: If $v = e_0 + e_1 y + e_2 y^2 + \cdots$ with $e_i \in E$ then we define

$$h(v) = h_0(e_0) + h_0(e_1)\,y + h_0(e_2)\,y^2 + \cdots.$$

The $h_0(e_i)$ are power series in y but, after multiplying out the right hand side, the coefficient of y^p for each $p \geqq 0$ is a finite sum. Therefore the right hand side is a power series in y and the homomorphism h is well defined. It is easy to see that h is a d-homomorphism which extends h_0. Conversely suppose that h' is a d-homomorphism which extends h_0. Then I) and II) imply that h and h' agree on any terminating power series of $E\{y\}$ and hence that $h = h'$. Q. E. D.

Given a d-homomorphism $h: E\{y\} \to \mathbf{Z}\{y\}$ and a fixed element $t \in E\{y\}$, there is a d-homomorphism h_t defined by

$$h_t(u) = h(t\,u)\,.$$

An immediate corollary of Lemma 17.1.1 is

Lemma 17.1.2. *Let h and h' be d-homomorphisms from $E\{y\}$ to $\mathbf{Z}\{y\}$. If $t \in E\{y\}$ is an element such that*

$$h'(u) = h(t\,u) \quad \textit{for all} \quad u \in E$$

then $h_t = h'$, that is the equation $h'(u) = h(t\,u)$ holds for all $u \in E\{y\}$.

In our applications the ring E will be of a particular form. Let $f_1, \ldots, f_r, w$ be indeterminates and let E be the ring generated over $\mathbf{Z}$ by these indeterminates together with $f_1^{-1}, \ldots, f_r^{-1}$. The products $w^\mu f_1^{\lambda_1} f_2^{\lambda_2} \ldots f_r^{\lambda_r}$ form an additive basis of E (here $\mu, \lambda_1, \ldots, \lambda_r$ are integers, μ is non-negative, and the element $1 \in \mathbf{Z}$ is regarded as a product with $\mu = \lambda_1 = \cdots = \lambda_r = 0$). Suppose that to each such product there is associated an element of $\mathbf{Z}\{y\}$. Then there is a unique additive homomorphism $E \to \mathbf{Z}\{y\}$, and hence by Lemma 17.1.1 a unique d-homomorphism $E\{y\} \to \mathbf{Z}\{y\}$, which takes the given values on the basis of E.

Now let V be a compact complex manifold, W a complex analytic vector bundle over V and $F_1, \ldots, F_r$ complex analytic line bundles over V. If E is defined as above there are two d-homomorphisms h and $\hat{h}$ from $E\{y\}$ to $\mathbf{Z}\{y\}$ defined by associating the following values to the basis elements of E:

$$\begin{aligned} h(w^\mu f_1^{\lambda_1} \ldots f_r^{\lambda_r}) &= \chi(V, W^\mu \otimes F_1^{\lambda_1} \otimes \cdots \otimes F_r^{\lambda_r}), \quad h(1) = \chi(V) \\ \hat{h}(w^\mu f_1^{\lambda_1} \ldots f_r^{\lambda_r}) &= \chi_y(V, W^\mu \otimes F_1^{\lambda_1} \otimes \cdots \otimes F_r^{\lambda_r}), \quad \hat{h}(1) = \chi_y(V)\,. \end{aligned} \tag{1}$$

On the right hand side powers of vector bundles are to be understood as tensor products. For line bundles negative powers are well defined.

Let $u \in E\{y\}$ be a power series with constant term u_0. Then

$$\hat{h}(u) \in \mathbf{Z}\{y\} \tag{2}$$

is a power series with constant term $h(u_0)$.

Convention: Let V be a compact complex manifold. If we are given a complex analytic vector bundle and a finite number of complex analytic line bundles over V, we denote the vector bundles by capital letters and associate to each vector bundle an indeterminate denoted by the corresponding lower case letter. If there is any possibility of confusion we write h_V, $\hat{h}_V$ for the homomorphisms h, $\hat{h}$ obtained in the above manner. If S is a non-singular divisor of V, the given vector bundles over V can be restricted to S. We denote these bundles over S (and the associated indeterminates) by the same letters as the corresponding bundles over V. Applying (1) to the complex manifold S we obtain d-homomorphisms h_S, $\hat{h}_S$ defined by

$$\begin{aligned} h_S(w^\mu f_1^{\lambda_1} \ldots f_r^{\lambda_r}) &= \chi(S, W^\mu \otimes F_1^{\lambda_1} \otimes \cdots \otimes F_r^{\lambda_r}), \quad h_S(1) = \chi(S) \\ \hat{h}_S(w^\mu f_1^{\lambda_1} \ldots f_r^{\lambda_r}) &= \chi_y(S, W^\mu \otimes F_1^{\lambda_1} \otimes \cdots \otimes F_r^{\lambda_r}), \quad \hat{h}_S(1) = \chi_y(S)\,. \end{aligned} \tag{3}$$

We associate to the line bundle $\{S\}$ over V the indeterminate s in accordance with the convention. Then formula 16.5 (11*) can be written

$$\chi_y(S, W) = \hat{h}_V\left(w \frac{1 - s^{-1}}{1 + y s^{-1}}\right). \tag{4}$$

Note that in $E\{y\}$ every element with constant term 1 has a unique multiplicative inverse. Particular cases of (4) are

$$\chi_y(S) = \hat{h}_V\left(\frac{1 - s^{-1}}{1 + y s^{-1}}\right)$$

and, by 16.2 (6), (6'),

$$\chi(S, W) = h_V(w(1 - s^{-1})), \quad \chi(S) = h_V(1 - s^{-1})\,.$$

17.2. We are now in a position to define the virtual χ_y-characteristic. Let V be a compact complex manifold of complex dimension n. Let W be a complex analytic vector bundle over V and $F_1, \ldots, F_r$ complex analytic line bundles over V. The r-ple $(F_1, \ldots, F_r)$ is called a virtual submanifold of V of (complex) dimension $n-r$. We allow the case $r > n$.

Definition [compare 17.1 (4)]:

$$\chi_y(F_1, \ldots, F_r|, W)_V = \hat{h}_V \left(w \prod_{i=1}^{r} \frac{1-f_i^{-1}}{1+y f_i^{-1}} \right).$$

$\chi_y(F_1, \ldots, F_r|, W)_V$ is an infinite power series on y with integer coefficients. It will be called the virtual χ_y-characteristic of the "restriction to the virtual submanifold $(F_1, \ldots, F_r)$" of the vector bundle W. Clearly it does not depend on the order in which the line bundles F_i appear. If W is the trivial line bundle we denote the virtual χ_y-characteristic by $\chi_y(F_1, \ldots, F_r)_V$ and call it the virtual χ_y-genus of the virtual submanifold $(F_1, \ldots, F_r)$. We write

$$\chi_y(F_1, \ldots, F_r|, W)_V = \sum_{p=0}^{\infty} \chi^p(F_1, \ldots, F_r|, W)_V \, y^p$$

and

$$\chi_y(F_1, \ldots, F_r)_V = \sum_{p=0}^{\infty} \chi^p(F_1, \ldots, F_r)_V \, y^p .$$

We shall always write χ for χ^0. Then by 17.1 (2)

$$\chi(F_1, \ldots, F_r|, W)_V = h_V \left(w \prod_{i=1}^{r} (1 - f_i^{-1}) \right).$$

The integer $\chi(F_1, \ldots, F_r|, W)_V$ is called the virtual χ-characteristic of the restriction to the virtual submanifold $(F_1, \ldots, F_r)$ of the vector bundle W. The integer $\chi(F_1, \ldots, F_r)_V$ is called the virtual arithmetic genus of the virtual submanifold $(F_1, \ldots, F_r)$.

In particular, the virtual arithmetic genus $\chi(F)_V$ of a line bundle F over V is defined by

$$\chi(F)_V = \chi(V) - \chi(V, F^{-1}) .$$

Now let S be a non-singular divisor of V. Then $\chi_y(S, W)$ is defined and is a polynomial of degree $\leq n-1$. By 17.1 (4)

$$\chi_y(S, W) = \chi_y(\{S\}|, W)_V . \tag{4'}$$

In this case the virtual χ_y-characteristic is a polynomial of finite degree. The fact that $\chi_y(F_1, \ldots, F_r|, W)_V$ is a polynomial of degree $\leq n-r$, and in particular that $\chi_y(F_1, \ldots, F_r|, W)_V$ is identically zero for $r > n$, is proved in Theorem 19.2.1 for the case that V is an algebraic manifold.

We prove the following generalisation of (4′) which justifies the above definitions.

Theorem 17.2.1. *Let $V, W, F_1, \ldots, F_r$ be as at the beginning of this section. Let S be a non-singular divisor of V and $\{S\} = F_1$. Then*

$$\chi_y(F_1, \ldots, F_r|, W)_V = \chi_y((F_2)_S, \ldots, (F_r)_S|, W_S)_S .$$

Proof: We write

$$\hat{R}(x) = \frac{1 - x^{-1}}{1 + y x^{-1}} . \tag{5}$$

Then by definition

$$\chi_y((F_2)_S, \ldots, (F_r)_S|, W_S)_S = \hat{h}_S\left(w \prod_{i=2}^{r} \hat{R}(f_i)\right).$$

It follows from (1), (3) and (4) that

$$\hat{h}_S(w^\mu f_1^{\lambda_1} \ldots f_r^{\lambda_r}) = \hat{h}_V(w^\mu f_1^{\lambda_1} \ldots f_r^{\lambda_r} \hat{R}(f_1)) .$$

Hence by Lemma 17.1.2 with $t = \hat{R}(f_1)$

$$\hat{h}_S\left(w \prod_{i=2}^{r} \hat{R}(f_i)\right) = \hat{h}_V\left(w \prod_{i=1}^{r} \hat{R}(f_i)\right) = \chi_y(F_1, \ldots, F_r|, W)_V . \qquad \text{Q.E.D.}$$

From the definition of the virtual χ_y-characteristic we obtain

Lemma 17.2.2. *If some F_i is the trivial line bundle* **1** *then*

$$\chi_y(F_1, \ldots, F_r|, W)_V = 0 .$$

17.3. We now show that the functional equation, established in 11.3 for the virtual T_y-characteristic, is also satisfied by the χ_y-characteristic.

Theorem 17.3.1. *Let V be a compact complex manifold, W a complex analytic vector bundle over V and $F_1, \ldots, F_r, A, B$ complex analytic line bundles over V. Then*

$$\begin{aligned}\chi_y(F_1, \ldots, F_r, A \otimes B|, W)_V \\ = \chi_y(F_1, \ldots, F_r, A|, W)_V + \chi_y(F_1, \ldots, F_r, B|, W)_V + \\ + (y-1)\,\chi_y(F_1, \ldots, F_r, A, B|, W)_V - y\,\chi_y(F_1, \ldots, F_r, A, B, A \otimes B|, W)_V .\end{aligned} \tag{6}$$

Proof: For brevity let $u = w \prod_{i=1}^{r} \hat{R}(f_i)$. Then by (5) it is necessary to prove the equation

$$\begin{aligned}&\hat{h}(u\,\hat{R}(a\,b)) \\ &= \hat{h}(u\,\hat{R}(a)) + \hat{h}(u\,\hat{R}(b)) + (y-1)\,\hat{h}(u\,\hat{R}(a)\,\hat{R}(b)) - y\,\hat{h}(u\,\hat{R}(a)\,\hat{R}(b)\,\hat{R}(ab)).\end{aligned}$$

Since $\hat{h}$ is a d-homomorphism it is sufficient, using 17.1 I) and 17.1 II) to prove that

$$\hat{R}(a\,b) = \hat{R}(a) + \hat{R}(b) + (y-1)\,\hat{R}(a)\,\hat{R}(b) - y\,\hat{R}(a)\,\hat{R}(b)\,\hat{R}(a\,b) .$$

But this is precisely the functional equation which occurs in 11.3.

Remark: The functional equation (6) is a relation between five formal power series. It cannot be assumed that these power series terminate or converge, and therefore it is not permissible to substitute particular numerical values for y. It is however permissible to equate coefficients in (6). The result is a relation between the $\chi^p(\ldots|, W)$ of the five virtual manifolds involved. For $\chi^0 = \chi$ this gives

$$\begin{aligned}&\chi(F_1, \ldots, F_r, A \otimes B|, W)_V \qquad (6')\\ &= \chi(F_1, \ldots, F_r, A|, W)_V + \chi(F_1, \ldots, F_r, B|, W)_V - \chi(F_1, \ldots, F_r, A, B|, W)_V.\end{aligned}$$

This is an equation for the virtual genus well known in algebraic geometry. In our formalism it arises from the identity

$$1 - (a\,b)^{-1} = (1 - a^{-1}) + (1 - b^{-1}) - (1 - a^{-1})\,(1 - b^{-1})\,.$$

17.4. Let V_m be a compact complex analytic split manifold [see 13.5 b)]. By definition the group of the tangent $\mathbf{GL}(m, \mathbf{C})$-bundle of V_m can be complex analytically reduced to the group $\varDelta(m, \mathbf{C})$ of triangular matrices. Let $A_1, \ldots, A_m$ be the m diagonal complex analytic line bundles [see 4.1 e)]. The complex analytic vector bundle $\lambda^p\,\boldsymbol{T}$ of covariant p-vectors on V_m admits the group $\varDelta\left(\binom{m}{p}, \mathbf{C}\right)$ as structure group; the corresponding $\binom{m}{p}$ diagonal complex analytic line bundles are (compare Theorem 4.1.1)

$$A_{i_1}^{-1} \otimes A_{i_2}^{-1} \otimes \cdots \otimes A_{i_p}^{-1} \quad (i_1 < i_2 < \cdots < i_p)\,.$$

Therefore, by Theorem 16.1.2 for $p \geqq 0$,

$$\begin{aligned}\chi^p(V_m, W) &= \chi(V_m, W \otimes \lambda^p\,\boldsymbol{T})\\ &= \sum_{i_1 < i_2 < \cdots < i_p} \chi(V_m, W \otimes A_{i_1}^{-1} \otimes A_{i_2}^{-1} \otimes \cdots \otimes A_{i_p}^{-1})\end{aligned} \qquad (7)$$

and, applying the notation of 17.1,

$$\chi_y(V_m, W) = h\left(w \prod_{i=1}^{m} (1 + y a_i^{-1})\right). \qquad (8)$$

We proved in 13.6 (13) a formula for the Todd genus of an almost complex split manifold, and will now obtain the corresponding formula for the arithmetic genus $\chi(V_m)$ of a complex analytic split manifold V_m.

Theorem 17.4.1. *Let V_m be a complex analytic split manifold with diagonal complex analytic line bundles $A_1, \ldots, A_m$. Let W be a complex analytic vector bundle over V_m. Then*

$$(1 + y)^m\,\chi(V_m, W) = \sum_{l=0}^{m} y^l \sum_{i_1 < i_2 < \cdots < i_l} \chi_y(A_{i_1}, \ldots, A_{i_l}|, W)_V\,. \qquad (9)$$

Proof: Note that (9) is a relation between formal power series. In the notation of 17.1 the right hand side can be written

$$\sum_{l=0}^{m} y^l \sum_{i_1 < i_2 < \cdots < i_l} \hat{h}(w \hat{R}(a_{i_1}) \ldots \hat{R}(a_{i_l})) \qquad \text{[Definition of } \hat{R} \text{ in (5)]}$$

$$= \hat{h}\left(\sum_{l=0}^{m} y^l \sum_{i_1 < i_2 < \cdots < i_l} w \hat{R}(a_{i_1}) \ldots \hat{R}(a_{i_l})\right) \qquad \text{(17.1 II))}$$

$$= \hat{h}\left(w \prod_{i=1}^{m} (1 + y \hat{R}(a_i))\right)$$

$$= \hat{h}\left(w \prod_{i=1}^{m} (1 + y)(1 + y\, a_i^{-1})^{-1}\right)$$

$$= (1+y)^m \hat{h}\left(w \prod_{i=1}^{m} (1 + y\, a_i^{-1})^{-1}\right).$$

A straightforward application of (8) shows that

$$\hat{h}(w^\mu a_1^{\lambda_1} a_2^{\lambda_2} \ldots a_r^{\lambda_r}) = h\left(w^\mu a_1^{\lambda_1} a_2^{\lambda_2} \ldots a_r^{\lambda_r} \prod_{i=1}^{m} (1 + y a_i^{-1})\right).$$

Now Lemma 17.1.2 with

$$t = \prod_{i=1}^{m} (1 + y a_i^{-1})$$

gives

$$(1+y)^m \hat{h}\left(w \prod_{i=1}^{m} (1 + y\, a_i^{-1})^{-1}\right) = (1+y)^m h\left(w \prod_{i=1}^{m} (1 + y\, a_i^{-1})^{-1} (1 + y\, a_i^{-1})\right)$$

$$= (1+y)^m h(w) = (1+y)^m \chi(V, W). \qquad \text{Q. E. D.}$$

§ 18. Some fundamental theorems of KODAIRA

18.1. Let V be a KÄHLER manifold. By definition (15.6) V is compact. Let $H^{1,1}(V, \mathbf{R})$ be the subgroup (see 15.7) of $H^2(V, \mathbf{R})$ which consists of elements of type (1, 1), and let $H^{1,1}(V, \mathbf{Z})$ be the corresponding subgroup of $H^2(V, \mathbf{Z})$. We introduce an "archimedean partial ordering" of $H^{1,1}(V, \mathbf{R})$:

Definition: An element $x \in H^{1,1}(V, \mathbf{R})$ is *positive* $(x > 0)$ if x can be chosen as the fundamental class of a KÄHLER metric on V.

If $x, y \in H^{1,1}(V, \mathbf{R})$ then the following rules hold:

(0) At least one element of $H^{1,1}(V, \mathbf{R})$ is positive.

(1) The zero element of $H^{1,1}(V, \mathbf{R})$ is not positive.

(2) If $x > 0$ and $y > 0$ then $x + y > 0$.

(3) If $x > 0$ and $r > 0$ $(r \in \mathbf{R})$ then $r\,x > 0$.

(4) If $x, y \in H^{1,1}(V, \mathbf{R})$ and $x > 0$ then there is a positive integer g (which depends on x, y) such that $g\,x - y > 0$.

Definition: An element $x \in H^{1,1}(V, \mathbf{Z})$ is *positive* if it is positive when regarded as an element of $H^{1,1}(V, \mathbf{R})$. A complex analytic line bundle F over V is *positive* if $c_1(F)$, which by Theorem 15.9.1 is an element of $H^{1,1}(V, \mathbf{Z})$, is positive.

A KÄHLER manifold V is called a HODGE manifold (see HODGE [2]) if $H^{1,1}(V, \mathbf{Z})$ contains at least one positive element, that is, if V admits a KÄHLER metric whose fundamental class is in the image of the natural homomorphism $H^2(V, \mathbf{Z}) \to H^2(V, \mathbf{R})$.

Examples are known of compact complex manifolds which are not KÄHLER manifolds, and of KÄHLER manifolds which are not HODGE manifolds.

Complex projective space $\mathbf{P}_n(\mathbf{C})$ is a KÄHLER manifold [and so automatically a HODGE manifold: $H^{1,1}(\mathbf{P}_n(\mathbf{C}), \mathbf{Z}) = H^2(\mathbf{P}_n(\mathbf{C}), \mathbf{Z}) \cong \mathbf{Z}$ implies that if $x \in H^{1,1}(\mathbf{P}_n(\mathbf{C}), \mathbf{R}) = H^2(\mathbf{P}_n(\mathbf{C}), \mathbf{R})$ there is a real number $r > 0$ such that $r\,x$ lies in the image of the homomorphism $H^2(\mathbf{P}_n(\mathbf{C}), \mathbf{Z}) \to H^2(\mathbf{P}_n(\mathbf{C}), \mathbf{R})$]. The positive elements of $H^2(\mathbf{P}_n(\mathbf{C}), \mathbf{Z})$ are the positive integral multiples of h_n [= the cohomology class of the oriented hyperplane $\mathbf{P}_{n-1}(\mathbf{C})$ in the oriented manifold $\mathbf{P}_n(\mathbf{C})$; see 4.2].

An algebraic manifold V (see 0.1) *is a* HODGE *manifold because V* can be regarded as a submanifold of $\mathbf{P}_m(\mathbf{C})$, m sufficiently large, and the restriction of $h_m \in H^2(\mathbf{P}_m(\mathbf{C}), \mathbf{Z})$ to V gives a positive element of $H^{1,1}(V, \mathbf{Z})$.

A complex analytic line bundle F over V is said to be *projectively induced* if, for some embedding of V in a projective space $\mathbf{P}_m(\mathbf{C})$, F is the restriction to V of the line bundle H with cohomology class h_m. [H is associated to the $\mathbf{C}^*$-bundle η_m of 4.2 with $c_1(\eta_m) = h_m$, and is determined by the hyperplane $\mathbf{P}_{m-1}(\mathbf{C})$ of $\mathbf{P}_m(\mathbf{C})$.] A projectively induced line bundle is positive, but in general there exist positive line bundles which are not projectively induced. The projectively induced line bundles can always be given by divisors (the hyperplane sections). More precisely we have

Theorem 18.1.1 (BERTINI). *Let F be a projectively induced line bundle over the algebraic manifold V. There is a non-singular divisor S of V with $F = \{S\}$.*

Remark: The theorem of BERTINI is often stated in the form: *A "general" hyperplane section S of a connected non-singular algebraic manifold V_n in $\mathbf{P}_m(\mathbf{C})$ is itself non-singular and, for $n \geqq 2$, connected.*

For proofs see AKIZUKI [1] and ZARISKI [2, 3]. It is easy to prove that S is non-singular; the fact that S is connected for $n \geqq 2$ is not needed in the sequel.

The following fundamental theorem is due to KODAIRA [6]. Another proof, which applies more generally to normal complex spaces, has been given by GRAUERT [3].

Theorem 18.1.2. *A compact complex manifold is algebraic if (and only if) it is a* HODGE *manifold.*

KODAIRA's proof makes essential use of a theorem on the vanishing of certain cohomology groups which is itself of considerable importance and is described in the next section. We then summarise the applications of Theorem 18.1.2 which are important for the present work.

18.2. In 15.2 we formulated the generalised RIEMANN-ROCH problem. Examples show that $H^0(V, W)$ does not depend only on the continuous vector bundle W: it is possible to find an algebraic manifold V and two complex analytic vector bundles W, W' over V which are isomorphic as continuous vector bundles but for which $\dim H^0(V, W) \neq \dim H^0(V, W')$. Nevertheless it turns out that $\chi(V, W)$ does depend only on the continuous vector bundle W. In fact it depends only on the CHERN classes of W. In many important cases it is moreover possible to prove that the cohomology groups $H^i(V, W)$ vanish for all $i > 0$. In such cases $\dim H^0(V, W) = \chi(V, W)$, and the calculation of $\chi(V, W)$ by means of CHERN classes gives a solution of the RIEMANN-ROCH problem.

Theorem 18.2.1. *Let F be a complex analytic line bundle over the compact complex manifold V. If F^{-1} is positive then the cohomology groups $H^i(V, F)$ vanish for all $i \neq n$.*

This theorem is proved by KODAIRA [4]. He uses a technique from differential geometry due to BOCHNER. Another proof has been given by AKIZUKI-NAKANO [1], who actually prove that if F^{-1} is positive the groups $H^{p,q}(V, F)$ [see 15.3 a)] vanish for $p + q < n$.

The SERRE duality theorem 15.4.3 shows that Theorem 18.2.1 is equivalent to

Theorem 18.2.2 (KODAIRA). *If $F \otimes K^{-1}$ is positive then the cohomology groups $H^i(V, F)$ vanish for all $i > 0$. In this case*

$$\dim H^0(V, F) = \chi(V, F).$$

Of course these theorems are non-vacuous only if V is a HODGE manifold. Theorem 18.2.2 and rule (4) of 18.1 imply immediately (see also GRIFFITHS [3])

Theorem 18.2.3 (KODAIRA). *Let F be a complex analytic line bundle over a* HODGE *manifold V, and let E be a positive line bundle over V. Then the cohomology groups $H^i(V, F \otimes E^k)$ vanish for all $i > 0$ and k sufficiently large.*

Theorem 18.2.2 is an essential preliminary in KODAIRA's proof of Theorem 18.1.2 (HODGE manifold $\rightarrow$ algebraic manifold). In the process KODAIRA [6] proves

Theorem 18.2.4. *Let V be a* HODGE *manifold. There is a positive element $x_0 \in H^{1,1}(V, \mathbf{Z})$ with the property:*

Every complex analytic line bundle F with $c_1(F) - x_0 > 0$ is projectively induced.

It is now possible to deduce

Theorem 18.2.5. *Let V be an algebraic manifold and F a complex analytic line bundle over V. There are projectively induced line bundles A, B with $F = A \otimes B^{-1}$.*

As a corollary, F can be written in the form

$$F = \{S\} \otimes \{T\}^{-1}$$

where S and T are non-singular divisors of V.

Proof: Let E be a projectively induced line bundle over V with $c_1(E) - x_0 > 0$. It is elementary that $A = E^k$ is projectively induced for $k > 0$. For k sufficiently large $k\, c_1(E) - c_1(F) - x_0 > 0$, and by 18.2.4 $B = E^k \otimes F^{-1}$ is projectively induced. Then $F = A \otimes B^{-1}$. The corollary follows from the theorem of BERTINI (18.1.1) with $A = \{S\}$ and $B = \{T\}$.

Remarks: The corollary shows that F can be represented by a divisor. It follows that the group of divisor classes of V is naturally isomorphic to the cohomology group $H^1(V, \mathbf{C}_\omega^*)$ (see 15.2 and KODAIRA-SPENCER [2]). The fact that every *divisor* D of an algebraic manifold is linearly equivalent (see 15.2) to a divisor of the form $S - T$, where S and T are non-singular, is elementary to prove (see for instance ZARISKI [4]).

From now on we make no distinction between HODGE manifolds and algebraic manifolds. In many cases (for instance in the next section) it is possible to show that a given compact complex manifold V admits a HODGE metric. V is then automatically algebraic.

18.3. Let L be a complex analytic fibre bundle over the algebraic manifold V with complex projective space $\mathbf{P}_r(\mathbf{C})$ as fibre and the projective group $\mathbf{PGL}(r+1, \mathbf{C})$ as structure group. Clearly L is a compact complex manifold. It is possible to construct a HODGE metric on L by using a HODGE metric on V and the usual HODGE metric on $\mathbf{P}_r(\mathbf{C})$. Hence

Theorem 18.3.1 (KODAIRA). *A complex analytic fibre bundle L over the algebraic manifold V with $\mathbf{P}_r(\mathbf{C})$ as fibre and $\mathbf{PGL}(r+1, \mathbf{C})$ as structure group is itself an algebraic manifold.*

The details of the proof can be found in KODAIRA [6], Theorem 8. A. BOREL (also using Theorem 18.1.2 of KODAIRA) has generalised the above theorem as follows:

Theorem 18.3.1* (A. BOREL). *Let L be a complex analytic fibre bundle over the algebraic manifold V with an algebraic manifold as fibre and a connected structure group. Assume that the first BETTI number of F is zero. Then L is itself algebraic.*

We shall apply the BOREL theorem only in the case where F is the flag manifold $\mathbf{F}(q) = \mathbf{GL}(q, \mathbf{C})/\varDelta(q, \mathbf{C})$ and L is associated to a $\mathbf{GL}(q, \mathbf{C})$-bundle ξ over V. In this case it is easy to prove that L is algebraic directly from Theorem 18.3.1 by induction over q:

Consider the fibre bundle L' associated to L but with $\mathbf{P}_{q-1}(\mathbf{C})$ as fibre. Then L is a complex analytic fibre bundle over L' with $\mathbf{F}(q-1) = \mathbf{GL}(q-1, \mathbf{C})/\varDelta(q-1, \mathbf{C})$ as fibre. By Theorem 18.3.1 L' is algebraic. By the induction hypothesis L is algebraic.

The fact that $\mathbf{F}(q)$ is an algebraic manifold was not used in this induction proof. It can be deduced by taking V as a point. In this case $L = \mathbf{F}(q)$.

Remark: Theorem 18.3.1* remains true if "algebraic" is replaced by "KÄHLER" throughout. It is then a special case of a theorem of BLANCHARD [2].

§ 19. The virtual χ_y-characteristic for algebraic manifolds

In § 17 we defined the virtual χ_y-characteristic $\chi_y(F_1, \ldots, F_r|, W)_V$ associated to a compact complex manifold V, a complex analytic vector bundle W over V and complex analytic line bundles $F_1, \ldots, F_r$ over V. By definition $\chi_y(F_1, \ldots, F_r|, W)_V$ is a formal power series in the indeterminate y with integer coefficients. We omit the suffix V when there is no danger of ambiguity. If V is an algebraic manifold, it is possible to obtain more detailed information about the χ_y-characteristic with the help of Theorem 18.2.5.

19.1. A 0-dimensional compact complex manifold is a finite number of isolated points.

Lemma 19.1.1. *Let V be a 0-dimensional complex manifold consisting of k points. Let W be a vector bundle over V with fibre $\mathbf{C}_q$ and $F_1, \ldots, F_r$ line bundles over V.*

Then I) $\chi_y(V, W) = q\,k$

II) $\chi_y(F_1, \ldots, F_r|, W) = 0$ for $r \geqq 1$.

Proof:

I) $\chi_y(V, W) = \chi(V, W) = \dim H^0(V, W) = q\,k$.

II) Every line bundle over V is trivial. Apply Lemma 17.2.2.

19.2. By definition $\chi_y(V, W)$ is a polynomial (terminating power series) with integer coefficients. We now prove by induction on the dimension n of V that the *virtual* χ_y-characteristic is also a polynomial in the case that V is an algebraic manifold.

Theorem 19.2.1. *Let V be an algebraic manifold of complex dimension n. Let W be a complex analytic vector bundle over V with fibre $\mathbf{C}_q$ and let $F_1, \ldots, F_r$ $(r \geqq 1)$ be complex analytic line bundles over V. Then*

a) *the virtual* χ_y*-characteristic* $\chi_y(F_1, \ldots, F_r|, W)$ *is zero for* $r > n$. *For* $r \leqq n$ *it is a polynomial of degree* $\leqq n - r$ *in* y *with integer coefficients.*

b) *if* $r = n \geqq 1$ *the virtual* χ_y*-characteristic* $\chi_y(F_1, \ldots, F_n|, W) = \chi(F_1, \ldots, F_n|, W)$ *is the integer* $q \cdot f_1 f_2 \ldots f_n[V]$, *where* $f_i \in H^2(V, \mathbf{Z})$ *is the cohomology class of* F_i.

Proof of a): Lemma 19.1.1 shows that a) is true for $\dim V = 0$. Now suppose that a) is proved for $\dim V < n$. By Theorem 18.2.5 there are non-singular divisors S and T of V such that $\{S\} = F_1 \otimes \{T\}$. The functional equation (6) in Theorem 17.3.1 then becomes

$$\begin{aligned}\chi_y(\{S\}, F_2, \ldots, F_r|, W)& \\ = \chi_y(F_1, \ldots, F_r|, W) + \chi_y(\{T\}, F_2, \ldots, F_r|, W)& \qquad (*)\\ + (y-1)\,\chi_y(\{T\}, F_1, \ldots, F_r|, W) - y\,\chi_y(\{S\}, \{T\}, F_1, \ldots, F_r|, W)&\,.\end{aligned}$$

The functional equation contains five terms, and we have to prove that term 2 has degree $\leqq n - r$. The induction hypothesis, together with Theorem 17.2.1 shows that terms 1, 3, 4, 5 all have degree $\leqq n - r$ and vanish for $r > n$. [If $r = 1$ then term 1 is just $\chi_y(S, W_S)$ and term 3 is $\chi_y(T, W_T)$; these terms are polynomials of degree $\leqq n - 1$ by the definition of the (non-virtual) χ_y-characteristic.] Therefore term 2 is a polynomial of degree $\leqq n - r$ and zero for $r > n$. Q. E. D.

Proof of b): Again by Theorem 18.2.5 there are non-singular divisors S and T of V such that $\{S\} = F_1 \otimes \{T\}$.

Then a) gives, for $n \geqq 2$,

$$\begin{aligned}\chi(\{S\}, F_2, \ldots, F_n|, W)& \\ = \chi(F_1, F_2, \ldots, F_n|, W) + \chi(\{T\}, F_2, \ldots, F_n|, W)& \qquad (1)\end{aligned}$$

and for $n = 1$

$$\chi(\{S\}|, W) = \chi(F_1|, W) + \chi(\{T\}|, W)\,. \qquad (2)$$

By (2) and Lemma 19.1.1 the case $n = 1$ gives

$$\chi(F_1|, W) = q\,s - q\,t = q\,f_1[V_1]$$

where s, t are the number of points of S, T. Therefore b) is true for $\dim V = 1$. Now suppose that b) is proved for $1 \leqq \dim V < n$. We now apply the induction hypothesis to (1), and use Theorem 17.2.1, Theorem 4.9.1 and 9.2 (3) to obtain

$$\begin{aligned}\chi(F_1, F_2, \ldots, F_n|, W) &= q \cdot (f_2 \ldots f_n)_S[S] - q \cdot (f_2 \ldots f_n)_T[T] \\ &= q \cdot c_1(\{S\})\, f_2 \ldots f_n[V] - q \cdot c_1(\{T\})\, f_2 \ldots f_n[V] \\ &= q \cdot f_1 f_2 \ldots f_n[V].\end{aligned}$$

Remark: The case $n = 1$ of Theorem 19.2.1 b) implies the RIEMANN-ROCH theorem for (connected) algebraic curves (see 0.5). Let F be a complex analytic line bundle over the algebraic curve V with cohomology class $f \in H^2(V, \mathbf{Z})$. Choose W to be the trivial line bundle. Then the virtual χ-genus of F^{-1} is given by

$$\chi(F^{-1}) = -f[V].$$

But, by 17.2, $\chi(F) = \chi(V) - \chi(V, F^{-1})$ and therefore, substituting F^{-1} for F,

$$\chi(V, F) = \chi(V) + f[V]. \tag{3}$$

By Theorem 15.7.1, $\chi(V) = 1 - g_1 = 1 - p$ where p = half first Betti number = genus of V. The integer $f[V]$ is called the degree of F. If F is represented by a divisor, which is always possible, then $\deg(F)$ is the algebraic number of points (number of zeros minus number of poles) of the divisor. The duality theorem 15.4.3 implies that

$$\begin{aligned}\chi(V, F) &= \dim H^0(V, F) - \dim H^1(V, F)\\ &= \dim H^0(V, F) - \dim H^0(V, K \otimes F^{-1})\end{aligned}$$

and therefore (3) becomes

$$\dim H^0(V, F) - \dim H^0(V, K \otimes F^{-1}) = 1 - p + \deg(F).$$

19.3. Let $(F_1, \ldots, F_r|, W)_V$ denote a set consisting of an algebraic manifold V, a complex analytic vector bundle W over V and complex analytic line bundles $F_1, \ldots, F_r$ over V. We allow the case $r = 0$, but in this case we also write (V, W) for $(\ldots|, W)_V$.

Theorem 19.3.1. *Let G be a function which associates to each set $(F_1, \ldots, F_r|, W)_V$ a power series in the indeterminate y with rational coefficients. Suppose that $G(F_1, \ldots, F_r|, W)_V$ is independent of the order in which the F_i appear and that*

I) $G(V, W) = \chi_y(V, W)$.

II) *G satisfies the functional equation*

$$\begin{aligned}&G(F_1, \ldots, F_r, A \otimes B|, W)_V = G(F_1, \ldots, F_r, A|, W)_V +\\ &+ G(F_1, \ldots, F_r, B|, W)_V + (y-1)\, G(F_1, \ldots, F_r, A, B|, W)_V -\\ &- G(F_1, \ldots, F_r, A, B, A \otimes B|, W)_V.\end{aligned}$$

III) *If S is a non-singular divisor of V and $F_1 = \{S\}$ then*

$$G(F_1, \ldots, F_r|, W)_V = G((F_2)_S, \ldots, (F_r)_S|, W_S)_S$$

(for $r = 1$ this means that $G(F_1|, W)_V = G(S, W_S)$). *If $F_1 = \{0\} = 1$ then $G(F_1, \ldots, F_r|, W)_V = 0$.*

Conclusion: *For all $(F_1, \ldots, F_r|, W)_V$ with $r \geqq 1$*

$$\chi_y(F_1, \ldots, F_r|, W)_V = G(F_1, \ldots, F_r|, W)_V.$$

Proof: χ_y has properties II) and III) and therefore the function $\chi_y - G$ has properties II) and III). It is therefore sufficient to show that *any function* G' *which satisfies* II), III) *and*

$$\text{I}') \quad G'(V, W) = 0$$

is identically zero. This will be proved by induction on the dimension n of V.

By Theorem 18.2.5 there are non-singular divisors S and T of V such that $\{S\} = F_1 \otimes \{T\}$. Then II) implies equation (*) of the proof of Theorem 19.2.1 a) with χ_y replaced by G'. This equation has five terms, and we have to prove that the second term is zero. The induction hypothesis and III) imply that terms 1, 3, 4, 5 are zero. [For $r = 1$ it is necessary to use I') to prove that terms 1, 3 are zero.] Therefore term 2 is zero. Q. E. D.

In the next theorem we consider only the virtual χ_y-genus, that is, the vector bundle W is always the trivial line bundle. A virtual submanifold (see 17.2) of V is denoted by $(F_1, \ldots, F_r)_V$. We allow the case $r = 0$ and in this case also write V for $(\ldots)_V$. By Theorem 19.2.1 the power series $\chi_y(F_1, \ldots, F_r)_V$ is actually a polynomial in y with integral coefficients. It is therefore permissible to substitute a particular number y_0 for the indeterminate y. If y_0 is a rational number then $\chi_{y_0}(F_1, \ldots, F_r)_V$ is a rational number; if y_0 is an integer then so is $\chi_{y_0}(F_1, \ldots, F_r)_V$.

Theorem 19.3.2. *Let* G *be a function which associates to each* $(F_1, \ldots, F_r)_V$ *a rational number which is independent of the order in which the* F_i *appear. Suppose that, for some fixed rational number* y_0,

I) $G(V) = \chi_{y_0}(V)$.

II) *G satisfies the functional equation*

$$G(F_1, \ldots, F_r, A \otimes B)_V = G(F_1, \ldots, F_r, A)_V + G(F_1, \ldots, F_r, B)_V + \\ + (y_0 - 1)\, G(F_1, \ldots, F_r, A, B)_V - y_0\, G(F_1, \ldots, F_r, A, B, A \otimes B)_V .$$

III) *If S is a non-singular divisor of V and $F_1 = \{S\}$ then*

$$G(F_1, \ldots, F_r)_V = G((F_2)_S, \ldots, (F_r)_S)_S$$

(for $r = 1$ this means that $G(F_1)_V = G(S)$). *If* $F_1 = \{0\} = 1$ *then* $G(F_1, \ldots, F_r)_V = 0$.

Conclusion: *For all* $(F_1, \ldots, F_r)_V$ *with* $r \geqq 1$

$$\chi_{y_0}(F_1, \ldots, F_r)_V = G(F_1, \ldots, F_r)_V .$$

Proof: Exactly as for Theorem 19.3.1.

Remark: The reason for choosing W to be the trivial line bundle is simply that we apply Theorem 19.3.2 in the above form. The same proof gives a formulation of Theorem 19.3.2 for arbitrary $(F_1, \ldots, F_r |, W)_V$. This is not more general, however, because the hypothesis is strengthened as well as the conclusion. The induction method of the proofs of Theorems 19.2.1, 19.3.1 and 19.3.2 is used frequently in algebraic geometry.

Certain results need only be proved for algebraic manifolds (in the non-virtual case) and then Theorem 18.2.5 allows them to be extended to apply to virtual manifolds. We have stated this induction principle only in the generality needed for the results in the present work, but at the cost of some repetition in statements of theorems and proofs.

19.4. Let $(F_1, \ldots, F_r|, W)_V$ be as at the beginning of this section, and let $f_i \in H^2(V, \mathbf{Z})$ be the cohomology class of F_i. The complex analytic vector bundle W is associated to a complex analytic $\mathbf{GL}(q, \mathbf{C})$-bundle which can be regarded as a continuous $\mathbf{GL}(q, \mathbf{C})$-bundle ξ. Then the virtual (TODD) T_y-characteristic $T_y(f_1, \ldots, f_r|, \xi)_V$ was defined in 12.3. We write

$$\begin{aligned} T_y(F_1, \ldots, F_r|, W)_V &= T_y(f_1, \ldots, f_r|, \xi)_V \\ T_y(V, W) &= T_y(V, \xi) . \end{aligned} \tag{4}$$

By Theorems 12.3.1 and 12.3.2 the T_y-characteristic has all the properties required of the function G in Theorem 19.3.1 except for the property

$$\text{I)} \quad T_y(V, W) = \chi_y(V, W)$$

which is not yet proved. We note already, however, that *it is only necessary to prove* I) *for all* V, W *in order to prove that* χ_y *and* T_y *agree on all* $(F_1, \ldots, F_r|, W)_V$.

19.5. The virtual T_y-genus is a polynomial in y with rational coefficients, and so it is permissible to substitute a particular value y_0 for y. If y_0 is an arbitrary (but fixed) rational number then, by Theorems 11.2.1 and 11.3.1, $T_{y_0}(F_1, \ldots, F_r)_V$ has all the properties required of the function G in Theorem 19.3.2 except for the property

$$\text{I)} \quad T_{y_0}(V) = \chi_{y_0}(V)$$

which is not yet proved. Note, however, that it is only necessary to prove I) for all algebraic manifolds V in order to prove that χ_{y_0} and T_{y_0} agree on all $(F_1, \ldots, F_r)_V$.

We now show that I) does hold for $y_0 = 1$ and $y_0 = -1$. The index theorem 8.2.2 implies [see 10.2 (6)] that for $y_0 = 1$

$$T_1(V) = \tau(V) = \text{index of } V.$$

Theorem 4.10.1 implies [see 10.2 (5)] that for $y_0 = -1$

$$T_{-1}(V) = E(V) = \text{EULER-POINCARÉ characteristic of } V.$$

Theorem 15.8.1 and the HODGE index theorem 15.8.2 imply the corresponding results for the χ_y-genus:

$$\chi_1(V) = \tau(V)$$

and

$$\chi_{-1}(V) = E(V) .$$

For $y_0 = 1$ or $y_0 = -1$ the function T_{y_0} satisfies all the properties required of the function G in Theorem 19.3.2 and therefore

Theorem 19.5.1. *The virtual T_y-genus and the virtual χ_y-genus agree for $y_0 = 1$ and $y_0 = -1$:*

Let V be an algebraic manifold and let $F_1, \ldots, F_r$, be complex analytic line bundles over V with cohomology classes $f_1, \ldots, f_r \in H^2(V, \mathbf{Z})$. Then

$$\chi_1(F_1, \ldots, F_r)_V = T_1(F_1, \ldots, F_r)_V = \tau(f_1, \ldots, f_r)_V ,$$

and

$$\chi_{-1}(F_1, \ldots, F_r)_V = T_{-1}(F_1, \ldots, F_r)_V = T_{-1}(f_1, \ldots, f_r)_V .$$

§ 20. The RIEMANN-ROCH theorem for algebraic manifolds and complex analytic line bundles

We are now in a position to prove that if V is an algebraic manifold then the TODD genus $T(V)$ and the arithmetic genus $\chi(V)$ agree. This will then imply the RIEMANN-ROCH theorem for a complex analytic line bundle over V.

20.1. The first step is to prove that $T(V)$ and $\chi(V)$ agree if the algebraic manifold V is also a complex analytic split manifold [see 13.5.b)].

Theorem 20.1.1. *Let V be an algebraic manifold which is also a complex analytic split manifold. Then $\chi(V) = T(V)$.*

Proof. Let $m = \dim V$ and let $A_1, \ldots, A_m$ be the complex analytic diagonal line bundles defined over V. By 13.6 (13) and Theorem 17.4.1 with $W = \mathbf{1}$,

$$(1+y)^m \, T(V) = \sum_{l=0}^{m} y^l \sum_{i_1 < \cdots < i_l} T_y(A_{i_1}, \ldots, A_{i_l})_V ,$$

$$(1+y)^m \, \chi(V) = \sum_{l=0}^{m} y^l \sum_{i_1 < \cdots < i_l} \chi_y(A_{i_1}, \ldots, A_{i_l})_V .$$

The T_y are polynomials. Since V is algebraic the χ_y are also polynomials (Theorem 19.2.1). The two equations show that $T(V) = \chi(V)$ provided that, for some $y_0 \neq -1$, T_{y_0} and χ_{y_0} agree for algebraic manifolds V and their virtual submanifolds $(A_{i_1}, \ldots, A_{i_l})_V$. Theorem 19.5.1 shows that T_{y_0} and χ_{y_0} agree in this sense for $y_0 = 1$. This completes the proof.

Remark: It is interesting to note that the agreement of χ_{y_0} and T_{y_0} for $y_0 = -1$ is not sufficient to prove the above theorem. The proof that χ_{y_0} and T_{y_0} agree for $y_0 = 1$ is based on the fact that the index of a differentiable manifold can be represented as a "polynomial in the PONTRJAGIN classes" (Theorem 8.2.2). This theorem was proved with the help of the cobordism theory of THOM.

20.2. The construction of 13.4 associates, to each compact complex manifold V_n, a compact complex split manifold V^{Δ} which is a complex analytic fibre bundle over V with the flag manifold $\mathbf{GL}(n, \mathbf{C})/\Delta(n, \mathbf{C})$ as fibre. This construction makes it possible to reduce the proof that $\chi(V) = T(V)$ for arbitrary algebraic manifolds V to the case considered in Theorem 20.1.1. By Theorem 14.3.1 $T(V) = T(V^\Delta)$. The corresponding result for the arithmetic genus is contained in

Theorem 20.2.1. *Let ξ be a complex analytic $\mathbf{GL}(q, \mathbf{C})$-bundle over the algebraic manifold V. Let V' be the fibre bundle associated to ξ with the flag manifold $\mathbf{F}(q) = \mathbf{GL}(q, \mathbf{C})/\Delta(q, \mathbf{C})$ as fibre. Then V' is an algebraic manifold and $\chi(V') = \chi(V)$.*

Proof: Theorem 18.3.1* implies that V' is algebraic. Let φ be the projection from V' on to V. The bundle $\varphi^* \xi$ over V' admits the group $\Delta(q, \mathbf{C})$ of diagonal matrices as structure group. Let $\xi_1, \ldots, \xi_q$ be the corresponding diagonal $\mathbf{C}^*$-bundles and let γ_i be the image of $c_1(\xi_i)$ under the natural homomorphism $H^2(V', \mathbf{Z}) \to H^2(V', \mathbf{C})$. The bundles ξ_i are complex analytic and therefore by Theorem 15.9.1 the cohomology classes γ_i are of type (1, 1). The cohomology homomorphism φ^* maps $H^*(V, \mathbf{C})$ monomorphically into $H^*(V', \mathbf{C})$ (BOREL [2]). Since φ is a complex analytic map, φ^* maps cohomology classes of type (p, q) to cohomology classes of type (p, q). It is known that $H^*(V', \mathbf{C})$ is generated by $\varphi^* H^*(V, \mathbf{C})$ and the γ_i (BOREL [2]). Since each γ_i is of type (1, 1), all cohomology classes of type $(0, p)$ in $H^*(V', \mathbf{C})$ must lie in $\varphi^* H^*(V, \mathbf{C})$. Therefore $h^{0,p}(V') = h^{0,p}(V)$ and hence $\chi(V) = \chi(V')$.

Theorem 20.2.2. *Let V be an algebraic manifold. The arithmetic genus $\chi(V)$ and the* TODD *genus $T(V)$ agree.*

Proof: Let V^Δ be the split manifold constructed from V. The previous theorem shows that V^Δ is an algebraic manifold and that

$$\chi(V) = \chi(V^\Delta) . \tag{1}$$

By Theorem 14.3.1

$$T(V) = T(V^\Delta) . \tag{2}$$

By Theorem 20.1.1

$$\chi(V^\Delta) = T(V^\Delta) . \tag{3}$$

The conclusion follows from (1)–(3).

20.3. Theorem 20.2.2 states that the χ_y-genus and the T_y-genus agree for algebraic manifolds when $y = 0$. The argument of 19.5 then implies

Theorem 20.3.1. *Let V be an algebraic manifold, and let $F_1, \ldots, F_r$ be complex analytic line bundles over V. Then*

$$\chi(F_1, \ldots, F_r)_V = T(F_1, \ldots, F_r)_V .$$

The theorem states for $r = 1$

$$\chi(F)_V = T(F)_V .$$

The virtual genus of a line bundle can be expressed in terms of the (non-virtual) genus of V. By 17.2 and 12.1 (4)

$$\chi(F)_V = \chi(V) - \chi(V, F^{-1}) ,$$

and

$$T(F)_V = T(V) - T(V, F^{-1}) .$$

Therefore, replacing F by F^{-1}, Theorems 20.2.2 and 20.3.1 imply the formula:

$$\chi(V, F) = T(V, F) . \tag{4}$$

Formula (4) is the RIEMANN-ROCH theorem for an algebraic manifold V and a complex analytic line bundle F over V.

Recall (15.9) that the cohomology class of F is the first CHERN class of the $\mathbf{C}^*$-bundle associated to F. The definitions of χ and T and the result just obtained can be collected together as

Theorem 20.3.2. *Let V be an algebraic manifold of dimension n and let F be a complex analytic line bundle over V with cohomology class $f \in H^2(V, \mathbf{Z})$. The cohomology groups $H^i(V, F)$ of V with coefficients in the sheaf of germs of local holomorphic sections of F are finite dimensional complex vector spaces which vanish for $i > n$. The* EULER-POINCARÉ *characteristic*

$$\chi(V, F) = \sum_{i=0}^{n} (-1)^i \dim H^i(V, F)$$

can be expressed as a "polynomial" $T(V, F)$ in the cohomology class f and the CHERN *classes c_i of V:*

$$\chi(V, F) = \varkappa_n \left[e^f \prod_{i=1}^{n} \frac{\gamma_i}{1 - e^{-\gamma_i}} \right] . \tag{4*}$$

Formula (4*) is to be understood as follows:

$c_i \in H^{2i}(V, \mathbf{Z})$ and there is a formal factorisation

$$1 + c_1 x + \cdots + c_n x^n = (1 + \gamma_1 x) \ldots (1 + \gamma_n x) . \tag{5}$$

Consider the term of degree n in f and the γ_i of the expression in square brackets. It is a symmetric function in the γ_i and is therefore a polynomial in f and the c_i with rational coefficients. If the multiplication is interpreted as the cup product in $H^*(V, \mathbf{Z})$, this polynomial defines an element of $H^{2n}(V, \mathbf{Z}) \otimes \mathbf{Q}$. The value of this element on the $2n$-dimensional cycle of V determined by the natural orientation is equal to $\chi(V, F)$.

If $F = \mathbf{1}$, so that f is 0, the above theorem gives Theorem 20.2.2.

Formula (4*) can also be written in the form (see 1.7):

$$\chi(V, F) = \varkappa_n \left[e^{f + \frac{1}{2} c_1} \prod_{i=1}^{n} \frac{\gamma_i/2}{\sinh \gamma_i/2} \right] . \tag{6}$$

The power series $\frac{x}{\sinh x}$ is a power series in x^2. The elementary symmetric functions of the γ_i^2 are the PONTRJAGIN classes $p_1, p_2, \ldots,$ of V, which depend only on the differentiable structure of V, not on the almost complex structure (see 4.6). Therefore

$\chi(V, F)$ is a polynomial in $f + \frac{1}{2} c_1$ and the PONTRJAGIN *classes of V.*

In terms of the polynomials A_i defined in 1.6 we deduce from (6) that

$$\chi(V, F) = \sum \frac{1}{2^{4s} r!} \left(f + \frac{1}{2} c_1\right)^r A_s(p_1, \ldots, p_s) [V], \tag{6*}$$

where the summation is over all r, s with $r + 2s = n = \dim V$.

The equation $\chi(V, F) = T(V, F)$ has an immediate corollary which was first proved by SERRE and KODAIRA-SPENCER [4]:

Let V be an algebraic manifold. The integer $\chi(V, F)$ depends only on the cohomology class f of F.

20.4. In the remaining sections of this paragraph we make some remarks on the connection between the present results and the classical theory (see 0.1–0.5). Let F and G be two fixed complex analytic line bundles over the algebraic manifold V. Then $\chi(V, F \otimes G^k)$ is an integer which depends on k. By Theorem 20.3.2

$$\chi(V, F \otimes G^k) = T(V, F \otimes G^k).$$

It is then clear from the definition of T that $\chi(V, F \otimes G^k)$ is a polynomial in k of degree $\leqq n = \dim V$. If f, g are the cohomology classes of F, G respectively, the coefficient of k^n in this polynomial is $\frac{1}{n!} g^n [V]$. The constant term of the polynomial is of course $\chi(V, F)$. Collecting these facts together we have

$$\begin{gathered} \chi(V, F \otimes G^k) = a_0 + a_1 k + \cdots + a_n k^n \\ \text{with} \quad a_0 = \chi(V, F) \quad \text{and} \quad n!\, a_n = g^n [V]. \end{gathered} \tag{7}$$

The a_i are rational numbers which by (4*) can be expressed as "polynomials" in f, g and the CHERN classes of V.

Remark: The fact that $\chi(V, F \otimes G^k)$ is a polynomial in k, and the formula for a_n, can easily be deduced from Theorem 19.2.1, so that it is not necessary to use (4*). On the other hand we obtain in this way very precise information about all the coefficients a_i. Another proof that $\chi(V, F \otimes G^k)$ is a polynomial in k was found by J.-P. SERRE who deduced, with the help of the PICARD manifold of V, that $\chi(V, F)$ depends only on the cohomology class f (see the end of the previous section).

Now let G be a positive line bundle (see 18.1). Then there is an integer k_0, depending on F and G, such that the line bundle $F \otimes G^k \otimes K^{-1}$ is positive for $k \geqq k_0$. [K is the canonical line bundle defined in 15.3. a).]

Therefore by Theorem 18.2.2

$$\dim H^0(V, F \otimes G^k) = \chi(V, F \otimes G^k) \quad \text{for} \quad k \geqq k_0 .$$

It follows that for any line bundle F and any positive line bundle G

$$\dim H^0(V, F \otimes G^k) = a_0 + a_1 k + \cdots + a_n k^n \tag{8}$$

for sufficiently large k, and that the coefficient $a_0 = \chi(V, F)$ is independent of G.

In the case in which G is projectively induced (that is, associated to a hyperplane section of some embedding of V in a complex projective space) these are known features of the classical theory (HILBERT characteristic function, postulation formula). It is then well known that a_0 in (8) does not depend on G. If we write $a_0 = a_0(F)$ we can define $\chi(V, F)$ in terms of the classical theory by

$$\chi(V, F) = a_0(F) . \tag{9}$$

By SERRE duality [15.5 (14)] or, alternatively, by 12.2 (12),

$$a_0(F) = (-1)^n a_0(K \otimes F^{-1}) . \tag{10}$$

In the classical theory the arithmetic genus is defined in two alternative ways

$$p_a(V) = (-1)^n(-1 + a_0(1)) \quad \text{and} \quad P_a(V) = -(-1)^n + a_0(K) . \tag{11}$$

The conjecture that $p_a(V) = P_a(V)$ was a long outstanding problem. SEVERI conjectured (see for instance SEVERI [1]) that for connected algebraic manifolds V

$$p_a(V) = P_a(V) = g_n - g_{n-1} + \cdots + (-1)^{n-1} g_1 ,$$

that is,

$$p_a(V) = P_a(V) = (-1)^n(-1 + \chi(V)) . \tag{12}$$

This equation follows from (10) for $F = 1$. Equations (12) state the equivalence of three definitions of the arithmetic genus of an algebraic manifold, all of which appear in the classical theory. This result was obtained by KODAIRA-SPENCER [1] in the manner described. For further information on the history of the arithmetic genus we refer to the work of KODAIRA [1, 2, 5]. Theorem 20.2.2 can be interpreted as stating that a fourth possible definition, namely the TODD genus, agrees with the three definitions just given (see 0.2).

20.5. Let V be a n-dimensional algebraic manifold and K the canonical line bundle of V. If c_1 is the first CHERN class of V then by Theorem 4.4.3 (see also 12.2 and 15.9) K has cohomology class $-c_1$. The i-th plurigenus of V is defined by

$$P^i = \dim H^0(V, K^i)$$

where K^i denotes the i-fold tensor product of the line bundle K. (The i in P^i is a suffix, not a power.) Then

$$P^1 = \dim H^0(V, K) = \dim H^n(V, \mathbf{1}) = g_n = \text{geometric genus of } V.$$

There is an interesting case in which the P^i can be calculated by means of the RIEMANN-ROCH theorem (20.3.2) and Theorem 18.2.2: Suppose that K is positive. Then

$$\chi(V, K^i) = P^i \quad \text{for} \quad i \geqq 2\,,$$

$$\chi(V, K^i) = \varkappa_n \left[\exp\left(-\frac{1}{2}(2i-1)\,c_1\right) \prod_{j=1}^{n} \frac{\gamma_j/2}{\sinh \gamma_j/2}\right], \tag{13}$$

$$\lim_{i\to\infty} \frac{P^i}{i^n} = \frac{1}{n!}(-c_1)^n[V] \neq 0\,.$$

The above hypothesis is for instance satisfied if V is the quotient space defined by a discontinuous group of automorphisms acting freely (*i. e.* no element other than the identity has fixed points) on a bounded domain of $\mathbf{C}_n$ (KODAIRA [6], p. 41). In this case the plurigenus P^i is equal to the number of linearly independent (over $\mathbf{C}$) automorphic forms of weight i.

20.6. Let F be a complex analytic line bundle over the n-dimensional algebraic variety V with cohomology class $f \in H^2(V, \mathbf{Z})$. Then $\chi(V, F)$ can be calculated by formula (4*) of 20.3. In (4*) there is a "multiplier" e^f in front of the product $\prod_{i=1}^{n}$. The identity

$$e^f = (1-(1-e^{-f}))^{-1} = \sum_{j=0}^{n} (1-e^{-f})^j$$

in the cohomology ring of V now implies (by the definition of the virtual TODD genus and the fact that the virtual T-genus and χ-genus agree) the following formula which was conjectured by SEVERI [1]:

$$\chi(V, F) = \chi(V) + \chi(F)_V + \chi(F, F)_V + \cdots + \underset{(n \text{ times})}{\chi(F, \ldots, F)_V}\,. \tag{14}$$

Associate to the n-dimensional algebraic manifold V the integers

$$\psi_j = \underset{(j \text{ times})}{\chi(K, \ldots, K)_V}\,, \quad \psi_0 = \chi(V)\,.$$

These are given in terms of the classical invariants Ω_i by the relations

$$\psi_n = \Omega_0 = (-c_1)^n[V]\,, \quad \psi_j = (-1)^{n-j}\,\Omega_{n-j} + 1\,.$$

Formula (14), with F replaced by K, becomes

$$(-1)^n \psi_0 = \sum_{j=0}^{n} \psi_j \quad (\text{SEVERI})\,. \tag{14'}$$

MAXWELL-TODD [1] obtained all other general relations between the integers ψ_j. All such relations are easily proved by using the virtual TODD genus and we give another example:

The definition of the virtual TODD genus shows that, if $c_1 \in H^2(V, \mathbf{Z})$ is the first CHERN class of V,

$$\psi_j = \varkappa_n \left[(1 - e^{c_1})^j \prod_{i=1}^{n} \frac{\gamma_i}{1 - e^{-\gamma_i}} \right].$$

This expression for ψ_j contains the "multiplier" $(1 - e^{c_1})^j$, and so the corresponding expression for $\sum_{j=0}^{n} 2^{n-j} \psi_j$ contains the multiplier

$$\sum_{j=0}^{n} 2^{n-j} (1 - e^{c_1})^j = 2^{n+1}/(1 + e^{c_1}).$$

Therefore

$$\sum_{j=0}^{n} 2^{n-j} \psi_j = \varkappa_n \left[2^{n+1} \frac{e^{c_1/2}}{1 + e^{c_1}} \prod_{i=1}^{n} \frac{\gamma_i/2}{\sinh \gamma_i/2} \right].$$

Since $\dfrac{e^{c_1/2}}{1 + e^{c_1}}$ is an even function of c_1, the expression in [ ] contains no terms of odd degree in c_1 and the γ_i. Therefore

$$\sum_{j=0}^{n} 2^{n-j} \psi_j = 0 \text{ for } n \text{ odd}. \tag{15}$$

Remark: The calculations of this section can be carried out without using the fact that the T-genus and the χ-genus agree. It is then necessary to use the formalism of § 17 together with Theorem 19.2.1 and the duality formula 15.5 (14). All the relations of MAXWELL-TODD can be obtained in this way. Nevertheless once T and χ are identified it is much easier to work with the T-genus. The calculations in the formalism of § 17 are precisely analogous to those using the T-genus. The reader will notice that the power series of § 17 correspond to the multipliers which occur in front of $\prod_{i=1}^{n} \frac{\gamma_i}{1 - e^{-\gamma_i}}$ in the calculations with the T-genus (and, similarly, to the multipliers which occur in front of $\prod_{i=1}^{n} Q(y; \gamma_i)$ for the T_y-genus; see 1.8). If for instance a complex analytic line bundle F with cohomology class f is involved, the formalism of § 17 will contain an indeterminate f and the calculations with the T-genus will contain a multiplier e^f.

20.7. We conclude this paragraph with some remarks on the RIEMANN-ROCH theorem for algebraic surfaces. Let V be an algebraic manifold of complex dimension two and F a complex analytic line bundle over V with cohomology class $f \in H^2(V, \mathbf{Z})$. Then formula (4*) of 20.3 and SERRE duality give

$$\begin{aligned}\dim H^0(V,F) - \dim H^1(V,F) + \dim H^0(V, K \otimes F^{-1}) \\ = \frac{1}{2}(f^2 + f\,c_1)\,[V] + \frac{1}{12}(c_1^2 + c_2)\,[V]\,.\end{aligned} \tag{16}$$

To express this in the classical notation we assume that V is connected. The superabundance (see ZARISKI [1], p. 68) of F is defined by

$$\dim H^1(V,F) = \sup(F)\,.$$

The integer $f^2[V]$ is called the virtual degree $g(F)$ of F. It is now easy to express (16) in the usual form of the RIEMANN-ROCH theorem. Alternatively it is possible to use formula 20.6 (14) and obtain

$$\dim H^0(V,F) + \dim H^0(V, K \otimes F^{-1}) = \chi(V) + \chi(F)_V + g(F) + \sup(F)\,. \tag{17}$$

In the classical terminology $\chi(F) = 1 - \pi(F)$, where $\pi(F)$ = "virtual genus of F", and $\chi(V) = 1 + p_a(V)$. Recall that, by the remark following Theorem 15.2.1,

$$\dim|F| + 1 = \dim H^0(V,F)\,,\ \dim|K \otimes F^{-1}| + 1 = \dim H^0(V, K \otimes F^{-1}),$$

so that (17) can be written in the classical form

$$\dim|F| + \dim|K \otimes F^{-1}| = p_a(V) - \pi(F) + g(F) + \sup(F)\,. \tag{18}$$

Unlike formula (16), formula (18) does not contain the fact that $\chi(V) = T(V)$. This equation arises in the classical theory in the following form:

Define the linear genus

$$p^{(1)} = g(K) + 1 = c_1^2[V] + 1\,.$$

Formula (18), with F replaced by K, gives an alternative definition

$$1 - \pi(K) = 1 - p^{(1)} = \chi(K)\,.$$

The ZEUTHEN-SEGRE invariant I of V is given by $c_2[V] = I + 4$, and the arithmetic genus $p_a(V)$ by $\chi(V) = 1 + p_a(V)$. Therefore the equation

$$\chi(V) = \frac{1}{12}(c_1^2 + c_2)\,[V] = T(V)$$

becomes

$$12 p_a + 9 = p^{(1)} + I\,. \tag{19}$$

This relation is due to M. NOETHER (see ZARISKI [1], p. 62).

§ 21. The RIEMANN-ROCH theorem for algebraic manifolds and complex analytic vector bundles

21.1. In this section we prove the main theorem

$$\chi(V, W) = T(V, W) \tag{1}$$

for an algebraic manifold V and a complex analytic vector bundle W over V. This theorem will be called the RIEMANN-ROCH theorem for vector bundles (or simply R-R). We recall the definitions of χ and T and summarise the situation in

Theorem 21.1.1. *Let V be an algebraic manifold of dimension n and let W be a complex analytic vector bundle over V with fibre $\mathbf{C}_q$. Let $c_0, c_1, \ldots, c_n$ be the* CHERN *classes of V and $d_0, d_1, \ldots, d_q$ the* CHERN *classes of W ($c_0 = d_0 = 1$; $c_i, d_i \in H^{2i}(V, \mathbf{Z})$). The cohomology groups $H^i(V, W)$ are finite dimensional vector spaces which vanish for $i > n$. The* EULER-POINCARÉ *characteristic*

$$\chi(V, W) = \sum_{i=0}^{n} (-1)^i \dim H^i(V, W)$$

can be expressed as a "polynomial" $T(V, W)$ in the CHERN *classes c_i and d_i:*

$$\begin{aligned} \chi(V, W) &= \varkappa_n \left[(e^{\delta_1} + \cdots + e^{\delta_q}) \prod_{i=1}^{n} \frac{\gamma_i}{1 - e^{-\gamma_i}}\right] \\ &= \varkappa_n \left[e^{c_1/2}(e^{\delta_1} + \cdots + e^{\delta_q}) \prod_{i=1}^{n} \frac{\gamma_i/2}{\sinh \gamma_i/2}\right] = T(V, W)\,. \end{aligned} \tag{1*}$$

Equation (1*) is to be understood as follows: *there are formal factorisations*

$$\sum_{i=0}^{n} c_i x^i = \prod_{i=1}^{n} (1 + \gamma_i x) \quad \text{and} \quad \sum_{i=0}^{q} d_i x^i = \prod_{i=1}^{q} (1 + \delta_i x)$$

and the term of degree n of the expression in square brackets is a polynomial in the c_i and d_i. This term determines an element of $H^{2n}(V, \mathbf{Z}) \otimes \mathbf{Q}$ which is to be evaluated on the fundamental $2n$-dimensional cycle of V.

Before giving the proof we make some remarks and discuss a special case. Of course R-R contains Theorem 20.3.2. R-R also implies that, for fixed V, the integer $\chi(V, W)$ depends only on the CHERN classes of W and therefore only on the continuous vector bundle W. This fact does not seem to have been proved without using R-R, except in the case that W is a line bundle (see the remark in 20.4). This may be connected with the fact that there is in general no algebraic manifold whose points represent (for fixed V and $q > 1$) the complex analytic $\mathbf{GL}(q, \mathbf{C})$-bundles over V which are trivial as continuous bundles. For $q = 1$ such an algebraic manifold does exist; it is called the PICARD manifold of V (see SERRE [1], KODAIRA-SPENCER [2]).

Theorem 21.1.1 is known for $n = 1$; in this case V is an algebraic curve and WEIL proved

Theorem 21.1.2 (WEIL [1], p. 63). *Let V be a connected algebraic curve and let W, W' be complex analytic vector bundles with* $\mathbf{C}_r$, $\mathbf{C}_{r'}$ *as typical fibres. Let* $d_1 \in H^2(V, \mathbf{Z})$ *be the first* CHERN *class of W and d_1' the first* CHERN *class of W'. Then*

$$\chi(V, W \otimes W'^*) = \dim H^0(V, W \otimes W'^*) - \dim H^0(V, K \otimes W^* \otimes W')$$
$$= r' d_1[V] - r d_1'[V] + r r'(1-p)$$

where p is the genus of V.

Proof with help of R-R: Let δ_i, δ_i' denote the formal roots of W, W'. Then by Theorem 4.4.3 or formula 10.1 (4)

$$\chi(V, W \otimes W'^*) = \varkappa_1\left[e^{c_1/2}(e^{\delta_1} + \cdots + e^{\delta_r})(e^{-\delta_1'} + \cdots + e^{-\delta_{r'}'})\right]$$
$$= \varkappa_1[(1 + c_1/2)(r + d_1)(r' - d_1')]. \qquad \text{Q.E.D.}$$

We now come to the proof of R-R. Let ξ be the complex analytic $\mathbf{GL}(q, \mathbf{C})$-bundle over V associated to W. Consider a fibre bundle E associated to ξ with the flag manifold $\mathbf{F}(q) = \mathbf{GL}(q, \mathbf{C})/\varDelta(q, \mathbf{C})$ as fibre and denote the projection of E on to V by φ. By Theorem 14.3.1

$$T(V, W) = T(E, \varphi^* W), \tag{2}$$

and, by the theorem of BOREL stated in the next section,

$$\chi(V, W) = \chi(E, \varphi^* W)\, \chi(\mathbf{F}(q)).$$

Since the arithmetic genus $\chi(\mathbf{F}(q))$ is 1 (see 15.10)

$$\chi(V, W) = \chi(E, \varphi^* W). \tag{3}$$

The group of the bundle reduces complex analytically to the group $\varDelta(q, \mathbf{C})$. Therefore there are q diagonal complex analytic line bundles $A_1, \ldots, A_q$ over E and by 12.1 (5) and Theorem 16.1.2

$$T(E, \varphi^* W) = \sum_{i=1}^{q} T(E, A_i) \quad \text{and} \quad \chi(E, \varphi^* W) = \sum_{i=1}^{q} \chi(E, A_i). \tag{4}$$

E is an algebraic manifold (Theorem 18.3.1*), and so by Theorem 20.3.2,

$$\chi(E, A_i) = T(E, A_i) \quad (1 \leq i \leq q). \tag{5}$$

Equations (2), (3), (4) and (5) now give

$$\chi(V, W) = T(V, W). \qquad \text{Q.E.D.}$$

21.2. The following previously unpublished theorem of BOREL was used in the proof of Theorem 21.1.1 (R-R).

Theorem 21.2.1. *Let E be a complex analytic fibre bundle over a compact complex manifold V with a (compact) connected* KÄHLER *manifold F as fibre and with a connected structure group. E is then automatically* *a compact complex manifold. Let φ be the projection from E on to V, and let W be a complex analytic vector bundle over V. Then*

$$\chi_y(E, \varphi^* W) = \chi_y(V, W)\, \chi_y(F)\,. \tag{6}$$

In particular, when $y = 0$,

$$\chi(E, \varphi^* W) = \chi(V, W)\, \chi(F)\,, \quad \chi(E) = \chi(V)\, \chi(F)\,. \tag{7}$$

Corollary: *If E, V and F are* KÄHLER *manifolds then the index τ satisfies*

$$\tau(E) = \tau(V)\, \tau(F)\,. \tag{8}$$

By Theorem 15.8.2, the corollary is the case $y = 1$, W trivial. The proof of Theorem 21.2.1 due to BOREL uses the spectral sequence for the $\bar{\partial}$-cohomology of a complex analytic fibre bundle, and is included in Appendix Two.

Remarks: (1) When F is a flag manifold, (6) implies formulae (10), (10*) of 14.3, 14.4. It is shown in CHERN-HIRZEBRUCH-SERRE [1] that (8) is true when E, V, F are compact connected oriented manifolds, provided that the orientation of E is induced by those of V and F and that the fundamental group $\pi_1(V)$ acts trivially on the cohomology ring $H^*(F)$ of F.

(2) Theorem 20.2.1 is a special case of Theorem 21.2.1. In 20.2.1 we proved only as much as was necessary for the application to Theorem 20.2.2.

(3) For the proof of R-R in the previous section it is enough to know formula (7) for F a flag manifold. The induction method used at the end of 18.3 shows that it is sufficient to know (7) for F a complex projective space. In this case $\chi(F) = 1$ and it is possible to prove that

$$\dim H^i(V, W) = \dim H^i(E, \varphi^* W)\,, \tag{9}$$

which implies the equation $\chi(V, W) = \chi(E, \varphi^* W)$. A direct proof of (9) is given in the appendix [23.2 (2)].

21.3. Theorem 21.1.1 (R-R) makes possible the complete identification of the χ-theory and the T-theory. By R-R

$$\chi(V, W \otimes \lambda^p \boldsymbol{T}) = T(V, W \otimes \lambda^p \boldsymbol{T})$$

and therefore

$$\chi^p(V, W) = T^p(V, W)\,.$$

Since χ^p, T^p are the coefficients of y^p in the polynomials χ_y, T_y this implies that

$$\chi_y(V, W) = T_y(V, W)\,.$$

Note that $\chi^p(V, W)$ depends only on the continuous vector bundle W. If W is a line bundle then this fact can be proved directly (KODAIRA-SPENCER [4]).

The explicit formula in the case $W = \mathbf{1}$ is [see 12.2 (9)]:

$$\begin{aligned}\chi^p(V) &= \sum_{q=0}^{n} (-1)^q h^{p,q}(V) \\ &= \varkappa_n \left[\sum e^{-(\gamma_{i_1} + \cdots + \gamma_{i_p})} \prod_{i=1}^{n} \frac{\gamma_i}{1 - e^{-\gamma_i}} \right] \\ &= \varkappa_n \left[\sum e^{\frac{1}{2}(\pm \gamma_{i_1} \pm \cdots \pm \gamma_{i_n})} \prod_{i=1}^{n} \frac{\gamma_i/2}{\sinh \gamma_i/2} \right]. \end{aligned} \tag{10}$$

(The summation in the last line is over all combinations of signs which contain exactly p minus signs.)

It now follows from 19.4 that χ_y and T_y also agree in the virtual case. We therefore obtain

Theorem 21.3.1. *Let V be an algebraic manifold, W a complex analytic vector bundle over V and $F_1, \ldots, F_r$ complex analytic line bundles over V. Then*

$$\chi_y(F_1, \ldots, F_r |, W)_V = T_y(F_1, \ldots, F_r |, W)_V . \tag{11}$$

Remark: The case $r = 0$ (see 19.3) and $y = 0$ of this formula is just R-R. Although formula (11) is the most general result of this chapter, it is not an essential generalisation. R-R is the central theorem.

Bibliographical note

At least four other proofs of the RIEMANN-ROCH theorem are now available. A proof that $\chi(V, W) = T(V, W)$ for a complex analytic vector bundle W over an arbitrary compact complex manifold V has been given by ATIYAH-SINGER [1]. Their method is based partly on the proof of the index theorem (8.2.2) in Chapter Two and is described in § 25. The argument of 21.3 then implies that the χ-theory and the T-theory agree on compact complex manifolds *i. e.* Theorem 21.3.1 holds for V a compact complex manifold. In particular $\chi_1(V) = T_1(V) = \tau(V)$ so that the HODGE index theorem (15.8.2) is true for V a compact complex manifold.

A direct proof that $\chi(V) = T(V)$ which avoids the index theorem is due to WASHNITZER. The proof holds for V an algebraic manifold and more generally for V a non-singular projective variety defined over an algebraically closed field $\mathbf{K}$. By results of CHOW and SERRE, $\chi(V)$ and $T(V)$ can still be defined in this case (SERRE [2, 4], BOREL-SERRE [2], GROTHENDIECK [4]). The published version (WASHNITZER [2]) contains an axiomatic characterisation of the arithmetic genus $\chi(V)$ but unfortunately omits the proof that $T(V)$ satisfies the axioms.

The GROTHENDIECK-RIEMANN-ROCH theorem for a proper map $f: V \to X$ of algebraic varieties (BOREL-SERRE [2]) is described in § 23. When X is a point the theorem becomes R-R for an algebraic vector bundle W over a non-singular projective variety V (both defined over an algebraically closed field $\mathbf{K}$). By results of SERRE [4] on the relation between analytic and algebraic sheaves when $\mathbf{K} = \mathbf{C}$, this implies R-R for a complex analytic vector bundle W over an algebraic manifold V.

Another proof of the GROTHENDIECK-RIEMANN-ROCH theorem when $\mathbf{K} = \mathbf{C}$ is due to ATIYAH-HIRZEBRUCH [8]. For $f: V \to X$ an embedding the proof includes the case that V, X are arbitrary compact complex manifolds. For general f it is necessary to assume that V, X are algebraic manifolds. This approach yields the shortest available proof of R-R but, as in this book, only for V a (complex) algebraic manifold.

Appendix One

by R. L. E. Schwarzenberger

§ 22. Applications of the Riemann-Roch theorem

Three typical applications of the Riemann-Roch theorem are summarised. The first uses the theorem to calculate invariants of complete intersections in projective space (22.1). The second uses the theorem to calculate invariants of algebraic manifolds which arise from the bounded homogeneous symmetric domains of E. Cartan (22. 2.—22.3). The third application of R-R is to the study of complex vector bundles over complex projective space (22.4).

22.1. Consider r non-singular hypersurfaces $F^{(a_1)}, \ldots, F^{(a_r)}$ of degrees $a_1, \ldots, a_r$ in complex projective space $\mathbf{P}_{n+r}(\mathbf{C})$. The intersection $V_n^{(a_1,\ldots,a_r)} = F^{(a_1)} \cap \cdots \cap F^{(a_r)}$ is an algebraic manifold of dimension n if the hypersurfaces $F^{(a_1)}, \ldots, F^{(a_r)}$ are in general position. The problem is to calculate the χ_y-characteristic of the algebraic manifold $V_n^{(a_1,\ldots,a_r)}$. It will appear that this depends only on the integers $a_1, \ldots, a_r$, n and not on the particular choice of hypersurfaces $F^{(a_1)}, \ldots, F^{(a_r)}$.

Let H be a line bundle over $\mathbf{P}_{n+r}(\mathbf{C})$ associated to the $\mathbf{C}^*$-bundle η_{n+r} (see 4.2). Then H corresponds to the divisor class of a hyperplane $\mathbf{P}_{n+r-1}(\mathbf{C})$ and has cohomology class $c_1(\eta_{n+r}) = h \in H^2(\mathbf{P}_{n+r}(\mathbf{C}), \mathbf{Z})$. The line bundle H^{a_i} corresponds to the divisor class of the hypersurface $F^{(a_i)}$. If $j: V_n^{(a_1,\ldots,a_r)} \to \mathbf{P}_{n+r}(\mathbf{C})$ is the embedding we write $\tilde{h}$ for $j^* h$ and $\tilde{H}$ for $j^* H$.

Consider the case $r = 1$. By 4.8.1 there is an exact sequence of vector bundles over $F^{(a_1)}$

$$0 \to \mathfrak{T}(F) \to j^* \mathfrak{T}(P) \to j^* H^{a_1} \to 0$$

where $\mathfrak{T}(F)$ and $\mathfrak{T}(P)$ are the tangent bundles of $F^{(a_1)}$ and $\mathbf{P}_{n+1}(\mathbf{C})$. Therefore

$$c(\mathfrak{T}(F)) = j^*\big(c(\mathfrak{T}(P)) \cdot c(H^{a_1})^{-1}\big) = (1 + \tilde{h})^{n+2} (1 + a_1 \tilde{h})^{-1} .$$

Theorem 4.8.1 can be applied r times to give the total Chern class of the algebraic manifold $V_n^{(a_1,\ldots,a_r)}$:

$$c(\mathfrak{T}(V_n)) = (1 + \tilde{h})^{n+r+1}(1 + a_1 \tilde{h})^{-1} \ldots (1 + a_r \tilde{h})^{-1} . \tag{1}$$

Theorem 22.1.1. *Let V_n be a complete intersection of r hypersurfaces of degrees $a_1, \ldots, a_r$ in general position in $\mathbf{P}_{n+r}(\mathbf{C})$, and let z be an in-*

determinate. The χ_y-characteristic of the line bundle $\tilde{H}^k$ over V_n is given by

$$\sum_{n=0}^{\infty} \chi_y(V_n, \tilde{H}^k)\, z^{n+r} = \frac{(1+zy)^{k-1}}{(1-z)^{k+1}} \prod_{i=1}^{r} \frac{(1+zy)^{a_i}-(1-z)^{a_i}}{(1+zy)^{a_i}+y(1-z)^{a_i}}\,. \tag{2}$$

Proof (HIRZEBRUCH [3], § 2.1): By the RIEMANN-ROCH theorem (21.3.1)

$$\chi_y(V_n, \tilde{H}^k) = T_y(V_n, \tilde{H}^k)\,.$$

Let $R(x) = [(1-e^{-x(y+1)})^{-1}(y+1) - y]^{-1}$. Then (1) implies

$$T_y(V_n, \tilde{H}^k) = \varkappa_n \left[e^{(1+y)k\tilde{h}} (\tilde{h}\, R(\tilde{h})^{-1})^{n+r+1} \prod_{i=1}^{r} (a_i \tilde{h})^{-1} R(a_i \tilde{h})\right]$$

$$= \varkappa_n \left[e^{(1+y)k\tilde{h}}\, \tilde{h}^{n+1} R(\tilde{h})^{-n-r-1} \prod_{i=1}^{r} a_i^{-1} R(a_i \tilde{h})\right].$$

The term of degree n is a multiple of $\tilde{h}^n$ and $\tilde{h}^n[V_n] = a_1 a_2 \ldots a_r$. Therefore $T_y(V_n, \tilde{H}^k)$ is the coefficient of x^{-1} in

$$e^{(1+y)kx} R(x)^{-n-r-1} \prod_{i=1}^{r} R(a_i x)\,.$$

This coefficient can be computed as a residue at $x=0$. The substitution $z = R(x)$ gives

$$e^{(1+y)x} = \frac{1+zy}{1-z}\,, \quad dz = (1+zy)(1-z)\,dx$$

$$R(ax) = \frac{(1+zy)^a-(1-z)^a}{(1+zy)^a+y(1-z)^a}\,.$$

Therefore $T_y(V_n, \tilde{H}^k)$ is the residue at $z=0$ of

$$z^{-n-r-1} \frac{(1+zy)^{k-1}}{(1-z)^{k+1}} \prod_{i=1}^{r} \frac{(1+zy)^{a_i}-(1-z)^{a_i}}{(1+zy)^{a_i}+y(1-z)^{a_i}}$$

as required.

Corollary: When $y=0$, equation (2) becomes

$$\sum_{n=0}^{\infty} \chi(V_n, \tilde{H}^k)\, z^{n+r} = (1-z)^{-k-1} \prod_{i=1}^{r} (1-(1-z)^{a_i})\,.$$

Similarly the cases $y=-1$, $y=+1$ give equations for the EULER-POINCARÉ characteristic and index of $V_n^{(a_1,\ldots,a_r)}$.

Remark: Theorem 22.1.1 can be proved directly from the "four term formula" [16.3 (10)] and this proof preceded the proof of the RIEMANN-ROCH theorem. It can be shown easily that the theorem holds also for $r=0$. The corollary gives for $r=0$ and $r=1$ respectively the well known formulae for $\chi(\mathbf{P}_n(\mathbf{C}), H^k)$ and $\chi(V_n^{(2)}, \tilde{H}^k)$ which were used for example in HIRZEBRUCH-KODAIRA [1] and BRIESKORN [1].

Theorem 22.1.1 can be used to calculate the integers $h^{p,q}$ for V_n (see 15.4). This is possible because of

Theorem 22.1.2. *Let* $V_n = V_n^{(a_1, \ldots, a_r)}$ *be a complete intersection. Then* $h^{p,q}(V_n) = \delta_{p,q}$ *for* $p + q \neq n$ *and*

$$\chi^p(V_n) = (-1)^{n-p}\, h^{p,n-p}(V_n) + (-1)^p \quad \textit{for} \quad 2p \neq n ,$$

$$\chi^m(V_n) = (-1)^m\, h^{m,m}(V_n) \quad \textit{for} \quad 2m = n .$$

The proof can be found in HIRZEBRUCH [3], § 2.2. It is by induction on r and uses the theorem of LEFSCHETZ on hyperplane sections (BOTT [4]).

22.2. Let M be a bounded domain in $\mathbf{C}_n$ endowed with the BERGMANN hermitian metric (KODAIRA [6], p. 42). This is a KÄHLER metric which is invariant under complex analytic homeomorphisms of M. Let $I(M)$ be the group of all such homeomorphisms and $Y = M/\Delta$ the quotient space defined by the action of a subgroup Δ of $I(M)$. The identification map $p: M \to Y$ is a complex analytic covering map of a compact complex manifold Y if

(a) Δ *is properly discontinuous, i. e. any compact set in* M *intersects only a finite number of its images under* Δ.

(b) M/Δ *is compact.*

(c) Δ *acts freely, i. e. only the identity element of* Δ *has fixed points.*

Properties *(a)*, *(b)*, *(c)* imply (see KODAIRA [6], p. 41) that the canonical line bundle K_Y is a positive line bundle over Y (see 18.1). Therefore Theorem 18.1.2 implies that Y is an algebraic manifold. A holomorphic function f on M is an *automorphic form of weight* r *with respect to* Δ if for all $x \in M$, $\gamma \in \Delta$,

$$f(\gamma x) = J_\gamma^{-r}(x)\, f(x)$$

where $J_\gamma(x)$ is the jacobian of γ at the point x. The complex vector space of all automorphic forms of weight r is isomorphic to $H^0(Y, K_Y^r)$. The dimension of this vector space, *i. e.* the "number" of linearly independent automorphic forms of weight r with respect to Δ, is denoted by $\Pi_r(M, \Delta)$. Since K_Y is positive, Theorems 18.2.1 and 18.2.2 imply that the cohomology groups of Y with coefficients in the sheaf of germs of holomorphic sections of K_Y^r are zero in all dimensions $\neq 0$ if $r \geqq 2$ and in all dimensions $\neq n$ if $r \leqq -1$. Therefore [see 20.5 (13)]

$$\begin{aligned} \Pi_r(M, \Delta) &= 0 \quad \text{for} \quad r \leqq -1 , \\ \Pi_0(M, \Delta) &= 1 , \\ \Pi_1(M, \Delta) &= g_n , \\ \Pi_r(M, \Delta) &= \chi(Y, K_Y^r) \quad \text{for} \quad r \geqq 2 . \end{aligned} \tag{3}$$

Here g_n is the "number" of holomorphic forms of degree n over $Y = M/\Delta$.

Now suppose that the bounded domain M is *homogeneous, i. e.* M admits a transitive group of complex analytic homeomorphisms. The CHERN classes c_i of Y can be represented by differential forms so that each partition $\pi = (j_1, \ldots, j_p)$ of n defines a differential form $P(\pi)$ of degree $2n$ and type (n, n) which represents the cohomology class $c_{j_1} \ldots c_{j_p}$. Since M is homogeneous $p^* P(\pi) = s(\pi).V$ where $s(\pi)$ is a real number which depends only on M and the partition π, and V is the invariant volume element of M with respect to the BERGMANN metric (see HIRZEBRUCH [5], § 2).

Theorem 22.2.1. *Let Δ_1, Δ_2 be two subgroups of $I(M)$ which satisfy (a), (b), (c). Let v_i be the volume of $Y_i = M/\Delta_i$ with respect to the* BERGMANN *metric on the bounded homogeneous domain M and $c = v_1/v_2$. Then*

$$\chi_y(Y_1) = c\, \chi_y(Y_2)\,, \quad \Pi_r(M, \Delta_1) = c\, \Pi_r(M, \Delta_2) \quad \textit{for} \quad r \geqq 2\,.$$

Proof: Let $s_i(\pi)$ be the CHERN number $c_{j_1} \ldots c_{j_p}[Y_i]$ of Y_i which corresponds to the partition π. Then $s_1(\pi) = s(\pi)\, v_1$, $s_2(\pi) = s(\pi)\, v_2$ and

$$s_1(\pi) = c\, s_2(\pi) \quad \text{for all} \quad \pi = (j_1, \ldots, j_p)\,. \tag{4}$$

Therefore (4) holds also for any linear combination of CHERN numbers. In particular (3) and the RIEMANN-ROCH theorem imply that (4) holds for $\chi_y(Y_i)$ and for $\Pi_r(M, \Delta_i)$, $r \geqq 2$.

Now suppose that the bounded homogeneous domain M is in addition *symmetric, i. e.* for each point $x \in M$ there is a complex analytic homeomorphism $\sigma_x : M \to M$ which has x as isolated fixed point and is an involution ($\sigma_x^2 =$ identity). The following special case of a theorem of BOREL shows that there always exist algebraic manifolds M/Δ.

Theorem 22.2.2 (BOREL [4]). *Let M be a bounded homogeneous symmetric domain, and $I(M)$ the group of complex analytic homeomorphisms of M. Then*

I) *$I(M)$ contains a subgroup Δ which satisfies (a), (b) and (c),*

II) *if Δ is a subgroup of $I(M)$ which satisfies (a) and (b), and which does not consist only of the identity element, then Δ has a proper normal subgroup of finite index which satisfies (a), (b) and (c).*

Remark: In the case considered, *(a)* holds if and only if Δ is a discrete subgroup of $I(M)$; property *(b)* holds if and only if $I(M)/\Delta$ is compact (BOREL [4], p. 112).

22.3. Let M be a bounded homogeneous symmetric domain in $\mathbf{C}_n$. Then M is a product $M = N_1 \times \cdots \times N_s$ of irreducible bounded homogeneous symmetric domains N_k. Each N_k is a quotient space $N = G/H$ with G a simple non-compact LIE group with centre the identity and H a maximal compact connected subgroup of G with centre of (real) dimension one. It is possible to associate to G a compact LIE group G' which also contains H. The quotient space $N' = G'/H$ is a compact irreducible homogeneous hermitian symmetric complex manifold which contains an open subset complex analytically homeomorphic to N (BOREL [1]). Full details of this construction can be found in HELGASON [1], p. 321.

It is shown in BOREL-HIRZEBRUCH [1], Part I, p. 520 that the canonical line bundle of N' is negative in the sense of KODAIRA (and hence that N' is an algebraic manifold). Let $e \in N \subset N'$ be the base point corresponding to the identity of G, G'. By a formula of E. CARTAN the curvature tensor at e associated to an invariant metric on N' is a negative multiple of the curvature tensor at e associated to an invariant metric on N (see HIRZEBRUCH [5]).

Let $M' = N_1' \times \cdots \times N_s'$ and $e = (e_1, \ldots, e_s) \in M$. Then M can be regarded as an open subset of M' and the invariant differential forms which represent given CHERN numbers on M, M' differ at e by a positive factor multiplied by $(-1)^n$. This is a consequence of the above-mentioned property of the curvature tensors. As in Theorem 22.2.1 an application of the RIEMANN-ROCH theorem gives

Theorem 22.3.1. *Let M be a bounded homogeneous symmetric domain in $\mathbf{C}_n$, and $I(M)$ the group of complex analytic homeomorphisms of M. Let $Y = M/\Delta$ be the quotient space defined by the action of a subgroup Δ of $I(M)$ which satisfies (a), (b) and (c) of 22.2, and M' the compact symmetric manifold corresponding to M. Then there is a real number c such that*

$$\chi_y(Y) = c\,\chi_y(M')\,, \quad \Pi_r(M, \Delta) = c\,\chi(M', (K_{M'})^r) \quad \text{for} \quad r \geqq 2\,.$$

If n is even $c > 0$. If n is odd $c < 0$.

In fact the manifolds M' have been classified directly (see HELGASON [1], p. 354). $M' = N_1' \times \cdots \times N_s'$ where each N' is one of the manifolds in the following list:

$$\begin{array}{llll} \text{I)} & \mathbf{U}(p+q)/\mathbf{U}(p) \times \mathbf{U}(q)\,, & \text{II)} & \mathbf{SO}(2p)/\mathbf{U}(p)\,, \\ \text{III)} & \mathbf{Sp}(p)/\mathbf{U}(p)\,, & \text{IV)} & \mathbf{SO}(p+2)/\mathbf{SO}(p) \times \mathbf{SO}(2), p \neq 2\,, \\ \text{V)} & \mathbf{E}_6/\mathbf{Spin}(10) \times \mathbf{T}^1 & \text{VI)} & \mathbf{E}_7/\mathbf{E}_6 \times \mathbf{T}^1\,. \end{array} \tag{4}$$

The fact that each such N' yields a bounded homogeneous symmetric domain N was proved by E. CARTAN by means of an explicit construction in each case. The first general proof is due to HARISH-CHANDRA [1], p. 591 (see HELGASON [1], p. 312). The BETTI numbers $b_r(N')$ of N' can be calculated by the formula of HIRSCH and it can be shown that the numbers $h^{p,q}(N')$ defined in 15.4 are zero for $p \neq q$ (see 15.10, BOREL [2] and BOREL-HIRZEBRUCH [1], § 14). It follows that $\chi(N') = 1$ (in fact N' is a rational algebraic manifold). Thus we see that *the constant c in* Theorem 22.3.1 *equals $\chi(Y)$*. It also follows that the index $\tau = \tau(N') = \sum (-1)^j b_{2j}(N')$ is zero except in the following cases:

I) *if $p = 2s$ and $q = 2t$, or if $p = 2s + 1$ and $q = 2t$, or if $p = 2t$ and $q = 2s + 1$, then* $\tau = \dfrac{(s+t)!}{s!\,t!}$, (5)

IV) *if $p = 4s$ then* $\tau = 2$,

V) $\tau = 3$.

Let Δ be a subgroup of $I(M)$ which satisfies (a), (b) and (c) of 22.2. Such a subgroup exists by Theorem 22.2.2. Also by this theorem there exists a proper normal subgroup Γ of index μ in Δ with μ arbitrarily large such that Γ acts freely on M. Then M/Γ is a finite covering of $Y = M/\Delta$ with μ sheets and the RIEMANN-ROCH theorem implies that $\chi_y(M/\Gamma) = \mu\, \chi_y(Y)$. By Theorem 22.3.1 and the equations $\chi(M') = 1$, $c = \chi(Y)$ mentioned above

$$\chi_y(M/\Gamma) = \mu\, \chi_y(Y) = \mu\, \chi(Y)\, \chi_y(M')$$

where $\chi(Y) > 0$ if n is even and $\chi(Y) < 0$ if n is odd. In particular if M' is a product of manifolds of the type listed in (5) then n is even and $\tau(M/\Gamma) = \mu\, \chi(Y)\, \tau(M')$, $\chi(Y) > 0$, $\tau(M') > 0$. In this way algebraic manifolds M/Γ can be constructed with arbitrarily large index.

The first example is the case $M' = \mathbf{U}(3)/\mathbf{U}(2) \times \mathbf{U}(1) = \mathbf{P}_2(\mathbf{C})$. Then $\chi(M') = 1$ and M is the open unit disc $\mathbf{B}_2 \subset \mathbf{C}_2$. Thus there exist algebraic surfaces M/Γ with arbitrarily large index. This contradicts a conjecture of ZAPPA [1]. Further details can be found in BOREL [4].

Theorem 22.3.1 can also be applied to calculate the integers $\Pi_r(M, \Delta)$. Because of (3) we suppose $r \geqq 2$. For simplicity let M be an irreducible bounded symmetric homogeneous domain. Then M' is one of the manifolds listed in (4). The values of $\Pi_r(M, \Delta)$ were calculated by HIRZEBRUCH [4], [5] using the values for $\chi(M', (K_{M'})^r)$. The latter can be calculated by R-R and are related to formulae of H. WEYL on degrees of representations (BOREL-HIRZEBRUCH [1], § 22). The results in each case are:

I) $\Pi_r(M, \Delta) = (-1)^{pq}\, \chi(M/\Delta) \prod \frac{r(p+q)-i-j}{p+q-i-j}$, *where the product is over all* $0 \leqq i \leqq p-1$, $1 \leqq j \leqq q$.

II) $\Pi_r(M, \Delta) = (-1)^{\frac{1}{2}p(p-1)}\, \chi(M/\Delta) \prod \frac{2(r-1)(p-1)+i+j}{i+j}$, *where the product is over all* $0 \leqq i < j \leqq p-1$.

III) $\Pi_r(M, \Delta) = (-1)^{\frac{1}{2}p(p+1)}\, \chi(M/\Delta) \prod \frac{2(r-1)(p+1)+i+j}{i+j}$, *where the product is over all* $0 \leqq i \leqq j \leqq p$.

IV) $\Pi_r(M, \Delta) = (-1)^{p}\, \chi(M/\Delta) \left(\binom{rp-1}{p} + \binom{rp}{p}\right)$.

V) $\Pi_r(M, \Delta) = \chi(M/\Delta) \prod \frac{12(r-1)+\mu_k}{\mu_k}$, *where the product is over* $k = 1, \ldots, 16$ *and the corresponding values of* μ_k *are* 1, 2, 3, 4, 4, 5, 5, 6, 6, 7, 7, 8, 8, 9, 10, 11.

VI) $\Pi_r(M, \Delta) = -\chi(M/\Delta) \prod \frac{18(r-1)+\mu_k}{\mu_k}$, *where the product is over* $k = 1, \ldots, 27$ *and the corresponding values of* μ_k *are* 1, 2, 3, 4, 5, 5, 6, 6, 7, 7, 8, 8, 9, 9, 9, 10, 10, 11, 11, 12, 12, 13, 13, 14, 15, 16, 17.

Remark: Another method of calculating the numbers $\Pi_r(M, \Delta)$, including also the case in which condition *(c)* is omitted, is due to SELBERG (Seminars on analytic functions, Vol. 2, p. 152–161. Institute for Advanced Study, Princeton 1957). Formulae I)–VI) have been generalised to a more general type of automorphic form by ISE [2]. He also uses the proportionality principle. LANGLANDS [1] has obtained these formulae, and corresponding formulae when condition *(c)* is omitted, using SELBERG's trace formula and HARISH-CHANDRA's work.

22.4. In this section the RIEMANN-ROCH theorem $\chi(V, W) = T(V, W)$ is applied with $V = \mathbf{P}_n(\mathbf{C})$. For each n we regard $\mathbf{P}_{n-1}(\mathbf{C})$ as a hyperplane in $\mathbf{P}_n(\mathbf{C})$. The corresponding divisor class defines a line bundle H over $\mathbf{P}_n(\mathbf{C})$ and a cohomology class $h \in H^2(\mathbf{P}_n(\mathbf{C}), \mathbf{Z})$.

Let W be a continuous complex vector bundle over $\mathbf{P}_n(\mathbf{C})$ with fibre $\mathbf{C}_q$ and CHERN class $1 + d_1 h + \cdots + d_s h^s$, $(d_j \in \mathbf{Z},\ s \leqq q,\ s \leqq n)$. In $H^*(\mathbf{P}_n(\mathbf{C}), \mathbf{C}) = H^*(\mathbf{P}_n(\mathbf{C}), \mathbf{Z}) \otimes \mathbf{C}$ there is a factorisation

$$1 + d_1 h + \cdots + d_s h^s = (1 + \delta_1 h) \ldots (1 + \delta_s h)$$

with $\delta_j \in \mathbf{C}$ and therefore by 10.1 and 4.4.3

$$T(\mathbf{P}_n(\mathbf{C}), W \otimes H^r) = \varkappa_n \left[\sum_{j=1}^{q} e^{(\delta_j + r) h} \left(\frac{h}{1 - e^{-h}} \right)^{n+1} \right]$$
$$= \sum_{j=1}^{q} \frac{1}{2\pi i} \int \frac{e^{(\delta_j + r) h}}{(1 - e^{-h})^{n+1}} \, dh$$

where $\delta_{s+1} = \cdots = \delta_q = 0$ and integration is over a small circle round the origin. The substitution $z = 1 - e^{-h}$ gives

$$T(\mathbf{P}_n(\mathbf{C}), W \otimes H^r) = \sum_{j=1}^{q} \binom{n + \delta_j + r}{n}.$$

If W is a complex analytic vector bundle the RIEMANN-ROCH theorem implies that $\sum_{j=1}^{q} \binom{n + \delta_j + r}{n}$, which a priori is a rational number with denominator $n!$, is an integer for all integers r. The same conclusion holds for W a continuous vector bundle by the integrality theorem of 26.1. This completes the proof of

Theorem 22.4.1. *Let W be a continuous complex vector bundle over $\mathbf{P}_n(\mathbf{C})$ with* CHERN *class*

$$1 + d_1 h + \cdots + d_s h^s = (1 + \delta_1 h) \ldots (1 + \delta_s h)$$

where $d_j \in \mathbf{Z}$, $\delta_j \in \mathbf{C}$ and $s \leqq n$. Let r be an integer. Then the symmetric function

$$\binom{n + r + \delta_1}{n} + \cdots + \binom{n + r + \delta_s}{n}$$

in the δ_j is an integer.

Examples: Consider the case $q=2$. Then $\binom{n+\delta_1}{n}+\binom{n+\delta_2}{n}$ is an integer. This implies the following restrictions on the integers $d_1=\delta_1+\delta_2$, $d_2=\delta_1\delta_2$:

$$
\begin{aligned}
n &= 2 \quad \text{no restriction} \\
n &= 3 \quad d_1 d_2 \equiv 0 \text{ modulo } 2 \\
n &= 4 \quad d_2(d_2+1-3d_1-2d_1^2) \equiv 0 \text{ modulo } 12 .
\end{aligned}
$$

Let W be the tangent bundle of $\mathbf{P}_2(\mathbf{C})$. Then $s=2$, $d_1=d_2=3$ and $d_1 d_2$ is odd. Therefore W is not the restriction to $\mathbf{P}_2(\mathbf{C})$ of any continuous vector bundle over $\mathbf{P}_3(\mathbf{C})$. It can be shown similarly that for all $n \geqq 3$ the tangent bundle of $\mathbf{P}_{n-1}(\mathbf{C})$ is not the restriction of any vector bundle over $\mathbf{P}_n(\mathbf{C})$. An example of a continuous vector bundle W over $\mathbf{P}_3(\mathbf{C})$ with fibre $\mathbf{C}_2$, which is not the restriction to $\mathbf{P}_3(\mathbf{C})$ of any vector bundle over $\mathbf{P}_4(\mathbf{C})$, is given by the following classical construction. Consider a linear complex in $\mathbf{P}_3(\mathbf{C})$, *i. e.* the set of lines satisfying an equation $\sum a_{ij} p_{ij} = 0$ where $p_{01}, p_{02}, p_{03}, p_{23}, p_{31}, p_{12}$ are Plücker coordinates. The lines of the linear complex which pass through a point $x \in \mathbf{P}_3(\mathbf{C})$ form a plane pencil. This defines an algebraic fibre bundle B over $\mathbf{P}_3(\mathbf{C})$ with fibre $\mathbf{P}_1(\mathbf{C})$. There is an associated vector bundle W over $\mathbf{P}_3(\mathbf{C})$ with fibre $\mathbf{C}_2$ and $d_1=d_2=2$. Thus $d_2(d_2+1-3d_1-2d_1^2) \equiv 2$ modulo 12 and W is not the restriction of any vector bundle over $\mathbf{P}_4(\mathbf{C})$.

In general Theorem 22.4.1 gives necessary conditions for integers $d_1, \ldots, d_s$ to appear as Chern classes of a continuous complex vector bundle over $\mathbf{P}_n(\mathbf{C})$ with fibre $\mathbf{C}_q$. These are hard to calculate for particular q, n but for fixed q they clearly become more restrictive as $n \to \infty$. In fact a lemma in algebraic number theory (which the author owes to J. W. S. Cassels) implies that if $\sum_{j=1}^{s}\binom{n+\delta_j}{n}$ is an integer for all n then each δ_j is an integer. This implies

Theorem 22.4.2. *Let W be a continuous vector bundle over $\mathbf{P}_n(\mathbf{C})$ with fibre $\mathbf{C}_q$ and suppose that W is the restriction to $\mathbf{P}_n(\mathbf{C})$ of a continuous vector bundle over $\mathbf{P}_N(\mathbf{C})$ for arbitrarily large N. Then there are integers $r_1, \ldots, r_q$ such that $c(W) = c(H^{r_1} \oplus \cdots \oplus H^{r_q})$.*

Further results on complex vector bundles over $\mathbf{P}_n(\mathbf{C})$ can be found in Horrocks [1], [2] and Schwarzenberger [1]. For the classification of complex analytic vector bundles over algebraic curves, which also makes use of R-R, see Atiyah [1], [2], Grothendieck [3], Narasimhan-Seshadri [1], [2], and Turin [1], [2].

§ 23. The Riemann-Roch theorem of Grothendieck

The generalisation of R-R due to Grothendieck depends on properties of coherent analytic sheaves over complex manifolds. These properties are summarised in 23.1—23.3 and are used to obtain an alternative proof of equation 21.2 (9). The GROTHENDIECK-RIEMANN-ROCH theorem itself is described in 23.4—23.6. Throughout this paragraph we shall for convenience assume that all algebraic manifolds are connected.

23.1. Let X_n be a complex manifold of dimension n, and Ω the sheaf over X_n of germs of local holomorphic functions (15.1). Each stalk Ω_x of Ω is a ring with identity $1 \in \Omega_x$.

Definition: A sheaf $\mathfrak{S} = (S, \pi, X_n)$ of abelian groups is an *analytic sheaf over* X_n if:

I) Every stalk S_x of $\mathfrak{S}$ is a module over the corresponding stalk Ω_x of Ω (the unit element $1 \in \Omega_x$ operates as the identity).

II) The map from $\bigcup_{x \in X} \Omega_x \times S_x$ (regarded as a subspace of $\Omega \times S$) to S defined by the module multiplication is continuous.

An essential role is played by the coherent analytic sheaves. Let Ω_p denote the sum $\Omega \oplus \cdots \oplus \Omega$ of p copies of Ω.

Definition: An analytic sheaf $\mathfrak{S}$ over X_n is *coherent* if for each point $x \in X_n$ there is an open neighbourhood U of x and an exact sequence of sheaves over U

$$\Omega_p|U \to \Omega_q|U \to \mathfrak{S}|U \to 0 .$$

For the basic properties of coherent analytic sheaves we refer to GRAUERT-REMMERT [1]. The definition given there is apparently more restrictive than the above definition. The theorem of OKA on the sheaf of relations determined by a system of holomorphic functions (see CARTAN [3], Exposé XIV) implies that Ω is coherent in the sense of GRAUERT-REMMERT [1]. It can then be deduced that the two definitions of coherence are equivalent (see SERRE [2], Chap. I, Prop. 7). Note that coherence is a purely local property.

The sheaf $\Omega(W)$ of germs of local holomorphic sections of a complex analytic vector bundle W over X_n with fibre $\mathbf{C}_q$ is locally isomorphic to Ω_q. Therefore $\Omega(W)$ is a coherent analytic sheaf.

If $\mathfrak{S}$ is an arbitrary sheaf over X_n the cohomology groups of X_n with coefficients in $\mathfrak{S}$ can be defined by "alternating" cochains (SERRE [3]). It follows from general considerations of dimension theory that $H^q(X_n, \mathfrak{S}) = 0$ for $q > 2n$. For coherent analytic sheaves a more precise result has been proved by MALGRANGE [Bull. Soc. Math. France **85**, 231—237 (1957)]:

Theorem 23.1.1. *Let $\mathfrak{S}$ be a coherent analytic sheaf over an n-dimensional complex manifold X_n. Then $H^q(X_n, \mathfrak{S}) = 0$ for $q > n$.*

The corresponding finiteness condition is due to CARTAN-SERRE [1] (see also CARTAN [4]):

Theorem 23.1.2. *Let $\mathfrak{S}$ be a coherent analytic sheaf over a compact complex manifold X. Then for all $q \geqq 0$ the complex vector space $H^q(X, \mathfrak{S})$ is finite dimensional.*

Theorems 23.1.1 and 23.1.2 generalise the results obtained, for the particular case $\mathfrak{S} = \Omega(W)$, in Theorem 15.4.2. The proof of Theorem 23.1.2 makes use of the theory of holomorphically complete manifolds (STEIN manifolds). Theorem B (SERRE [1] and CARTAN [3], Exposé XIX) implies that if $\mathfrak{S}$ is a coherent analytic sheaf over a holomorphically complete manifold X then $H^q(X, \mathfrak{S}) = 0$ for $q > 0$. If now X is a compact complex manifold there is a finite covering $\mathfrak{U} = \{U_i\}_{i \in I}$ of X such that each intersection $U_{i_0} \cap \cdots \cap U_{i_p}$ is holomorphically complete. This is for instance automatically the case if each U_i is a unit disc with respect to some complex analytic chart on X. The LERAY spectral sequence (GODEMENT [1], Chap. II, 5.2.4) can then be used to prove that $H^q(X, \mathfrak{S}) = H^q(\mathfrak{U}, \mathfrak{S})$ for $q \geqq 0$. This is one of the basic facts required for the proof of CARTAN-SERRE.

23.2. Let $f: X \to Y$ be a holomorphic map of complex manifolds and $\mathfrak{S}$ an analytic sheaf over X. The q-th direct image of $\mathfrak{S}$ is an analytic sheaf $f^q_* \mathfrak{S}$ over Y which is defined by means of a presheaf. For any open set U of Y the cohomology group $H^q(f^{-1}(U), \mathfrak{S})$ is a module over the ring of holomorphic functions defined on $f^{-1}(U)$. A holomorphic function $g: U \to \mathbf{C}$ can be lifted to a holomorphic function $g f: f^{-1}(U) \to \mathbf{C}$ and so $H^q(f^{-1}(U), \mathfrak{S})$ can also be regarded as a module over the ring of holomorphic functions defined on U. These modules define a presheaf for $f^q_* \mathfrak{S}$. The definition implies that $f^q_* \mathfrak{S}$ is an analytic sheaf over Y.

Consider an exact sequence of analytic sheaves over X

$$0 \to \mathfrak{S}' \to \mathfrak{S} \to \mathfrak{S}'' \to 0 .$$

By Theorem 2.8.2 the open set $f^{-1}(U)$ is paracompact for every open set U of Y. By Theorem 2.10.1 there is an exact sequence

$$0 \to H^0(f^{-1}(U), \mathfrak{S}') \to H^0(f^{-1}(U), \mathfrak{S}) \to H^0(f^{-1}(U), \mathfrak{S}'') \to H^1(f^{-1}(U), \mathfrak{S}') \to \cdots$$
$$\cdots \to H^q(f^{-1}(U), \mathfrak{S}') \to H^q(f^{-1}(U), \mathfrak{S}) \to H^q(f^{-1}(U), \mathfrak{S}'') \to$$
$$\to H^{q+1}(f^{-1}(U), \mathfrak{S}') \to \cdots$$

and hence an exact sequence of analytic sheaves over Y

$$\begin{aligned} &0 \to f^0_* \mathfrak{S}' \to f^0_* \mathfrak{S} \to f^0_* \mathfrak{S}'' \to f^1_* \mathfrak{S}' \to \cdots \\ &\cdots \to f^q_* \mathfrak{S}' \to f^q_* \mathfrak{S} \to f^q_* \mathfrak{S}'' \to f^{q+1}_* \mathfrak{S}' \to \cdots . \end{aligned} \tag{1}$$

Theorem 23.2.1. *Let $f: X \to Y$ be a holomorphic map of complex manifolds and $\mathfrak{S}$ an analytic sheaf over X. Suppose that $f^i_* \mathfrak{S} = 0$ for all $i > 0$. Then the complex vector spaces $H^q(Y, f^0_* \mathfrak{S})$ and $H^q(X, \mathfrak{S})$ are isomorphic for all $q \geqq 0$.*

The direct image sheaves figure already in the fundamental work of LERAY [1], [2]. The exact sequence (1) and Theorem 23.2.1 are reformulations, for holomorphic maps and analytic sheaves, of results of LERAY on continuous maps. Theorem 23.2.1 follows immediately from the LERAY spectral sequence (see CARTAN [2] and GODEMENT [1], Chap. II, 4.17.1). A direct proof can be found in GRAUERT-REMMERT [1] (p. 417, Satz 6).

The proof of the RIEMANN-ROCH theorem in 21.1 depends on a result of BOREL (Theorem 21.2.1). As remarked in 21.2, in order to complete the proof of R-R directly it is sufficient to establish equation 21.2 (9). We first prove

Lemma 23.2.2. *Let X be a complex analytic fibre bundle over the complex manifold Y with fibre $\mathbf{P}_n(\mathbf{C})$ and projection map f. Let W be a complex analytic vector bundle over Y. There is a natural isomorphism between the analytic sheaves $\Omega(W)$ and $f^0_* \Omega(f^* W)$. The analytic sheaf $f^i_* \Omega(f^* W)$ is zero for $i > 0$.*

Proof: Let U be an open set of Y. A holomorphic section s of W over U determines a holomorphic section $s\,f$ of $f^* W$ over $f^{-1}(U)$. Since each fibre $\mathbf{P}_n(\mathbf{C})$ is compact and connected, this defines an isomorphism $H^0(U, \Omega(W)) \to H^0(f^{-1}(U), \Omega(f^* W))$ and proves the first part of the lemma. The second part is purely local, so we may choose U to be a holomorphically complete open set over which both W and X are trivial. We wish to prove that $H^i(f^{-1}(U), \Omega(f^* W)) = 0$ for $i > 0$. Since $f^* W | f^{-1}(U)$ is a sum of trivial bundles it is sufficient to prove that $H^i(f^{-1}(U), \mathbf{1}) = 0$ for $i > 0$. Now $H^r(U, \mathbf{1}) = 0$ for $r > 0$ (23.1) and $H^s(\mathbf{P}_n(\mathbf{C}), \mathbf{1}) = 0$ for $s > 0$ (15.10). Therefore (KAUP [1], § 7, Satz 1) the KÜNNETH formula for analytic sheaves can be applied in this case to give

$$H^i(f^{-1}(U), \mathbf{1}) = H^i(U \times \mathbf{P}_n(\mathbf{C}), \mathbf{1}) = \sum_{r+s=i} H^r(U, \mathbf{1}) \otimes H^s(\mathbf{P}_n(\mathbf{C}), \mathbf{1}) = 0 \text{ for } i > 0\,.$$

Remark: The KÜNNETH formula for sheaves is due originally to GROTHENDIECK (see BOTT [1] and BOREL-SERRE [2]). A proof of the formula for algebraic coherent sheaves was given by SAMPSON-WASHNITZER [3]. The formula for analytic coherent sheaves used here depends on finiteness assumptions on the higher dimensional cohomology groups involved; in the present case these groups are zero. For full details see KAUP [1].

Lemma 23.2.2 and Theorem 23.2.1 (with $\mathfrak{S} = \Omega(f^* W)$) together imply

Theorem 23.2.3. *Let X be a complex analytic fibre bundle over the complex manifold Y with fibre $\mathbf{P}_n(\mathbf{C})$ and projection map f. Let W be a complex analytic vector bundle over Y. Then the complex vector spaces $H^q(Y, W)$ and $H^q(X, f^* W)$ are isomorphic for all $q \geqq 0$.*

As a corollary we obtain the equation 21.2 (9) required to complete the proof of the RIEMANN-ROCH theorem:

$$\dim H^q(Y, W) = \dim H^q(X, f^* W)\,. \tag{2}$$

The direct image sheaves $f^q_* \mathfrak{S}$ have special properties when $\mathfrak{S}$ is coherent. Let X be a complex manifold of dimension n and $\mathfrak{S}$ a coherent analytic sheaf over X. Let $f: X \to Y$ be a holomorphic map of complex manifolds. The following theorems reduce to 23.1.1 and 23.1.2 when Y is a point.

Theorem 23.2.4. *Under the above hypotheses $f^q_* \mathfrak{S} = 0$ for $q > n$.*

Theorem 23.2.5. *Under the above hypotheses, if f is a proper map then $f^q_* \mathfrak{S}$ is coherent for all $q \geqq 0$.*

Theorem 23.2.4 is an immediate consequence of 23.1.1. Theorem 23.2.5 is a deep theorem of GRAUERT [2]. If X and Y are both algebraic manifolds then Theorem 23.2.5 can be proved algebraically (BOREL-SERRE [2], Théorème 1) by using the correspondence between coherent analytic sheaves and coherent algebraic sheaves (SERRE [4]).

23.3. Let X be a complex manifold, $C(X)$ the set of isomorphism classes of coherent analytic sheaves over X and $F(X)$ the free abelian group generated by $C(X)$. An element of $F(X)$ is a finite linear combination $\sum_i n_i \mathfrak{S}_i$, $n_i \in \mathbf{Z}$, $\mathfrak{S}_i$ a coherent analytic sheaf over X. Let $R(X)$ be the subgroup generated by all elements $\mathfrak{S} - \mathfrak{S}' - \mathfrak{S}''$ where

$$0 \to \mathfrak{S}' \to \mathfrak{S} \to \mathfrak{S}'' \to 0$$

is an exact sequence of coherent analytic sheaves over X. The GROTHENDIECK group "of coherent analytic sheaves over X" is the quotient group $K_\omega(X) = F(X)/R(X)$.

Let X be a compact complex manifold and $b \in K_\omega(X)$ an element represented by a linear combination $\sum_i n_i \mathfrak{S}_i$ of coherent analytic sheaves $\mathfrak{S}_i$ on X. Theorems 23.1.1 and 23.1.2 show that $\mathfrak{S}_i$ is of type (F) and therefore the integer $\chi(X, \mathfrak{S}_i)$ is defined (see 2.10). The integer

$$\chi(X, b) = \sum_i n_i \chi(X, \mathfrak{S}_i)$$

depends only on the element $b \in K_\omega(X)$.

Let $f: X \to Y$ be a holomorphic map of complex manifolds which is also proper. If $\mathfrak{S} \in C(X)$ then, by 23.2.4 and 23.2.5, $f^q_* \mathfrak{S} \in C(Y)$ for $q \geqq 0$ and $f^q_* \mathfrak{S} = 0$ for $q > \dim_{\mathbf{C}} X$. Consider the homomorphism $f_!: F(X) \to F(Y)$ defined on generators of $F(X)$ by

$$f_!(\mathfrak{S}) = \sum_{q=0}^{n} (-1)^q f^q_* \mathfrak{S}, \quad n = \dim_{\mathbf{C}} X .$$

The exact sequence (1) shows that $f_!$ maps the subgroup $R(X)$ to $R(Y)$. Therefore $f_!$ induces a homomorphism

$$f_!: K_\omega(X) \to K_\omega(Y) .$$

The LERAY spectral sequence can be used (BOREL-SERRE [2], p. 111) to prove that if $f: X \to Y$ and $g: Y \to Z$ are proper holomorphic maps of complex manifolds X, Y, Z then

$$(gf)_! = g_! f_! . \tag{3}$$

Consider the special case in which Y is a point and $f: X \to Y$ is the constant map. Then f is proper if and only if X is compact. A coherent analytic sheaf over Y is a finite dimensional complex vector space and therefore $K_\omega(Y) = \mathbf{Z}$. In this case

$$f_!(b) = \chi(X, b) . \tag{4}$$

The homomorphism $f_!$ is analogous to the GYSIN homomorphism f_* for cohomology. If X, Y are compact connected oriented manifolds (not necessarily complex), and $f: X \to Y$ is a continuous map, there is a homomorphism of $H^*(Y, \mathbf{Z})$-modules

$$f_* : H^*(X, \mathbf{Z}) \to H^*(Y, \mathbf{Z})$$

which maps classes of codimension q to classes of codimension q. As in 4.3, $f_*(x) = D_Y^{-1}(f_* D_X(x))$ for $x \in H^*(X, \mathbf{Z})$ where D_X, D_Y denote the duality isomorphism from cohomology to homology. The homomorphism $f_* : H^*(X, \mathbf{Q}) \to H^*(Y, \mathbf{Q})$ is defined in the same way. If $g: Y \to Z$ is another continuous map of compact connected oriented manifolds then

$$(g f)_* = g_* f_* . \tag{5}$$

Consider the special case in which Y is a point, f is the constant map and X is a compact connected oriented manifold of (real) dimension m. In this case

$$f_*(v) = \varkappa^m[v] \cdot 1 , \quad v \in H^*(X) \tag{6}$$

where $1 \in H^0(Y)$ is the identity element and $\varkappa^m[\;]$ is defined as in 9.2.

23.4. Let X be a complex manifold, $C'(X)$ the set of isomorphism classes of complex analytic vector bundles over X and $F'(X)$ the free abelian group generated by $C'(X)$. Exactly as in 23.3 we can define the GROTHENDIECK group $K'_\omega(X)$ "of complex analytic vector bundles over X". There is a natural homomorphism $h: K'_\omega(X) \to K_\omega(X)$ induced by $h(W) = \Omega(W)$.

Theorem 23.4.1. *Let X be an algebraic manifold. Then $h: K'_\omega(X) \to$ $\to K_\omega(X)$ is an isomorphism.*

The main step in the proof of Theorem 23.4.1 is

Lemma 23.4.2. *Let $\mathfrak{S}$ be a coherent analytic sheaf over an n-dimensional algebraic manifold X. Then there are complex analytic vector bundles*

$W_0, W_1, \ldots, W_n$ over X and an exact sequence

$$0 \to \Omega(W_n) \to \Omega(W_{n-1}) \to \cdots \to \Omega(W_0) \to \mathfrak{S} \to 0 \tag{7}$$

of analytic sheaves over X.

Lemma 23.4.2 shows that the homomorphism h is surjective. It must then be shown that the element $\sum_{i=0}^{n} (-1)^i W_i$ of $K'_\omega(X)$ determined by (7) depends only on $\mathfrak{S}$. Proofs for $\mathfrak{S}$ a coherent algebraic sheaf over X are given in BOREL-SERRE [2]. The above statements then follow from the correspondence between coherent analytic sheaves and coherent algebraic sheaves over an algebraic manifold (SERRE [4]). A similar remark applies to all the other results mentioned in this section including the RIEMANN-ROCH theorem of GROTHENDIECK (23.4.3). The proofs are purely algebraic and apply to non-singular irreducible projective varieties defined over an arbitrary algebraically closed field $\mathbf{K}$. They are formulated in terms of the ZARISKI topology, coherent algebraic sheaves and algebraic fibre bundles with fibre $\mathbf{K}_q$. The cohomology ring $H^*(X, \mathbf{Z})$ is replaced by the CHOW ring $A(X)$ of rational equivalence classes of algebraic cycles on X. When $\mathbf{K} = \mathbf{C}$ the results of SERRE mentioned above allow algebraic statements to be reformulated in the complex analytic terminology used in this book.

Let $\mathfrak{S}$ be a coherent analytic sheaf over X with a "resolution by vector bundles" as in (7). Then the CHERN character of $\mathfrak{S}$ can be defined by $\mathrm{ch}(\mathfrak{S}) = \sum_{i=0}^{n} (-1)^i \mathrm{ch}(W_i)$. By 23.4.1 this is independent of the choice of resolution. If

$$0 \to \mathfrak{S}' \to \mathfrak{S} \to \mathfrak{S}'' \to 0$$

is an exact sequence of coherent analytic sheaves then (see 10.1)

$$\mathrm{ch}(\mathfrak{S}) = \mathrm{ch}(\mathfrak{S}') + \mathrm{ch}(\mathfrak{S}'') .$$

Therefore the CHERN character defines a homomorphism

$$\mathrm{ch}: K_\omega(X) \to H^*(X, \mathbf{Q})$$

for every algebraic manifold X.

Let $\mathrm{td}(X)$, $\mathrm{td}(Y)$ be the total TODD class of the tangent bundle of X, Y defined in 10.1. The RIEMANN-ROCH theorem of GROTHENDIECK can now be stated.

Theorem 23.4.3 (G-R-R). *Let X, Y be algebraic manifolds and $f: X \to Y$ a holomorphic map. Then the equation*

$$\mathrm{ch}(f_! b) \cdot \mathrm{td}(Y) = f_*(\mathrm{ch}(b) \cdot \mathrm{td}(X)) \tag{8}$$

holds in $H^*(Y, \mathbf{Q})$ for all $b \in K_\omega(X)$.

Let $f: X \to Y$, $g: Y \to Z$ be holomorphic maps of algebraic manifolds. It follows from (3) and (5) that if G-R-R is true for both f and g then it is true for $g f: X \to Z$. Since X is an algebraic manifold there is a holomorphic embedding $X \to \mathbf{P}_N(\mathbf{C})$ for some integer N. The map $f: X \to Y$ is then the composition of an embedding $X \to Y \times \mathbf{P}_N(\mathbf{C})$ and a product projection $Y \times \mathbf{P}_N(\mathbf{C}) \to Y$. It is therefore sufficient to prove G-R-R in the two cases:

I) $f: X \to Y$ is an embedding. There is an algebraic proof in BOREL-SERRE [2] and a complex analytic proof in ATIYAH-HIRZEBRUCH [8]. The special case in which X is a non-singular divisor of Y and $b \in K_\omega(X)$ arises from the restriction to X of a vector bundle over Y is proved in 23.5.

II) $f: Y \times \mathbf{P}_N(\mathbf{C}) \to Y$ is a product projection. An algebraic proof is given in BOREL-SERRE [2].

We have formulated the RIEMANN-ROCH theorem of GROTHENDIECK only for algebraic manifolds. It is possible to formulate it for a proper holomorphic map $f: X \to Y$ of complex manifolds: the problem is to define $\operatorname{ch}(\mathfrak{S})$ for an arbitrary analytic coherent sheaf $\mathfrak{S}$ over a compact complex manifold X, and this can be done by considering resolutions by real analytic, and by differentiable, vector bundles. At the time of writing this version of G-R-R has been proved only for f an embedding (ATIYAH-HIRZEBRUCH [8]). Two special cases of G-R-R are discussed in 23.5; two applications are described in 23.6.

23.5. Suppose first that Y is an algebraic manifold with complex analytic tangent bundle θ and that $j: X \to Y$ is an embedding of X as a submanifold of Y. Then $j^* \theta$ has the tangent bundle of X as subbundle and the complex analytic normal bundle ν as quotient bundle (4.9). Thus $\operatorname{td}(X) = (\operatorname{td}(\nu))^{-1} \cdot j^* \operatorname{td}(Y)$ by 10.1 and (8) becomes

$$\operatorname{ch}(j_! b) \cdot \operatorname{td}(Y) = j_*\big(\operatorname{ch}(b) \cdot (\operatorname{td}(\nu))^{-1} \cdot j^* \operatorname{td}(Y)\big).$$

Now j_* is a $H^*(Y, \mathbf{Q})$-module homomorphism and $\operatorname{td}(Y)$ is invertible in $H^*(Y, \mathbf{Q})$. Therefore G-R-R implies the RIEMANN-ROCH theorem for an embedding:

$$\operatorname{ch}(j_! b) = j_* \operatorname{ch}(b) \cdot (\operatorname{td}(\nu))^{-1} \textit{ for all } b \in K_\omega(X). \tag{9}$$

We prove the following special case of (9). Let X be a non-singular divisor S of Y and $\{S\}$ the corresponding line bundle (15.2). Let W be a complex analytic vector bundle over Y and $b \in K_\omega(S)$ the element represented by the coherent analytic sheaf $\Omega(j^*(W \otimes \{S\}))$ over S. Let U be an open set on Y such that $V = U \cap S$ is holomorphically complete. Then, in the notation of 16.2,

$$j_*^q \, \Omega(j^*(W \otimes \{S\}))(U) = H^q(V, j^*(W \otimes \{S\})) = 0 \text{ for } q > 0$$

and so $j_! b$ is represented by the trivial extension $j^0_* \Omega(j^*(W \otimes \{S\})) = \hat{\Omega}((W \otimes \{S\})_S)$ of $\Omega(j^*(W \otimes \{S\}))$ from S to Y. By 16.2 (4) there is a resolution of $\hat{\Omega}((W \otimes \{S\})_S)$ by vector bundles over Y

$$0 \to \Omega(W) \to \Omega(W \otimes \{S\}) \to \hat{\Omega}((W \otimes \{S\})_S) \to 0$$

and therefore $\mathrm{ch}(j_! b) = \mathrm{ch}(W \otimes \{S\}) - \mathrm{ch}(W) = (e^h - 1)\,\mathrm{ch}(W)$ where $h \in H^2(Y, \mathbf{Z})$ is the cohomology class of S.

On the other hand $c_1(\nu) = j^* h$ by 4.8.1 and $j_* 1 = h$ by 4.9.1. Therefore the right hand side of (9) is

$$j_*\left(j^* \mathrm{ch}(W \otimes \{S\}) \cdot (\mathrm{td}(\nu))^{-1}\right) = j_* j^* \left(\mathrm{ch}(W) \cdot e^h \cdot \left(\frac{h}{1 - e^{-h}}\right)^{-1}\right)$$
$$= (e^h - 1)\,\mathrm{ch}(W) .$$

This proves (9) in the special case and helps to explain why the Todd class arises in G-R-R.

Now consider the special case of G-R-R in which Y is a point and f is the constant map. Let $b \in K_\omega(X)$ be the element represented by the coherent analytic sheaf $\Omega(W)$ of germs of local holomorphic sections of a complex analytic vector bundle W over X. By (4) the left hand side of (8) becomes $\chi(X, W)$. By (6) the right hand side of (8) becomes $\varkappa_n[\mathrm{ch}(W)\,\mathrm{td}(X)] = T(X, W)$. Therefore the Riemann-Roch theorem of Grothendieck implies Theorem 21.1.1 (R – R):

$$\chi(X, W) = T(X, W) .$$

23.6. Let E, F, V be algebraic manifolds and let $\varphi: E \to V$ be a holomorphic fibre bundle with fibre F and connected structure group (see Theorem 18.3.1*). As in 23.2.2 let U be a holomorphically complete open set of V over which E is trivial. Then, by the Künneth theorem for coherent analytic sheaves used in the proof of 23.2.2,

$$H^i(\varphi^{-1}(U), \mathbf{1}) = H^0(U, \mathbf{1}) \otimes H^i(F, \mathbf{1}) .$$

Therefore $\varphi^i_* \Omega(\mathbf{1}) = \Omega(W_i)$ for some complex analytic vector bundle W_i over V with fibre dimension $\dim H^i(F, \mathbf{1})$. The fact that the structure group of E is connected implies that W_i is trivial. Hence

$$\begin{aligned} &\mathrm{ch}_0(\varphi_! \Omega(\mathbf{1})) = \sum_i (-1)^i \dim H^i(F, \mathbf{1}) = \chi(F) = T(F) , \\ &\mathrm{ch}_j(\varphi_! \Omega(\mathbf{1})) = 0 \ \text{ for } j > 0 . \end{aligned} \tag{10}$$

On the other hand G-R-R applied to the map $\varphi: E \to V$ and the sheaf $\Omega(\mathbf{1})$ over E gives

$$\mathrm{ch}(\varphi_! \Omega(\mathbf{1})) \cdot \mathrm{td}(V) = \varphi_* \mathrm{td}(E) = \varphi_* \mathrm{td}(\theta) \cdot \mathrm{td}(V)$$

where θ is the bundle over E of tangent vectors "along the fibres".

Therefore (10) implies

$$T(F)\cdot \mathrm{td}(V) = \varphi_* \,\mathrm{td}(E)\,, \tag{11}$$

$$T(F)\cdot 1 = \varphi_* \,\mathrm{td}(\theta)\,, \tag{11*}$$

where $1 \in H^0(V, \mathbf{Q})$ denotes the unit element. Formula (11*) is the strict multiplicative property of BOREL-HIRZEBRUCH [1], § 21. If ζ is a continuous $\mathbf{GL}(q, \mathbf{C})$-bundle over V then multiplication of both sides of (11) by $\mathrm{ch}(\zeta)$ gives

$$T(F)\cdot(\mathrm{ch}(\zeta)\cdot \mathrm{td}(V)) = \varphi_*(\mathrm{ch}(\varphi^*\zeta)\cdot \mathrm{td}(E))\,.$$

Equating terms of top dimension we obtain the multiplicative property of the TODD genus (compare Theorem 14.3.1):

Theorem 23.6.1 (BOREL-SERRE [2], Prop. 16). *Let E, F, V be algebraic manifolds and $\varphi: E \to V$ a holomorphic fibre bundle with fibre F and connected structure group. Let ζ be a continuous $\mathbf{GL}(q, \mathbf{C})$-bundle over V. Then $T(F)\cdot T(V,\zeta) = T(E, \varphi^*\zeta)$.*

A second application of G-R-R is to monoidal transformations. Let X be a submanifold of codimension q of an algebraic manifold Y, $i: X \to Y$ the embedding, $\boldsymbol{\nu}$ the complex analytic normal $\mathbf{GL}(q, \mathbf{C})$-bundle of X, and $f: X' \to X$ the associated bundle over X with fibre $\mathbf{P}_{q-1}(\mathbf{C})$. There is an algebraic manifold Y', called the monoidal transform of Y along X, an embedding $j: X' \to Y'$ and a map $g: Y' \to Y$ such that the diagram

$$\begin{array}{ccc} X' & \xrightarrow{\ j\ } & Y' \\ {\scriptstyle f}\downarrow & & \downarrow{\scriptstyle g} \\ X & \xrightarrow{\ i\ } & Y \end{array} \tag{12}$$

is commutative. Let U be an open set of Y which admits local analytic coordinates. If U does not meet X then $g^{-1}(U)$ is biholomorphically equivalent to U. If U meets X there are holomorphic functions $f_1, \ldots, f_q$ on U such that $U \cap X$ is the submanifold $\{u \in U;\, f_1(u) = \cdots = f_q(u) = 0\}$ of U and such that the differentials $d f_1, \ldots, d f_q$ are linearly independent at each point of $U \cap X$. In terms of homogeneous coordinates $z = (z_1 : \cdots : z_q)$ for $\mathbf{P}_{q-1}(\mathbf{C})$, the open set $g^{-1}(U)$ is biholomorphically equivalent to the submanifold $\{(u, z) \in U \times \mathbf{P}_{q-1}(\mathbf{C});\, z_i f_j(u) = z_j f_i(u),\ 1 \leqq i < j \leqq q\}$ of $U \times \mathbf{P}_{q-1}(\mathbf{C})$.

Let $\mathfrak{T}$, $\mathfrak{T}'$ be the complex analytic tangent vector bundles of Y, Y' and $\mathfrak{N}$ the normal vector bundle of X in Y associated to $\boldsymbol{\nu}$. Let H be the line bundle over Y' determined by the non-singular divisor X' of Y'. A lemma of PORTEOUS [1] implies that in $K_\omega(Y')$ there is an equation

$$\Omega(g^*\mathfrak{T}) - \Omega(\mathfrak{T}') = j_!(\Omega(f^*\mathfrak{N}) - \Omega(j^* H))\,.$$

By the RIEMANN-ROCH theorem for an embedding (9) the CHERN character of the right hand side is

$$j_* \left((f^* \operatorname{ch}(\nu) - j^* e^h) \cdot j^* \left(\frac{1 - e^{-h}}{h} \right) \right)$$

where $h \in H^2(Y', \mathbf{Z})$ is the cohomology class of H. We obtain

Theorem 23.6.2 (PORTEOUS [1]). *Let X be a submanifold of an algebraic manifold Y. Let (12) be the diagram obtained from a monoidal transformation of Y along X, ν the normal bundle of X in Y, and $h \in H^2(Y', \mathbf{Z})$ the class represented by the cycle X'. Let θ, θ' be the tangent bundles of Y, Y'. Then*

$$g^* \operatorname{ch}(\theta) - \operatorname{ch}(\theta') = \left(\frac{1 - e^{-h}}{h} \right) \cdot j_* (f^* \operatorname{ch}(\nu) - j^* e^h) . \tag{13}$$

The CHERN character of Y' can be calculated in terms of the CHERN character of Y by (13). A refinement of the RIEMANN-ROCH theorem (involving integer cohomology; see PORTEOUS [1] and ATIYAH-HIRZEBRUCH [8]) gives the corresponding formula for the CHERN classes of Y and Y' which had been conjectured by TODD [5] and SEGRE [1]. The RIEMANN-ROCH theorem for an embedding is proved in ATIYAH-HIRZEBRUCH [8] for arbitrary compact complex manifolds. Therefore (13), and also the TODD-SEGRE formula, is true for a monoidal transformation of a compact complex manifold Y along a submanifold X. In certain cases this had been proved by VAN DE VEN [1]. A calculation due to HIRZEBRUCH (unpublished) shows that the TODD-SEGRE formula implies $T(Y') = T(Y)$, that is, the TODD genus is invariant under monoidal transformations. In the special case when X is a point (quadratic transform, HOPF σ-process) this can also be proved directly with the help of Lemma 1.7.2.

If Y is algebraic the invariance of the TODD genus can be obtained more easily either from the birational invariance of the arithmetic genus (see 0.1 and SAMPSON-WASHNITZER [2]) or by applying G-R-R to the map $g: Y' \to Y$. Then $g^q_* \,\Omega(\mathbf{1}) = 0$ for $q > 0$ and G-R-R gives $g_* \operatorname{td}(Y') = \operatorname{td}(Y)$; the equation $T(Y') = T(Y)$ follows by equating coefficients of the top dimension.

§ 24. The GROTHENDIECK ring of continuous vector bundles

The definition of the group $K'_\omega(X)$ "of complex analytic vector bundles over a complex manifold X" in 23.4 is due to GROTHENDIECK. His construction can be imitated in the continuous case to give a "GROTHENDIECK ring of continuous vector bundles" (ATIYAH-HIRZEBRUCH [1], [3]). Although the vector bundles themselves are not elements of the GROTHENDIECK ring, this abuse of language may be permitted. There is one slight simplification: by Theorem 4.1.4 a sequence

$$0 \to W' \to W \to W'' \to 0$$

of continuous complex vector bundles over a paracompact space X is exact if and only if $W = W' \oplus W''$. Throughout this paragraph we shall for convenience suppose that X is a compact space. This implies, if X is finite dimensional, that X is admissible in the sense of 4.2.

24.1. Let X be a compact space and $C(X)$ the set of isomorphism classes of continuous complex vector bundles over X (see 3.5). The WHITNEY sum $\oplus$ makes $C(X)$ a semi-group. Let $F(X)$ be the free abelian group generated by $C(X)$, and $R(X)$ the subgroup generated by all elements of the form $W - W' - W''$ where $W = W' \oplus W''$. Define $K(X) = F(X)/R(X)$. The tensor product of vector bundles defines a ring structure on $K(X)$. This is the GROTHENDIECK ring of continuous complex vector bundles over X. If X is a point then $K(X) = \mathbf{Z}$. If X is a complex manifold there is a homomorphism $K'_\omega(X) \to K(X)$ obtained by ignoring the complex analytic structure.

The natural map $C(X) \to F(X)$ defines a homomorphism of semi-groups $i: C(X) \to K(X)$. Let G be an additive group and $\tilde{f}: C(X) \to G$ a homomorphism of semi-groups. Then there is a unique homomorphism $f: K(X) \to G$ such that $\tilde{f} = f\, i$. This universal property allows homomorphisms defined on $C(X)$ to be extended to $K(X)$. If X is finite dimensional the CHERN class and TODD class give homomorphisms

$$c: K(X) \to G(X, \mathbf{Z})$$
$$\mathrm{td}: K(X) \to G(X, \mathbf{Q})$$

where $G(X, A)$ denotes the set of all sums $1 + h_1 + h_2 + \cdots$ with $h_i \in H^{2i}(X, A)$ and with group operation defined by cup product. Similarly the CHERN character defines a ring homomorphism

$$\mathrm{ch}: K(X) \to H^*(X, \mathbf{Q}) \tag{1}$$

and a map $f: X \to X'$ defines a ring homomorphism

$$f^!: K(X') \to K(X)$$

which depends only on the homotopy class of f. By 4.2 there is a commutative diagram

$$\begin{array}{ccc} K(X') & \xrightarrow{f^!} & K(X) \\ \downarrow{\scriptstyle \mathrm{ch}} & & \downarrow{\scriptstyle \mathrm{ch}} \\ H^*(X', \mathbf{Q}) & \xrightarrow{f^*} & H^*(X, \mathbf{Q}) . \end{array} \tag{2}$$

If X is infinite dimensional, $H^*(X, \mathbf{Q})$ must be replaced by the direct product (infinite sums allowed) $H^{**}(X, \mathbf{Q})$.

The GROTHENDIECK ring can also be defined for pairs (X, Y) where X is a compact space and Y is a closed subspace. If Y is empty define $K(X, \emptyset) = K(X)$. If Y consists of a single point define $K(X, \{x_0\})$ to be the kernel of the homomorphism $i^!: K(X) \to K(\{x_0\}) = \mathbf{Z}$ induced by the embedding $i: \{x_0\} \to X$. In general let $X \cup TY$ be the space obtained by attaching a cone on Y with vertex z_0 and define $K(X, Y) = K(X \cup TY, \{z_0\})$. There is a canonical map $X \cup TY \to X/Y$ which collapses the cone TY to a point y_0 and induces an isomorphism

$$K(X/Y, \{y_0\}) \to K(X, Y).$$

The CHERN character can be defined in the relative case. It is a ring homomorphism $\mathrm{ch}: K(X, Y) \to H^*(X, Y; \mathbf{Q})$. A map of compact pairs $f: (X, Y) \to (X', Y')$ defines a ring homomorphism

$$f^!: K(X', Y') \to K(X, Y)$$

which depends only on the homotopy class of f. In particular the embeddings $i: (Y, \emptyset) \to (X, \emptyset)$ and $j: (X, \emptyset) \to (X, Y)$ define a sequence

$$K(X, Y) \xrightarrow{j^!} K(X) \xrightarrow{i^!} K(Y) \tag{3}$$

which is an exact sequence of $K(X)$-modules. If Y is a retract of X, *i. e.* if there exists a map $f: X \to Y$ such that $f\,i(y) = y$ for all $y \in Y$, then it can be shown that one has a short exact sequence

$$0 \to K(X, Y) \xrightarrow{j^!} K(X) \underset{f^!}{\overset{i^!}{\rightleftarrows}} K(Y) \to 0$$

which splits by means of $f^!$.

The definition of the relative GROTHENDIECK ring $K(X, Y)$ is the first step in the construction of an "extraordinary cohomology theory" $K^*(X, Y)$ which satisfies all the axioms of EILENBERG-STEENROD except for the dimension axiom. Further details can be found in ATIYAH-HIRZEBRUCH [3].

24.2. Let X be a compact space, Y a closed subspace, E and F continuous complex vector bundles over X and $\alpha: E|Y \to F|Y$ an isomorphism between the restrictions of E and F to Y. In this section we construct an element $d(E, F, \alpha)$ of $K(X, Y)$ which can be regarded as a first obstruction to extending the isomorphism α to the whole of X. For the original (and slightly different) construction see ATIYAH-HIRZEBRUCH [7].

Let I be the unit interval and form the subspace $Z = X \times 0 \cup \cup X \times 1 \cup Y \times I$ of $X \times I$. On Z define a complex vector bundle L by putting E over $X \times 1$, putting F over $X \times 0$ and using α to "join" them along $Y \times I$. More precisely let

$$I_0 = I - \{0\}, \qquad I_1 = I - \{1\},$$
$$Z_0 = X \times 0 \cup Y \times I_1, \qquad Z_1 = X \times 1 \cup Y \times I_0,$$
$$E_0 = F, \qquad E_1 = E,$$

and let $f_0: Z_0 \to X$, $f_1: Z_1 \to X$, $f: Z \to X$ be induced by the product projection $X \times I \to X$. Then $f_i^*(E_i)$ is a bundle over the open set Z_i for $i = 0, 1$ and α induces an isomorphism $f_1^*(E_1) \to f_0^*(E_0)$ on the open set $Z_0 \cap Z_1 = Y \times (I_0 \cap I_1)$. This gives the required bundle L over Z. The element $L - f^* F$ of $K(Z)$ is trivial when restricted to $X \times 0$. Since $f: Z \to X = X \times 0$ is a retraction map, we get a short exact sequence

$$0 \to K(Z, X \times 0) \to K(Z) \underset{f^!}{\rightleftarrows} K(X \times 0) \to 0$$

which splits. Thus $L - f^* F$ and this splitting define an element $d(E, F, \alpha)$ in $K(Z, X \times 0) = K(X, Y)$. The element $d(E, F, \alpha)$ is called the difference bundle of the triple (E, F, α). The following properties of the difference bundle are easily checked (ATIYAH-HIRZEBRUCH [7], Prop. 3.3).

Theorem 24.2.1. I) *If* $f: (X, Y) \to (X', Y')$ *is a map then* $d(f^* E', f^* F', f^* \alpha') = f^! d(E', F', \alpha')$.

II) $d(E, F, \alpha)$ *depends only on the homotopy class of* α.

III) *If* $Y = \emptyset$ *then* $d(E, F, \alpha) = E - F$.

IV) *If* $j^!: K(X, Y) \to K(X)$ *is as in* (3) *then* $j^! d(E, F, \alpha) = E - F$.

V) $d(E, F, \alpha) = 0$ *if and only if there is a vector bundle* G *over* X *such that* $\alpha \oplus 1$ *extends to an isomorphism* $E \oplus G \to F \oplus G$ *over the whole of* X.

VI) $d(E_1 \oplus E_2, F_1 \oplus F_2, \alpha_1 \oplus \alpha_2) = d(E_1, F_1, \alpha_1) + d(E_2, F_2, \alpha_2)$.

VII) $d(E, F, \alpha) + d(F, E, \alpha^{-1}) = 0$.

VIII) *If* $\beta: F|Y \to G|Y$ *is an isomorphism over* Y *then* $d(E, G, \beta\alpha) = d(E, F, \alpha) + d(F, G, \beta)$.

24.3. There is an important special case in which the CHERN character of the difference bundle $d(E, F, \alpha)$ can be computed by 24.2.1 IV).

Let W be a real vector bundle with fibre $\mathbf{R}^{2q}$ and group $\mathbf{SO}(2q)$ over a compact space X. Let $B(W)$, $S(W)$ denote the unit disc and unit sphere bundles of W and $\pi: B(W) \to X$ the projection map. We shall consider difference bundles $d(\pi^* E, \pi^* F, \alpha)$ where E, F are continuous complex vector bundles over X and α is an isomorphism

$$\pi^* E|S(W) \to \pi^* F|S(W).$$

The CHERN character of the difference bundle is then a relative class

$$\operatorname{ch} d(\pi^* E, \pi^* F, \alpha) \in H^*(B(W), S(W); \mathbf{Q}). \tag{4}$$

The cohomology ring $H^*(B(W), S(W); \mathbf{Q})$ has been described by THOM [1]. It is a free module over $H^*(B(W), \mathbf{Q}) = H^*(X, \mathbf{Q})$ generated by a class

$$U \in H^{2q}(B(W), S(W); \mathbf{Q}) .$$

The THOM isomorphism $\varphi_*: H^i(X, \mathbf{Q}) \to H^{i+2q}(B(W), S(W); \mathbf{Q})$ is defined by $\varphi_*(x) = (\pi^* x) \cdot U$ and is an isomorphism for all i. Let j be the embedding $(B(W), \emptyset) \to (B(W), S(W))$. A comparison with 4.11 shows that the EULER class $e(W)$ of W can be defined (since π^* is an isomorphism) by

$$j^* U = \pi^* e(W) . \tag{5}$$

It follows that

$$j^* \varphi_*(x) = \pi^*(x \cdot e(W)) \text{ for } x \in H^*(X, \mathbf{Q}) . \tag{6}$$

Theorem 24.3.1. *Let E, F be complex vector bundles over X, and W a real oriented vector bundle over X. Let $B(W)$ and $S(W)$ be the unit disc and unit sphere bundles of W, $\pi: B(W) \to X$ the projection map and $\alpha: \pi^* E|S(W) \to \pi^* F|S(W)$ an isomorphism. Then*

$$e(W) \cdot \varphi_*^{-1} \operatorname{ch} d(\pi^* E, \pi^* F, \alpha) = \operatorname{ch} E - \operatorname{ch} F .$$

Proof: $j^* \operatorname{ch} d(\pi^* E, \pi^* F, \alpha) = \operatorname{ch} j^! d(\pi^* E, \pi^* F, \alpha)$
$= \operatorname{ch} \pi^* E - \operatorname{ch} \pi^* F$

by 24.2.1 IV), and therefore

$$j^* \varphi_* \varphi_*^{-1} \operatorname{ch} d(\pi^* E, \pi^* F, \alpha) = \pi^*(\operatorname{ch} E - \operatorname{ch} F) .$$

By (6) this gives

$$\pi^*(e(W) \cdot \varphi_*^{-1} \operatorname{ch} d(\pi^* E, \pi^* F, \alpha)) = \pi^*(\operatorname{ch} E - \operatorname{ch} F)$$

and the result follows from the fact that π^* is an isomorphism.

We consider a case in which 24.3.1 gives an explicit formula for $\varphi_*^{-1} \operatorname{ch} d(\pi^* E, \pi^* F, \alpha)$. Suppose that W is induced by a map $f: X \to \mathfrak{G}^+(2q, N; \mathbf{R})$ from the standard vector bundle W' with fibre $\mathbf{R}^{2q}$ over $\mathfrak{G}^+(2q, N; \mathbf{R})$ [see 4.1 a)]. Then f induces a map

$$g: (B(W), S(W)) \to (B(W'), S(W')) .$$

Suppose that E', F' are complex vector bundles over $\mathfrak{G}^+(2q, N; \mathbf{R})$ such that $E = f^* E'$, $F = f^* F'$ and that $\alpha': E'|S(W) \to F'|S(W')$ is an isomorphism such that $\alpha = g^* \alpha'$.
Then by 24.2.1 I)

$$\varphi_*^{-1} \operatorname{ch} d(\pi^* E, \pi^* F, \alpha) = f^* \varphi_*'^{-1} \operatorname{ch} d(\pi'^* E', \pi'^* F', \alpha') .$$

If N is sufficiently large the ring $H^*(\mathfrak{G}^+(2q, N; \mathbf{R}), \mathbf{Q})$ has no divisors of zero in dimensions $\leqq \dim X$ (BOREL [2]) and therefore Theorem 24.3.1

Theorem 24.5.2 is also due to BOTT who originally gave a direct proof using MORSE theory (BOTT [3]). It implies that $\mathbf{S}^{2n}$ does not admit an almost complex structure for $n \geqq 4$ (if θ were a tangent $\mathbf{GL}(n, \mathbf{C})$-bundle then 4.11 (16) would imply that $(c_n(\theta))[\mathbf{S}^{2n}] = 2$). KERVAIRE and MILNOR deduced from Theorem 24.5.2 that $\mathbf{S}^{2n-1}$ is parallelisable if and only if $n = 1, 2$ or 4 (KERVAIRE [1], MILNOR [2]; see also BOREL-HIRZEBRUCH [1], § 26.11 and ATIYAH-HIRZEBRUCH [5]).

Consider the homomorphism $\varphi_! : K(X) \to K(B(E), S(E))$ when X is a point. Then $K(B(E), S(E)) = K(\mathbf{S}^{2q}, y_0)$ for some base point $y_0 \in \mathbf{S}^{2q}$ and $\varphi_! : \mathbf{Z} \to K(\mathbf{S}^{2q}, y_0)$ is a homomorphism with $(\mathrm{ch}_q \varphi_!\, 1)[\mathbf{S}^{2q}] = 1$. In this case it can be shown that $p^! : K(\mathbf{S}^{2q}) \to K(\mathbf{S}^2 \times \cdots \times \mathbf{S}^2)$ is a monomorphism and that $\varphi_! : \mathbf{Z} \to K(\mathbf{S}^{2q}, y_0)$ is an isomorphism. It is often convenient to introduce the element $h = 1 + \varphi_!\, 1 \in K(\mathbf{S}^{2q})$. For $q = 1$ this coincides with the element used in the proof of Theorem 24.5.2. The same argument involving reduced products implies that Theorem 24.5.1 holds with $\mathbf{S}^2$ replaced by $\mathbf{S}^{2q}$ for any $q > 0$.

The BOTT periodicity theorem is the basic tool for the definition of the complete "extraordinary cohomology theory" $K^*(X, Y)$ (see 24.1) and hence also for the proof of the THOM isomorphism theorem mentioned above. We give one further application: to the proof of differentiable analogues of the RIEMAHH-ROCN theorem.

Let $j : X \to Y$ be an embedding of compact connected oriented differentiable manifolds such that the normal bundle E of X in Y admits a complex structure, *i. e.* E is associated to a $\mathbf{U}(q)$-bundle η as in 24.4.2. There is a map $r : Y \to B(E)/S(E)$ under which all points outside $B(E) \subset Y$ are collapsed to the base point and hence a homomorphism $r^! : K(B(E), S(E)) \to K(Y)$. Define $j_! : K(X) \to K(Y)$ by $j_!\, a = r^!\, \varphi_!\, a$ so that

$$\begin{aligned} \mathrm{ch}\, j_!\, a &= r^*\, \varphi_*((\mathrm{td}\,\eta)^{-1} \cdot \mathrm{ch}\, a) \\ &= j_*((\mathrm{td}\,\eta)^{-1} \cdot \mathrm{ch}\, a) \end{aligned}$$

where $j_* : H^*(X, \mathbf{Q}) \to H^*(Y, \mathbf{Q})$ is the GYSIN homomorphism. This is a differentiable analogue of the RIEMANN-ROCH theorem for an embedding [23.5 (9)]. We give two corollaries for the case when X is an almost complex manifold.

Theorem 24.5.3. *Let X be a connected almost complex manifold. There exists an embedding $j : X \to \mathbf{S}^{2N}$ and a homomorphism $j_! : K(X) \to K(\mathbf{S}^{2N})$ such that* $\mathrm{ch}\, j_!\, a = j_*(\mathrm{td}(X) \cdot \mathrm{ch}\, a)$.

Proof: Let θ be the tangent $\mathbf{U}(n)$-bundle of X. For q sufficiently large there is a $\mathbf{U}(q)$-bundle η over X such that $\theta \oplus \eta$ is a trivial $\mathbf{U}(n+q)$-bundle and such that η is the normal bundle of a differentiable embedding $X \to \mathbf{C}_{n+q}$. We regard $\mathbf{S}^{2N}$ as the one point compactification of $\mathbf{C}_N$ where $N = n + q$. The result follows from the equation $\mathrm{td}\,\theta \cdot \mathrm{td}\,\eta = 1$.

Theorem 24.5.4. *Let X be an almost complex manifold and η a $\mathbf{U}(q)$-bundle over X. Then $T(X, \eta)$ is an integer.*

Corollary: *The* Todd *genus of X is an integer.*

Proof: Let $j: X \to \mathbf{S}^{2N}$ be the embedding constructed in 24.5.3. Then η determines an element $a \in K(X)$ and

$$\begin{aligned} T(X, \eta) &= \varkappa_N [j_* (\mathrm{td}(X) \cdot \mathrm{ch}\, a)] \\ &= \varkappa_N [\mathrm{ch} j_! \, a] \end{aligned}$$

is an integer by Theorem 24.5.2.

Theorem 24.5.3 is due to Atiyah-Hirzebruch [1], [8]. It is a special case of a theorem on continuous maps of differentiable manifolds which is described in 26.5. Similarly Theorem 24.5.4 is a special case of more general integrality theorems on differentiable manifolds (26.1–26.2).

§ 25. The Atiyah-Singer index theorem

25.1. Let $x_1, \ldots, x_n$ be coordinates for $\mathbf{R}^n$ and define, for each n-ple $t = (t_1, \ldots, t_n)$ of non-negative integers,

$$\begin{aligned} |t| &= t_1 + \cdots + t_n \\ D^t &= (-i)^{|t|} \frac{\partial^{|t|}}{\partial x_1^{t_1} \ldots \partial x_n^{t_n}}, \quad i^2 = -1 \,. \end{aligned} \tag{1}$$

Let A, B be finite dimensional complex vector spaces and $C^\infty(U, A)$ the space of differentiable functions f from an open set $U \subset \mathbf{R}^n$ to A. The linear map

$$D : C^\infty(U, A) \to C^\infty(U, B)$$

is a *linear differential operator of order r* if there exist functions $g_t \in C^\infty(U, \mathrm{Hom}(A, B))$ such that

$$D f = \sum_{|t| \leqq r} g_t D^t f \,.$$

The differential operator D of order r defines a linear map $\sigma_r(D)(v) \in \mathrm{Hom}(A, B)$ for each $v = (u, (y_1, \ldots, y_n)) \in U \times \mathbf{R}^n$ by

$$\sigma_r(D)(v) = \sum_{|t| = r} g_t(u) \, y_1^{t_1} \ldots y_n^{t_n} \,. \tag{2}$$

D is *elliptic* of order r if, for all $u \in U$ and all non-zero

$$y = (y_1, \ldots, y_n) \in \mathbf{R}^n, \; v = (u, y),$$

the homomorphism $\sigma_r(D)(v)$ is invertible. The homomorphism $\sigma_r(D)$ is called the *symbol* of D. Note that the symbol depends on the choice of r: if D is regarded as a differential operator of order $r + 1$ then the symbol $\sigma_{r+1}(D)$ is zero.

Now let X be a differentiable manifold, ${}_{\mathbf{R}}\theta^*$ the dual tangent bundle of X (see 4.6), $B(X)$ and $S(X)$ the disc and sphere bundles associated to ${}_{\mathbf{R}}\theta^*$, and $\pi : B(X) \to X$ the projection map. Let E, F be differentiable complex vector bundles over X and $\Gamma(E)$, $\Gamma(F)$ the corresponding vector spaces of (global) differentiable sections. A linear map

$$D : \Gamma(E) \to \Gamma(F)$$

is a *differential operator of order* r if there is an open covering of X by coordinate neighbourhoods U_j such that, over each U_j, we have $E = U_j \times A$ and $F = U_j \times B$ and D is given by a differential operator $D_j : C^\infty(U_j, A) \to C^\infty(U_j, B)$ of order r.

Regard $\pi^* E$, $\pi^* F$ as subspaces of $B(X) \times E$, $B(X) \times F$ respectively and define a homomorphism

$$\sigma_r(D) : \pi^* E \to \pi^* F\,,$$

called the *symbol* of D by

$$\sigma_r(D)\,(v, s(x_0)) = \left(v, \frac{i^r}{r!}\, D(f^r s)\,(x_0)\right) \tag{3}$$

where $x_0 \in X$, $v \in B(X)$, $\pi(v) = x_0$, $s \in \Gamma(E)$ and f is a differentiable function with $f(x_0) = 0$ and $df = v$. In terms of local coordinates $x_1, \ldots, x_n$ at the point x_0 we have $D^t(f^r s)_{x_0} = 0$ for $|t| > r$ and

$$D^t(f^r s)_{x_0} = (-i)^r \left(\frac{\partial f}{\partial x_1}\right)_{x_0}^{t_1} \cdot \ldots \cdot \left(\frac{\partial f}{\partial x_n}\right)_{x_0}^{t_n} r!\, s(x_0)$$

for $|t| = r$. Therefore $\sigma_r(D)\,(v, s(x_0))$ depends only on the coordinates $\frac{\partial f}{\partial x_1}, \ldots, \frac{\partial f}{\partial x_n}$ of df and on the value $s(x_0)$ of s.

This proves that the bundle homomorphism $\sigma_r(D)$ is well defined and that it agrees at x_0 with the homomorphism defined by (2).

If E, F, G are complex vector bundles over X, and if $D_1 : \Gamma(E) \to \Gamma(F)$ and $D_2 : \Gamma(F) \to \Gamma(G)$ are differential operators of orders r_1 and r_2, then $D_2 D_1$ is a differential operator of order $r_1 + r_2$ and

$$\sigma_{r_1 + r_2}(D_2 D_1) = \sigma_{r_2}(D_2)\, \sigma_{r_1}(D_1)\,.$$

Definition: The differential operator D is *elliptic* of order r if $\sigma = \sigma_r(D) | S(X)$ is an isomorphism.

Remark: If D is elliptic then E, F have the same fibre dimension. A monomorphism between vector bundles of the same fibre dimension must be an isomorphism. Therefore D is elliptic provided that E, F have the same fibre dimension and that, if $s \in \Gamma(E)$ is a section with $s(x) \neq 0$ and f is a differentiable function with $f(x) = 0$, $df(x) \neq 0$, then $D(f^r s)\,(x) \neq 0$.

25.2. Now suppose that X is compact with a RIEMANN metric. The volume element makes it possible to define integration over X. Suppose that the complex vector bundles E, F are given hermitian metrics $H(\,,)$. A differential operator $D^* : \Gamma(F) \to \Gamma(E)$ is called a *formal* *adjoint for* D if for all $s \in \Gamma(E)$, $t \in \Gamma(F)$

$$\int_X H(D\,s, t) = \int_X H(s, D^*\,t)\,.$$

The hermitian metrics on E, F define metrics on $\pi^* E$, $\pi^* F$. Therefore the symbol $\sigma_r(D): \pi^* E \to \pi^* F$ defines an adjoint homomorphism $\sigma_r(D)^*: \pi^* F \to \pi^* E$.

Theorem 25.2.1. *Let X be a compact differentiable manifold with a* RIEMANN *metric, and E, F differentiable complex vector bundles over X with hermitian metrics. There exists a unique formal adjoint D^* for D and $\sigma_r(D^*) = \sigma_r(D)^*$.*

For the proof see PALAIS [1]. If D is a differential operator of order r then $D^* D: \Gamma(E) \to \Gamma(E)$ is a differential operator of order $2r$ by 25.1. With respect to the hermitian metrics in E, F

$$H(e, \sigma_{2r}(D^* D)\,e) = H(\sigma_r(D)\,e, \sigma_r(D)\,e)$$

for all $e \neq 0$ in $\pi^* E$. Therefore if D is elliptic, $D^* D$ is *strongly elliptic*, *i. e.* $H(e, \sigma_{2r}(D^* D)\,e) > 0$ for all $e \neq 0$ in $\pi^* E$. Conversely suppose that E, F have the same fibre dimension and that $D^* D$ is strongly elliptic. Then $\sigma_r(D) | S(X)$ is a monomorphism and hence D is elliptic.

Let ker D and coker D be the kernel and cokernel of the differential operator D. If D is elliptic then D^* is elliptic, ker D is finite dimensional and dim ker D^* = dim coker D (PALAIS [1], GELFAND [1]). The *index*, or *analytic index*, $\tau(D)$ of D is defined by

$$\tau(D) = \dim \ker D - \dim \operatorname{coker} D = \dim \ker D - \dim \ker D^*\,. \qquad (4)$$

VEKUA and GELFAND [1] conjectured that the integer $\tau(D)$ could be expressed in terms of topological invariants. This conjecture was checked in special cases by AGRANOVIC [1], DYNIN [1], VOLPERT [1], [2] and others.

25.3. Let X be a compact differentiable m-dimensional manifold, which need not be orientable, and ${}_{\mathbf{R}}\theta$ the tangent $\mathbf{GL}(m, \mathbf{R})$-bundle of X. Let T^* be the total space of the covariant tangent vector bundle ${}_{\mathbf{R}}\mathfrak{T}^*$ of X and $\pi: T^* \to X$ the projection map. Then T^* is a $2m$-dimensional manifold with tangent $\mathbf{GL}(2m, \mathbf{R})$-bundle $\pi^*\,{}_{\mathbf{R}}\theta \oplus \pi^*\,{}_{\mathbf{R}}\theta^*$. A RIEMANN metric on X defines an isomorphism ${}_{\mathbf{R}}\theta \cong {}_{\mathbf{R}}\theta^*$ and hence (in the notation of 4.5) an isomorphism

$$\pi^*_{\mathbf{R}}\theta \oplus \pi^*_{\mathbf{R}}\theta^* \cong \pi^*_{\mathbf{R}}\theta \oplus \pi^*_{\mathbf{R}}\theta \cong \varrho(\pi^*\,\psi({}_{\mathbf{R}}\theta))\,.$$

Therefore the $\mathbf{GL}(m, \mathbf{C})$-bundle $\eta = \pi^*\,\psi({}_{\mathbf{R}}\theta)$ gives an almost complex structure for the manifold T^*. For a detailed study of the almost complex structure on T^* see DOMBROWSKI [1].

In terms of local coordinates $x_1, \ldots, x_m$ an element v in the fibre of T^* over $(0, \ldots, 0)$ has the form $\sum_{j=1}^{m} v_j \, dx_j$. The ordering of coordinates $(x_1, v_1, \ldots, x_m, v_m)$ defines the orientation of T^* induced by η. This orientation induces orientations of the unit disc bundle $B(X)$ and the unit sphere bundle $S(X)$ and hence a fundamental cycle in

$$H_{2m}(B(X), S(X); \mathbf{Q}).$$

The value of a cohomology class $u \in H^*(B(X), S(X); \mathbf{Q})$ on the fundamental class is denoted by $\varkappa^{2m}[u]$.

Let $D: \Gamma(E) \to \Gamma(F)$ be an elliptic differential operator of order r with symbol $\sigma_r(D)$. By 24.2 the restriction $\sigma = \sigma_r(D)|S(X)$ defines a difference bundle $d(\pi^* E, \pi^* F, \sigma)$ in $K(B(X), S(X))$ with CHERN character

$$\mathrm{ch} D \in H^*(B(X), S(X); \mathbf{Q}).$$

The relative cohomology group $H^*(B(X), S(X); \mathbf{Q})$ can be regarded, using the relative cup product, as a module over $H^*(B(X), \mathbf{Q})$. The *topological index* $\gamma(D)$ of D is then defined by

$$\gamma(D) = \varkappa^{2m}[\mathrm{ch} D \cdot \mathrm{td}\eta]. \tag{5}$$

Theorem 25.3.1 (ATIYAH-SINGER [1]). *Let E, F be differentiable complex vector bundles over a compact differentiable manifold X and $D: \Gamma(E) \to \Gamma(F)$ an elliptic differential operator. Then $\tau(D) = \gamma(D)$.*

Corollary: *$\gamma(D)$ is an integer.*

The ATIYAH-SINGER index theorem $\tau(D) = \gamma(D)$ implies Theorem 21.1.1 (R-R) for an arbitrary compact complex manifold V. In addition it implies the index theorem of Chapter Two (Theorem 8.2.2). These implications are proved in 25.4. The proof of Theorem 25.3.1 is discussed very briefly in 25.5. In certain cases it can be proved directly that $\gamma(D) = 0$. Theorem 25.3.1 then implies that $\tau(D) = 0$. For example

Lemma 25.3.2. *Let D be an elliptic differential operator on a compact differentiable manifold of odd dimension. Then $\gamma(D) = 0$.*

Proof: Let $D: \Gamma(E) \to \Gamma(F)$ be elliptic of order r. If $v \in S(X)$, $\pi(v) = x$, then the symbol $\sigma_r(D)(x, v): E_x \to F_x$ is defined by a homogeneous polynomial of degree r in the local fibre coordinates $v_1, \ldots, v_m$ for $B(X)$. Therefore

$$\sigma_r(D)(x, -v) = (-1)^r \sigma_r(D)(x, v). \tag{6}$$

Let $f: B(X), S(X) \to B(X), S(X)$ be the antipodal map and $\beta: \pi^* F \to \pi^* F$ scalar multiplication by $(-1)^r$. Then (6) gives

$$\beta \sigma_r(D) = f^* \sigma_r(D): f^* \pi^* E \to f^* \pi^* F.$$

Therefore, since $\pi f = \pi$ and $f^* \pi^* = \pi^*$,

$$d(\pi^* E, \pi^* F, f^* \sigma) = d(\pi^* E, \pi^* F, \beta \sigma).$$

It follows from Theorem 24.2.1 that since β is homotopic to the identity

$$d(f^* \pi^* E, f^* \pi^* F, f^* \sigma) = d(\pi^* E, \pi^* F, \sigma),$$

$$f^* \operatorname{ch} D = \operatorname{ch} D.$$

On the other hand $\operatorname{td}\eta$ is a class in $H^*(B(X), \mathbf{Q}) = H^*(X, \mathbf{Q})$ and so $f^* \operatorname{td}\eta = \operatorname{td}\eta$. If X is odd dimensional the map f is orientation reversing and therefore $-\gamma(D) = \gamma(D)$.

If X is orientable, ${}_{\mathbf{R}}\mathfrak{T}$ is associated to the $\mathbf{SO}(m)$-bundle ${}_{\mathbf{R}}\theta$. There is a Thom isomorphism

$$\varphi_* : H^*(X, \mathbf{Q}) \to H^*(B(X), S(X); \mathbf{Q}) \tag{7}$$

defined by the orientation of X, and the orientation of $B(X)$ given by the ordering of coordinates $(x_1, \ldots, x_m, v_1, \ldots, v_m)$. This orientation differs from that used above by a factor $(-1)^{\frac{1}{2}m(m-1)}$. Therefore

$$\begin{aligned}\gamma(D) &= \varphi_*^{-1}\left((-1)^{\frac{1}{2}m(m-1)} \operatorname{ch} D \cdot \operatorname{td}\eta\right)[X] \\ &= \varkappa^m \left[\varphi_*^{-1}\left((-1)^{\frac{1}{2}m(m-1)} \operatorname{ch} D\right) \cdot \operatorname{td}\psi({}_{\mathbf{R}}\theta)\right].\end{aligned} \tag{8}$$

The Todd class $\operatorname{td}\psi({}_{\mathbf{R}}\theta)$ can be expressed as a polynomial in the Pontrjagin classes $p_j(X) = (-1)^j c_{2j}(\psi({}_{\mathbf{R}}\theta))$ of X: if

$$p(X) = \prod_j (1 + y_j^2) \in H^*(X, \mathbf{Q})$$

then (see 4.5)

$$c(\psi({}_{\mathbf{R}}\theta)) = \prod_j (1 - y_j^2) = \prod_j (1 + y_j)(1 - y_j)$$

and so

$$\operatorname{td}\psi({}_{\mathbf{R}}\theta) = \prod_j \left(\frac{y_j}{1 - e^{-y_j}}\right)\left(\frac{-y_j}{1 - e^{y_j}}\right) = \prod_j \left(\frac{\frac{1}{2} y_j}{\sinh \frac{1}{2} y_j}\right)^2. \tag{9}$$

The right hand side is a symmetric function of the y_j^2 and is therefore expressible as a polynomial in the $p_j(X)$ (compare the corresponding formula in 1.7).

25.4. In this section we outline two important applications of the Atiyah-Singer index theorem. Further details can be found in Palais [1], Cartan-Schwartz [1].

a) Let V_n be a compact complex manifold of dimension n and W a complex analytic vector bundle over V_n with fibre $\mathbf{C}_q$. We wish to show that Theorem 25.3.1 implies the Riemann-Roch theorem $\chi(V_n, W) = T(V_n, W)$. Let $\boldsymbol{T}$ be the complex covariant tangent vector bundle of V_n. In the notations of 15.4, $\Gamma(W \otimes \lambda^p \overline{\boldsymbol{T}}) = A^{0,p}(W)$. The differential operator

$$\bar{\partial} + \vartheta : \Gamma\Big(\sum_p W \otimes \lambda^p \overline{\boldsymbol{T}}\Big) \to \Gamma\Big(\sum_p W \otimes \lambda^p \overline{\boldsymbol{T}}\Big)$$

is self adjoint [15.4 (9)]. Since $\bar{\partial}$ is of degree $+1$, and ϑ is of degree -1, the differential operator $\bar{\partial} + \vartheta$ maps odd degree forms into even degree forms and conversely. Let

$$E = \sum_s W \otimes \lambda^{2s} \overline{\boldsymbol{T}}\,, \quad F = \sum_s W \otimes \lambda^{2s+1} \overline{\boldsymbol{T}}\,,$$

$$D = \bar{\partial} + \vartheta : \Gamma(E) \to \Gamma(F)\,.$$

Then D is a differential operator of order 1. The decomposition

$$A^{0,p}(W) = \bar{\partial}\, A^{0,p-1}(W) \oplus \vartheta\, A^{0,p+1}(W) \oplus B^{0,p}(V, W)$$

of 15.4 shows that if $\bar{\partial}\,\alpha + \vartheta\,\beta = 0$, $\alpha \in A^{0,p-1}(W)$, $\beta \in A^{0,p+1}(W)$, then $\bar{\partial}\,\alpha = \vartheta\,\beta = 0$. Therefore

$$\ker D \;= \sum_s B^{0,2s}(V, W)\,,$$

$$\ker D^* = \sum_s B^{0,2s+1}(V, W)\,.$$

By Theorem 15.4.1

$$\begin{aligned} \tau(D) &= \dim \ker D - \dim \ker D^* \\ &= \sum_p (-1)^p \dim H^p(V, W) = \chi(V, W)\,. \end{aligned} \tag{10}$$

Let ${}_{\mathbf{R}}\mathfrak{T}^*_{\mathbf{C}}$ be the complexification of the real dual tangent bundle ${}_{\mathbf{R}}\mathfrak{T}^*$ of X. The isomorphism ${}_{\mathbf{R}}\mathfrak{T}^*_{\mathbf{C}} = \boldsymbol{T} \oplus \overline{\boldsymbol{T}}$ [see 4.7 (12)] defines a projection map $p : {}_{\mathbf{R}}\mathfrak{T}^*_{\mathbf{C}} \to \overline{\boldsymbol{T}}$. The induced map ${}_{\mathbf{R}}\mathfrak{T}^* \to \overline{\boldsymbol{T}}$ can be used to identify the disc bundle $B(X) = B({}_{\mathbf{R}}\mathfrak{T}^*)$ with $B(\overline{\boldsymbol{T}})$. We assume this identification when calculating the symbol of the differential operator D. By (3) the symbol of $\bar{\partial} : \Gamma(\lambda^{r-1}\, \overline{\boldsymbol{T}}) \to \Gamma(\lambda^r\, \overline{\boldsymbol{T}})$ is defined at $d f \in B({}_{\mathbf{R}}\mathfrak{T}^*_{\mathbf{C}})$, $u_1 \wedge u_2 \wedge \cdots \wedge u_{r-1} \in \lambda^{r-1}\, \overline{\boldsymbol{T}}$ by

$$\sigma_1(\bar{\partial})\,(d f, u_1 \wedge \cdots \wedge u_{r-1}) = i\, \bar{\partial} f \wedge u_1 \cdots \wedge u_{r-1}$$

and at $p(d f) = \bar{\partial} f \in B(\overline{\boldsymbol{T}})$ by

$$\sigma_1(\bar{\partial})\,(\bar{\partial} f, u_1 \wedge \cdots \wedge u_{r-1}) = i\, \bar{\partial} f \wedge u_1 \wedge \cdots \wedge u_{r-1}\,.$$

The isomorphism $\lambda^r\, \overline{\boldsymbol{T}} \to \lambda^r\, \boldsymbol{T}^*$ [see 15.3 c)] induces hermitian metrics for each $\lambda^r\, \overline{\boldsymbol{T}}$ such that ϑ, defined by 15.4 (9), is a formal adjoint for $\bar{\partial}$ in the sense of 25.2. Therefore, in the notation of 24.4,

$$\sigma_1(\bar{\partial}) = i\, \beta_r$$

$$\sigma_1(D)|S(X) = i\,\beta : \pi^* \sum_s W \otimes \lambda^{2s}\, \overline{\boldsymbol{T}}|S(X) \to \pi^* \sum_s W \otimes \lambda^{2s+1}\, \overline{\boldsymbol{T}}\,|S(X).$$

By 24.4, β is an isomorphism and therefore D is elliptic. Alternatively an explicit calculation shows that $D^* D = \square$ is strongly elliptic and hence that D is elliptic.

As in 24.4.2 we have

$$\varphi_*^{-1}\operatorname{ch}(D) = \varphi_*^{-1}\operatorname{ch} d\left(\pi^* \sum_s W \otimes \lambda^{2s}\,\overline{\boldsymbol{T}},\, \pi^* \sum_s W \otimes \lambda^{2s+1}\,\overline{\boldsymbol{T}},\, \beta\right)$$
$$= (-1)^n \operatorname{ch} W \cdot (\operatorname{td}\theta^*)^{-1}$$

where θ is the tangent $\mathbf{U}(n)$-bundle of V_n. Then

$$\begin{aligned}\gamma(D) &= \varkappa_n[(-1)^{2n}\operatorname{ch} W \cdot (\operatorname{td}\theta^*)^{-1}\cdot \operatorname{td}\theta \cdot \operatorname{td}\bar{\theta}] \\ &= \varkappa_n[\operatorname{ch} W \cdot \operatorname{td}\theta] = T(V_n, W)\,. \end{aligned} \tag{11}$$

Equations (10) and (11) show that Theorem 25.3.1 implies R-R for an arbitrary compact complex manifold V_n. The same theorem applied to the vector bundle $W \otimes \lambda^p\,\boldsymbol{T}$ gives

$$\chi^p(V_n, W) = T^p(V_n, W) \text{ and } \chi_y(V_n, W) = T_y(V_n, W)\,.$$

In particular the case $y = 1$ shows that the HODGE index theorem (15.8.2) is valid for an arbitrary compact complex manifold.

b) Now let X be a compact oriented differentiable manifold of dimension $2n$. We wish to show that Theorem 4.11.4 and Theorem 8.2.2 are both consequences of Theorem 25.3.1.

Let ${}_{\mathbf{R}}\mathfrak{T}^*_{\mathbf{C}}$ be the complexification of the real vector bundle of covariant tangent vectors (see 4.6) and define

$$W = \sum_r \lambda^r\, {}_{\mathbf{R}}\mathfrak{T}^*_{\mathbf{C}}\,.$$

A section of W is a complex valued differential form on X. The exterior derivative

$$d : \Gamma(W) \to \Gamma(W)$$

is a differential operator of degree 1 (see 2.12). Equation (3) shows that if $v = df$, $\pi(v) = x$, $f(x) = 0$ and $\omega \in \Gamma(W)$ then the symbol of d is defined by

$$\sigma_1(d)\,(v, \omega(x)) = (v, i\, v \wedge \omega(x))\,.$$

A RIEMANN metric on X defines a homomorphism

$$* : \lambda^r\, {}_{\mathbf{R}}\mathfrak{T}^*_{\mathbf{C}} \to \lambda^{2n-r}\, {}_{\mathbf{R}}\mathfrak{T}^*_{\mathbf{C}}$$

and hence a homomorphism $* : \Gamma(W) \to \Gamma(W)$. Since X is even dimensional the formal adjoint δ for d in the sense of 25.2 is defined by $\delta = -* d *$. As in 15.4 we have $dd = \delta\delta = 0$ and $(d+\delta)(d+\delta) = d\delta + \delta d = \triangle$. A form $\omega \in \Gamma(W)$ is called harmonic if $\triangle\omega = 0$. ω is harmonic if and only if $d\omega = \delta\omega = 0$. If $B^r(X)$ denotes the vector space of harmonic forms of degree r there is a natural isomorphism (DE RHAM [1], HODGE [1]; compare 15.7)

$$H^r(X, \mathbf{C}) = B^r(X)$$

and therefore $\dim B^r(X) = b_r(X)$ is the r-th BETTI number of X.

The differential operator $d + \delta : \Gamma(W) \to \Gamma(W)$ is self adjoint. We therefore seek decompositions $W = E \oplus F$ such that

$$D = d + \delta : \Gamma(E) \to \Gamma(F)$$

is elliptic. Consider the endomorphisms of W defined by $*$ and $\alpha = i^{r(r+1)-n} * : \lambda^r\, {}_{\mathbf{R}}\mathfrak{T}^*_{\mathbf{C}} \to \lambda^{2n-r}\, {}_{\mathbf{R}}\mathfrak{T}^*_{\mathbf{C}}$. Since X is even dimensional $** = (-1)^r$ and $\alpha^2 = (-1)^{r^2} ** = 1$. The eigenspaces of the involutions $**$ and α provide decompositions of W of the required type.

1) Define $E = \sum_s \lambda^{2s}\, {}_{\mathbf{R}}\mathfrak{T}^*_{\mathbf{C}}$, $F = \sum_s \lambda^{2s+1}\, {}_{\mathbf{R}}\mathfrak{T}^*_{\mathbf{C}}$ and

$$D = d + \delta : \Gamma(E) \to \Gamma(F)\,.$$

In the notation of 24.4 the symbol of D is $i\beta$ and therefore D is elliptic. Alternatively an explicit calculation shows that $D^* D = \triangle$ is strongly elliptic and hence that D is elliptic. As in a)

$$\gamma(D) = \dim \ker D - \dim \ker D^* = \sum_r (-1)^r \dim B^r(X)\,.$$

Therefore $\gamma(D)$ is the EULER-POINCARÉ characteristic $E(X)$ of X. By Theorem 10.1.1, if ${}_{\mathbf{R}}\theta$ is the tangent $\mathbf{SO}(2n)$-bundle of X,

$$\begin{aligned} \operatorname{ch} E - \operatorname{ch} F &= c_{2n}(\psi({}_{\mathbf{R}}\theta)) \cdot (\operatorname{td} \psi({}_{\mathbf{R}}\theta))^{-1} \\ &= (-1)^n (e({}_{\mathbf{R}}\theta))^2 (\operatorname{td} \psi({}_{\mathbf{R}}\theta))^{-1}\,. \end{aligned}$$

By 24.3 (7) and 25.3 (8), $\gamma(D) = e({}_{\mathbf{R}}\theta)\,[X]$. Therefore Theorem **25.3.1** implies Theorem 4.11.4 for even dimensional X. The case of odd dimensional X is covered by Theorem **25.3.2**.

2) Now let E, F be the eigenspaces corresponding to the eigenvalues $+1, -1$ of α. The argument of **1**) shows that the differential operator $d + \delta : \Gamma(W) \to \Gamma(W)$ is elliptic. Now $\alpha(d + \delta) = -(d + \delta)\,\alpha$ and therefore there is a differential operator

$$D = d + \delta : \Gamma(E) \to \Gamma(F)\,.$$

The symbols of D and $d + \delta$ form a commutative diagram

$$\begin{array}{ccc} \pi^* E & \xrightarrow{\sigma_1(D)} & \pi^* F \\ \downarrow & & \downarrow \\ \pi^* W & \xrightarrow{\sigma_1(d+\delta)} & \pi^* W \end{array}$$

in which vertical arrows denote inclusions. Since $\sigma_1(d + \delta)$ is an isomorphism over $S(X)$ the symbol $\sigma_1(D)$ is a monomorphism. The same argument shows that $\sigma_1(D^*)$ is a monomorphism and hence (see **25.2.1**) that $\sigma_1(D)$ is an epimorphism. Therefore D is elliptic.

The kernel of D is the space of harmonic forms ω such that $\alpha\,\omega = \omega$. The kernel of D^* is the space of harmonic forms ω such that $\alpha\,\omega = -\omega$.

Thus (compare the proof of Theorem 15.8.2)

$$\tau(D) = \dim \ker D - \dim \ker D^* = \dim B^n_+(X) - \dim B^n_-(X)$$

where $B^n_\pm(X)$ is the subspace $\{\omega \in B^n(X); \alpha\,\omega = \pm\,\omega\}$ of $B^n(X)$. The homomorphism $\alpha: \lambda^n\,{}_{\mathbf{R}}\mathfrak{T}_{\mathbf{C}} \to \lambda^n\,{}_{\mathbf{R}}\mathfrak{T}_{\mathbf{C}}$ is defined by $\alpha = i\,*$ for n odd and $\alpha = *$ for n even. If n is odd the map $\omega \to \omega$ is an isomorphism $B^n_+(X) \to B^n_-(X)$ and therefore $\tau(D) = 0$. The direct sum

$$H^n(X, \mathbf{C}) = B^n_+(X) \oplus B^n_-(X)$$

induces a corresponding direct sum, when n is even,

$$H^n(X, \mathbf{R}) = B^n_{+,\mathbf{R}}(X) \oplus B^n_{-,\mathbf{R}}(X)$$

where $B^n_{\pm,\mathbf{R}}(X)$ is the subspace $\{\omega \in B^n_\pm(X); \omega = \bar{\omega}\}$ of $B^n_\pm(X)$. The inner product $(\omega_1, \omega_2) = \int_X \omega_1 \wedge * \,\omega_2$ on $B^n_{\mathbf{R}}(X) = H^n(X, \mathbf{R})$ is positive definite and $B^n_{+,\mathbf{R}}(X)$, $B^n_{-,\mathbf{R}}(X)$ are orthogonal with respect to this inner product if n is even. The quadratic form $Q(\omega_1, \omega_2) = \int_X \omega_1 \wedge \omega_2$ is positive definite on $B^n_{+,\mathbf{R}}(X)$ and negative definite on $B^n_{-,\mathbf{R}}(X)$. Therefore if n is even, $\dim B^n_+(X)$, $\dim B^n_-(X)$ is the number p_+, p_- of positive, negative eigenvalues of Q. Hence $\tau(D) = p_+ - p_-$ is the index of X as defined in 8.2.

A calculation on the classifying space $\mathfrak{G}^+(2n, N; \mathbf{R})$ which is similar to 1) and which is given with full details in Palais [1], shows that, in terms of the factorisation

$$p(X) = \prod_{j=1}^{n} (1 + y_j^2)\,, \quad e({}_{\mathbf{R}}\theta^*) = \prod_{j=1}^{n} y_j\,,$$

$$\operatorname{ch} E - \operatorname{ch} F = \prod_{j=1}^{n} (e^{-y_j} - e^{y_j})$$

and therefore by 24.3 (7) and 24.4 (7)

$$\varphi_*^{-1} \operatorname{ch} D = \prod_{j=1}^{n} \frac{e^{-y_j} - e^{y_j}}{y_j}\,,$$

$$\begin{aligned}
\gamma(D) &= \varkappa^{2n}\left[(-1)^n \prod_{j=1}^{n} \frac{e^{-y_j} - e^{y_j}}{y_j} \cdot \frac{y_j}{1 - e^{-y_j}} \cdot \frac{-y_j}{1 - e^{y_j}}\right] \\
&= \varkappa^{2n}\left[\prod_{j=1}^{n} y_j \frac{\cosh \frac{1}{2} y_j}{\sinh \frac{1}{2} y_j}\right] \\
&= \varkappa^{2n}\left[2^n \prod_{j=1}^{n} \frac{1}{2} y_j \left(\tanh \frac{1}{2} y_j\right)^{-1}\right] \\
&= \varkappa^{2n}\left[\prod_{j=1}^{n} y_j \big/ \tanh y_j\right] = L(X)\,.
\end{aligned}$$

Therefore Theorem 25.3.1 implies Theorem 8.2.2. The case of odd dimensional X is again covered by 25.3.2.

25.5. There are two proofs of the ATIYAH-SINGER index theorem. The first proof, which is modelled on that of Theorem 8.2.2, was outlined in ATIYAH-SINGER [1]. Details can be found in PALAIS [1] and CARTAN-SCHWARTZ [1]. The second proof is due to appear in a forthcoming paper of ATIYAH and SINGER.

The starting point for both proofs is the fact that formula (8) defines the index $\gamma(b)$ for any element $b \in K(B(X), S(X))$. We wish to extend the analytic index $\tau(D)$ similarly so that it becomes a function $\tau: K(B(X), S(X)) \to \mathbf{Q}$. It is known that homotopic operators have the same analytic index and that symbols which determine the same difference bundle are homotopic. Thus the index will depend only on the difference bundle if (i) a homotopy between symbols can be raised to a homotopy between operators. The function τ will be defined if (ii) every $b \in K(B(X), S(X))$ is the difference bundle determined by some elliptic operator. In general neither (i) nor (ii) is true for elliptic differential operators. It is necessary to introduce the elliptic integral operators of SEELEY [1]. This class includes the elliptic differential operators but is large enough for (i) and (ii) to hold. In this way the analytic index defines a homomorphism $\tau: K(B(X), S(X)) \to \mathbf{Q}$ which always takes integral values.

The remainder of the proof is devoted to showing that the two homomorphisms

$$\gamma: K(B(X), S(X)) \to \mathbf{Q},$$
$$\tau: K(B(X), S(X)) \to \mathbf{Q},$$

coincide. We summarise both methods very briefly. In the first it is assumed that X is oriented and even dimensional.

a) By 25.4 b) there is a differential operator D_0 over X whose topological index is the L-genus of X. Let $b_0 \in K(B(X), S(X))$ be the corresponding difference bundle. The ring $K(B(X), S(X))$ is a $K(X)$-module (24.5) and the function γ is determined completely by its values on the subgroup $K(X) \cdot b_0$ of finite index. Define a function $\gamma(X, \ ): K(X) \to \mathbf{Q}$ by $\gamma(X, b) = \gamma(b \cdot b_0)$. In fact $\gamma(X, b)$ is none other than the T_y-characteristic of b with $y = 1$ (this is defined also for differentiable manifolds and by 12.2 (13) the definition can be extended to elements $b \in K(X)$). The homomorphism has the properties

I) $\gamma(X + Y, b + c) = \gamma(X, b) + \gamma(Y, c)$ where on the left hand side $+$ denotes disjoint union (*not* direct sum);

II) $\gamma(X \times Y, b \otimes c) = \gamma(X, b)\, \gamma(Y, c)$ where $\otimes$ denotes tensor product; this follows from 12.2 (14) with $y = 1$;

III) $\gamma(X, b) = 0$ if there exists a manifold X' with boundary $\partial X' = X$ and an element $b' \in K(X')$ whose restriction to X is b; this is proved by a more complicated version of Theorem 7.2.1;

IV) $\gamma(\mathbf{S}^{2n}, h) = 2^n$ where $h \in K(\mathbf{S}^{2n})$ is the element with $\varkappa^{2n}[\mathrm{ch}\, h] = 1$ defined in 24.5; by 12.2 (10) we have, if $\mathrm{ch}\, h = \sum e^{\delta_i}$,

$$\gamma(\mathbf{S}^{2n}, h) = \varkappa^{2n}[\sum e^{2\delta_i}] = 2^n \varkappa^{2n}[\mathrm{ch}\, h] = 2^n;$$

V) $\gamma(\mathbf{P}_{2n}(\mathbf{C}), 1) = 1$; this follows from 1.5.1.

The next, and most difficult step in the proof is to show that the analytic index τ also satisfies properties I)—V). Finally it is shown that a function $K(X) \to \mathbf{Q}$ is uniquely determined by properties I)—V). As in 7.1 we consider a cobordism group Ω_n. For each $n \geqq 0$, Ω_n is obtained by considering pairs (X, b) with X a compact oriented n-dimensional differentiable manifold and $b \in K(X)$. A pair (X, b) bounds if there exists a manifold X' and element $b' \in K(X')$ such that $X = \partial X'$ and $b = b'|X$. The groups $\Omega_n \otimes \mathbf{Q}$ are determined as in 7.2.3: elements of $\Omega_n \otimes \mathbf{Q}$ are determined uniquely by mixed Pontrjagin-Chern numbers

$$p_{j_1}(X) \ldots p_{j_r}(X) \cdot \mathrm{ch}_{k_1}(b) \ldots \mathrm{ch}_{k_s}(b)\, [X]\,.$$

Properties I)—III) show that γ gives a function $\Omega = \sum_{n=0}^{\infty} \Omega_n \to \mathbf{Q}$. Properties IV), V) are sufficient to determine γ on the generators of $\Omega \otimes \mathbf{Q}$ and hence to determine γ uniquely. A general theory of such cobordism groups of pairs can be found in Conner-Floyd [1].

The theorem for X odd dimensional follows by considering $X \times \mathbf{S}^1$.

b) The second proof that the homomorphisms γ and τ coincide does not depend on cobordism theory. By 25.3 the unit disc bundle $B(X)$ is an almost complex manifold with boundary $S(X)$. For convenience we write $T^* X$ for $B(X) - S(X)$ and $K(T^* X)$ for $K(B(X), S(X))$. Let $V = \mathbf{R}^N$ so that $K(T^* V) = K(\mathbf{S}^{2N}, y_0) = \mathbf{Z}$. An embedding $X \subset V$ defines an embedding $j: T^*X \to T^*V$. Now Theorem 24.5.3, suitably modified to apply to manifolds with boundary, implies that there is a homomorphism

$$j_! : K(T^*X) \to K(T^*V) = \mathbf{Z}$$

such that $j_!\, a = \varkappa^{2m}[\mathrm{ch}\, a \cdot \mathrm{td}\eta]$ where m is the dimension of X and η is the tangent $\mathbf{GL}(m, \mathbf{C})$-bundle of T^*X. By 25.3 (5) the homomorphism $j_!$ coincides with the homomorphism $\gamma: K(T^*X) \to \mathbf{Q}$. Note in particular that γ always takes integer values, so that for applications of the Atiyah-Singer index theorem to integrality theorems (26.2; Mayer [1]) the full proof is not needed.

It remains to prove that the homomorphism $j_!$ coincides with the analytic index $\tau: K(T^*X) \to \mathbf{Z}$. The first part of the proof consists in extending the definition of τ to apply to operators on non-compact manifolds U, V. This is done by SEELEY [1]. The result is a diagram of homomorphisms

$$\begin{array}{ccccccc} K(T^*X) & \xrightarrow{\varphi_!} & K(T^*U) & \xrightarrow{r^!} & K(T^*V) & \xleftarrow{\varphi_!} & K(y_0) \\ \downarrow{\scriptstyle\tau} & & \downarrow{\scriptstyle\tau} & & \downarrow{\scriptstyle\tau} & & \downarrow{\scriptstyle\tau} \\ \mathbf{Z} & \xrightarrow{id} & \mathbf{Z} & \xrightarrow{id} & \mathbf{Z} & \xrightarrow{id} & \mathbf{Z} \end{array}$$

in which U is a tubular neighbourhood of X in V, the $\varphi_!$ are THOM isomorphisms (see 24.5) and $r^!$ is induced by the map $T^*V \to T^*U$ which collapses everything outside TU to a point. The second and difficult part of the proof consists in proving that each of the squares in this diagram is commutative.

The techniques involved in this proof have been extended by ATIYAH to give generalisations of the ATIYAH-SINGER index theorem which apply to manifolds with boundary (ATIYAH-BOTT [2]), to families of elliptic operators, and to actions of compact LIE groups on differentiable manifolds.

25.6. The latter development can be described briefly as follows. Let X be a compact space, and G a compact LIE group which acts on X. Then a *G-vector bundle* over X consists of a complex vector bundle E over X together with a G-action on E, commuting with the projection $E \to X$, given by linear maps $g: E_x \to E_{gx}$ for all $g \in G$, $x \in X$. The definitions of 24.1 can be imitated in this case to give a GROTHENDIECK ring $K_G(X)$ "of G-vector bundles over X". In the special case when G consists only of the identity element, $K_G(X)$ coincides with $K(X)$. When X is a point, $K_G(X)$ is the representation ring $R(G)$ of G. If Y is a G-stable closed subspace of X then the relative group $K_G(X, Y)$ is defined. Note that the groups $K_G(X)$, $K_G(X, Y)$ depend not only on G, X, Y but also on the particular action of G on X. The results in K-theory mentioned in § 24 all have analogues (due to ATIYAH and SEGAL) in K_G-theory.

Now suppose that X is a compact differentiable manifold, and that G acts differentiably on X. If E is a *differentiable* G-vector bundle over X (that is, both E and the action of G on E are differentiable) there is an action of G on the space $\Gamma(E)$ of differentiable sections of E defined by

$$(gs)(x) = g \cdot s(g^{-1}x), \qquad g \in G, s \in \Gamma(E), x \in X.$$

Let E, F be differentiable G-vector bundles over X and

$$D: \Gamma(E) \to \Gamma(F)$$

an elliptic differential operator compatible with the action of G. Then G acts linearly on the finite dimensional vector spaces $\ker D$ and $\operatorname{coker} D$.

The *analytic index* of D is the element $\tau(D)$ of the representation ring $R(G)$ of G defined by

$$\tau(D) = \ker D - \operatorname{coker} D .$$

When G consists of the identity element, $R(G) = \mathbf{Z}$ and this definition coincides with 25.2 (4).

On the other hand it is possible to define a *topological index* $\gamma(D) \in R(G)$ which reduces to that defined in 25.3 when G is the identity. The definition involves, not only the symbol of D and the PONTRJAGIN classes of X, but also the fixed point sets X^g of elements $g \in G$.

The second proof of the ATIYAH-SINGER index theorem (25.5) can be given in terms of K_G-theory, and then shows that $\tau(D) = \gamma(D)$ for every elliptic differential operator D compatible with the action of G. There is a G-invariant metric on X, and hence a G-action on the disc bundle $B(X)$ for which $S(X)$ is a G-stable subspace. The elliptic integral, or "pseudo-differential", operators of SEELEY are used to define homomorphisms

$$\tau : K_G(B(X), S(X)) \to R(G) ,$$

$$\gamma : K_G(B(X), S(X)) \to R(G) ,$$

which are then proved to coincide.

Consider the special case in which G is a cyclic group, and in which the generator $g : X \to X$ has only simple fixed points [a fixed point $x \in X$ is simple if $\det(1 - dg_x) \neq 0$, where dg_x is the induced map on the tangent space to X at x; this implies that x is an isolated fixed point]. In this special case the formula $\tau(D) = \gamma(D)$ is also given by a "LEFSCHETZ fixed point formula" due to ATIYAH and BOTT. The latter theorem, which is proved by quite different methods, applies to more general maps $f : X \to X$ (again with only simple fixed points, but not necessarily the generator of a cyclic group acting on X). As in 25.4, applications follow by considering particular differential operators D. Thus the operator of 25.4a) gives a theorem, on the fixed points of a holomorphic map $f : V \to V$ of a compact complex manifold V, which is analogous to R-R. The operator of 25.4b) gives 1) a theorem analogous to the HIRZEBRUCH index theorem, and 2) the original LEFSCHETZ fixed point formula, for a compact oriented differentiable manifold. Full details of these results, with a sketch of the proof of the general formula, can be found in ATIYAH-BOTT [3].

§ 26. Integrality theorems for differentiable manifolds

26.1. The ATIYAH-SINGER index theorem implies in particular [25.4 a)] that the T-characteristic $T(V_n, \eta)$ of a complex analytic $\mathbf{GL}(q, \mathbf{C})$-bundle η over a compact complex manifold V_n is an integer.

This is a special case of a more general theorem for continuous $\mathbf{GL}(q, \mathbf{C})$-bundles over compact oriented differentiable manifolds.

Let $\{A_k(p_1, \ldots, p_k)\}$ be the multiplicative sequence with characteristic power series $Q(z) = \frac{2\sqrt{z}}{\sinh 2\sqrt{z}}$ defined in 1.6. The power series $\frac{\frac{1}{2}\sqrt{z}}{\sinh \frac{1}{2}\sqrt{z}}$ defines a multiplicative sequence $\{\hat{A}_j(p_1, \ldots, p_j)\}$ with $A_j = 2^{4j}\hat{A}_j$.

Throughout this paragraph we assume that X is a compact oriented differentiable manifold of dimension m with PONTRJAGIN classes p_i.

Theorem 26.1.1. *Let d be an element of $H^2(X, \mathbf{Z})$ whose reduction* mod 2 *is the* WHITNEY *class $w_2(X)$, and η a continuous $\mathbf{GL}(q, \mathbf{C})$-bundle over X. Then*

$$\hat{A}\left(X, \frac{1}{2}d, \eta\right) = \varkappa^m \left[e^{\frac{1}{2}d} \cdot \mathrm{ch}\,\eta \cdot \sum_{j=0}^{\infty} \hat{A}_j(p_1, \ldots, p_j)\right]$$

is an integer.

Remark: Since X is oriented, $w_{2i+1}(X)$ is the reduction modulo 2 of an integral STIEFEL-WHITNEY class $W_{2i+1}(X)$. The exact sequence $0 \to \mathbf{Z} \to \mathbf{Z} \to \mathbf{Z}_2 \to 0$ defines a cohomology coboundary homomorphism δ such that $\delta w_{2i}(X) = W_{2i+1}(X)$. Hence there is an element $d \in H^{2i}(X, \mathbf{Z})$ whose restriction modulo 2 is $w_{2i}(X)$ if and only if $W_{2i+1}(X) = 0$. In particular Theorem 26.1.1 can be applied only if $W_3(X) = 0$.

If m is odd then $\hat{A}\left(X, \frac{1}{2}d, \eta\right) = 0$. It is therefore sufficient to prove Theorem 26.1.1 when m is even. In 26.3—26.5 we give references to three proofs of 26.1.1. We first note two important special cases which have been proved already in 24.5.

1) Let X be an almost complex manifold with tangent $\mathbf{GL}(n, \mathbf{C})$-bundle θ and η a continuous $\mathbf{U}(q)$-bundle over X. Let $d = c_1(\theta)$ and $p_i = p_i(\varrho(\theta))$. Then equation 1.7 (12) shows that

$$\mathrm{td}\,\theta = e^{\frac{1}{2}d} \sum_{j=0}^{\infty} \hat{A}_j(p_1, \ldots, p_j)$$

and therefore $\hat{A}\left(X, \frac{1}{2}d, \eta\right) = T(X, \eta)$. Since the reduction of $c_1(\theta)$ modulo 2 is $w_2(X)$, Theorem 26.1.1 implies Theorem 24.5.4: the TODD characteristic $T(X, \eta)$ is an integer.

2) Let η be a continuous $\mathbf{U}(q)$-bundle over the $2n$-dimensional sphere $\mathbf{S}^{2n}$. The PONTRJAGIN classes $p_i(\mathbf{S}^{2n})$ are zero for $i > 0$ (see 7.2.1) and therefore $\hat{A}(\mathbf{S}^{2n}, 0, \eta) = \varkappa^{2n}[\mathrm{ch}\,\eta] = (\mathrm{ch}_n \eta)[\mathbf{S}^{2n}]$. Thus 26.1.1 implies Theorem 24.5.2: $(\mathrm{ch}_n \eta)[\mathbf{S}^{2n}]$ is an integer.

26.2. The integrality theorem (26.1.1) is itself a special case of a "non-stable" integrality theorem due to MAYER [1]. Let ξ be a $\mathbf{SO}(k)$-bundle over X with $k = 2s$ or $2s+1$ and consider a formal factorisation $p(\xi) = \prod_{i=1}^{s}(1 + y_i^2)$.

Theorem 26.2.1 (MAYER [1]). *Let d be an element of $H^2(X, \mathbf{Z})$ whose reduction* mod 2 *is $w_2(X) + w_2(\xi)$, and η a continuous* $\mathbf{GL}(q, \mathbf{C})$*-bundle over X. Then*

$$2^s \varkappa^m \left[e^{\frac{1}{2}d} \cdot \mathrm{ch}\,\eta \cdot \prod_{i=1}^{s} \cosh\left(\frac{1}{2} y_i\right) \cdot \sum_{j=0}^{\infty} \hat{A}_j(p_1, \ldots, p_j) \right]$$

is an integer.

In certain cases Theorem 26.2.1 can be improved by a factor of two (MAYER [1]). The corollaries of 26.2.1 include

I) if ξ is the zero bundle, Theorem 26.1.1;

II) if $k = m$ and ξ is the tangent bundle of X, the integrality of the L-genus (see 1.5 and 8.2);

III) if ξ is the normal bundle of an embedding or immersion of X in $\mathbf{S}^{m+k}$, the non-embedding theorems of ATIYAH-HIRZEBRUCH [2] and the non-immersion theorems of SANDERSON-SCHWARZENBERGER [1].

The proof of 26.2.1 is by an application of the ATIYAH-SINGER index theorem and is outlined in 26.3.

26.3. Let X be a compact oriented differentiable manifold of dimension $m = 2n$, and W a complex vector bundle over X associated to a $\mathbf{U}(q)$-bundle η. If $\mathbf{Spin}(2n)$ is the universal covering group of $\mathbf{SO}(2n)$ there is an exact sequence

$$1 \to \mathbf{Z}_2 \to \mathbf{Spin}(2n) \xrightarrow{\lambda} \mathbf{SO}(2n) \to 1\,.$$

The tangent bundle of X is an element ${}_{\mathbf{R}}\theta \in H^1(X, \mathbf{SO}(2n)_c)$. It can be shown that there is an exact sequence of cohomology sets with distinguished elements with coboundary map $\delta : H^1(X, \mathbf{SO}(2n)_c) \to H^2(X, \mathbf{Z}_2)$ such that $\delta({}_{\mathbf{R}}\theta) = w_2(X)$. Therefore ${}_{\mathbf{R}}\theta$ is associated to a $\mathbf{Spin}(2n)$-bundle if and only if $w_2(X) = 0$ (BOREL-HIRZEBRUCH [1], § 26.3).

Suppose that $w_2(X) = 0$. Then it is possible, using the two irreducible spinor representations of $\mathbf{Spin}(2n)$, to construct complex vector bundles W^+, W^- and an elliptic differential operator (the DIRAC operator; see PALAIS [1]) $D : \Gamma(W^+) \to \Gamma(W^-)$ such that $\gamma(D) = \hat{A}(X, 0, \eta)$. By the ATIYAH-SINGER index theorem $\gamma(D)$ is an integer. This gives the following special case of Theorem 26.1.1.

Theorem 26.3.1. *Let X be a compact oriented differentiable manifold of dimension $2n$ with $w_2(X) = 0$. Let η be a continuous* $\mathbf{U}(q)$*-bundle over X and $d \in H^2(X, \mathbf{Z})$. Then $\hat{A}(X, d, \eta)$ is an integer.*

Proof: There is a $\mathbf{U}(1)$-bundle ξ with $c_1(\xi) = d$ (see 3.8). Then $\hat{A}(X, d, \eta) = \hat{A}(X, 0, \xi \otimes \eta)$ is an integer by the above argument.

Corollary: *If $w_2(X) = 0$ then the $\hat{A}$-genus of X is an integer.*

The proofs of 26.1.1 and 26.2.1 are similar. Let $\lambda_{2n} : \mathbf{Spin}(2n+2) \to$ $\to \mathbf{SO}(2n+2)$ and $\lambda_{2n+k} : \mathbf{Spin}(2n+k+2) \to \mathbf{SO}(2n+k+2)$ be the

2-fold covering maps and put

$$\mathbf{G}_{2n} = \lambda_{2n}^{-1}(\mathbf{SO}(2n) \times \mathbf{SO}(2)), \quad \mathbf{G}_{2n,k} = \lambda_{2n+k}^{-1}(\mathbf{SO}(2n) \times \mathbf{SO}(k) \times \mathbf{SO}(2)).$$

Then $\mathbf{G}_{2n}$ is isomorphic to the complex spinor group $\mathbf{Spin}^c(2n)$ defined in Atiyah-Bott-Shapiro [1] (see also Hirzebruch, A Riemann-Roch theorem for differentiable manifolds, Séminaire Bourbaki, **11** (1958/59) and Mayer [1]). There is an exact sequence

$$1 \to \mathbf{U}(1) \to \mathbf{G}_{2n} \to \mathbf{SO}(2n) \to 1$$

and the tangent bundle ${}_{\mathbf{R}}\theta$ of X is associated to a $\mathbf{G}_{2n}$-bundle if and only if $w_2(X)$ is the reduction mod 2 of an integral class $d \in H^2(X, \mathbf{Z})$. The proof of Theorem 26.1.1 now proceeds similarly to that of 26.3.1 but using the irreducible representations of $\mathbf{G}_{2n}$. Similarly ${}_{\mathbf{R}}\theta \oplus \xi$ is associated to a $\mathbf{G}_{2n,k}$-bundle if and only if $w_2(X) + w_2(\xi)$ is the reduction mod 2 of an integral class, and the proof of Theorem 26.2.1 proceeds using the irreducible representations of $\mathbf{G}_{2n,k}$.

Alternatively, 26.1.1 and 26.2.1 can be proved by a direct application of 26.3.1 to a certain fibre bundle over X (Roberts [1]).

In certain cases Theorems 26.2.1 and 26.3.1 can be improved by a factor of two. The following theorem, due originally to Atiyah-Hirzebruch [1], generalises a theorem of Rohlin [1]. A proof using complex spinor representations and the Atiyah-Singer index theorem has been given by Mayer [1] (see also Palais [1]).

Theorem 26.3.2. *Let X be a compact oriented differentiable manifold with* $\dim X \equiv 4 \bmod 8$ *and* $w_2(X) = 0$. *Let ξ be a continuous* $\mathbf{O}(k)$*-bundle over X. Then $\hat{A}(X, 0, \psi(\xi))$ is an even integer.*

26.4. A second proof of Theorem 26.1.1 can be found in Parts II and III of Borel-Hirzebruch [1]. In this approach Theorem 26.1.1 is deduced from the integrality of the Todd genus (Theorem 24.5.4). A proof of the integrality of the Todd genus except for powers of two is given in 14.3; it depends essentially on the index theorem (8.2.2) and hence on cobordism theory. Another direct proof of the integrality of the Todd genus has been given by Milnor [3]: it involves the complete determination of the complex cobordism ring (see the bibliographical note to Chapter Three) showing that for each almost complex manifold we can find an algebraic manifold with the same Chern numbers. By R-R the Todd genus is then an integer also for almost complex manifolds.

26.5. A more direct proof of the integrality theorems is due to Atiyah-Hirzebruch [1]. As was remarked in 25.5 it is not necessary to apply the full Atiyah-Singer theorem; the method of 24.5 is sufficient.

Every m-dimensional differentiable manifold can be embedded in $\mathbf{S}^{2m}$. Theorem 24.5.2 implies that $(\mathrm{ch}_m\, b)\,[\mathbf{S}^{2m}]$ is an integer for all $b \in K(\mathbf{S}^{2m})$. Therefore 26.1.1 is a consequence of 24.5.2 and the following generalisation of 24.5.3:

Theorem 26.5.1. *Let X, Y be compact connected oriented differentiable manifolds with* $\dim Y - \dim X = 2N$ *and let $j: X \to Y$ be an embedding. Let $d \in H^2(Y, \mathbf{Z})$ be an element whose reduction* mod 2 *is $w_2(X) - j^* w_2(Y)$. Then for each element $a \in K(X)$ there exists an element $j_! a$ such that*

$$\mathrm{ch} j_!\, a \cdot \sum_{i=0}^{\infty} \hat{A}_i(p_1(Y), \ldots, p_i(Y)) = j_* \left(\mathrm{ch}\, a \cdot e^{\frac{1}{2}d} \sum_{i=0}^{\infty} \hat{A}_i(p_1(X), \ldots, p_i(X))\right) \tag{1}$$

where $j_*: H^*(X, \mathbf{Q}) \to H^*(Y, \mathbf{Q})$ is the Gysin *homomorphism.*

Let ν be the normal $\mathbf{SO}(2N)$-bundle of X in Y. The reduction mod 2 of d is $w_2(\nu)$ and (1) can be written

$$\mathrm{ch} j_!\, a = j_* \left(\mathrm{ch}\, a \cdot \left(e^{-\frac{1}{2}d} \sum_{i=0}^{\infty} \hat{A}_i(p_1(\nu), \ldots, p_i(\nu))\right)^{-1}\right). \tag{1*}$$

Let B and S be the unit disc and unit sphere bundles associated to ν and identify B with a tubular neighbourhood of X in Y. There is a map $r: Y \to B/S$ obtained by collapsing the complement of $B - S$ in Y to a point, and hence a homomorphism $r^!: K(B, S) \to K(Y)$. To construct an element $j_!\, a \in K(Y)$ which satisfies (1*) it is sufficient to construct an element $b \in K(B, S)$ such that

$$\mathrm{ch}\, b = \varphi_* \left(\left(e^{-\frac{1}{2}d} \sum_{i=0}^{\infty} \hat{A}_i(p_1(\nu), \ldots, p_i(\nu))\right)^{-1}\right)$$

where $\varphi_*: H^i(X, \mathbf{Q}) \to H^{i+2N}(Y, \mathbf{Q})$ is the Thom isomorphism (24.3). The existence of b is proved by means of the representations of $\mathbf{Spin}^c(2N)$ mentioned already in 26.3.

The same method applied to the Grothendieck ring of real vector bundles yields the original proof of Theorem 26.3.2. Theorem 26.5.1 can also be generalised to give:

Theorem 26.5.2 (Atiyah-Hirzebruch [1]). *Let X, Y be compact connected oriented differentiable manifolds with* $\dim X \equiv \dim Y$ mod 2. *Let $f: X \to Y$ be a continuous map, and $d \in H^2(X, \mathbf{Z})$ an element whose reduction* mod 2 *is $w_2(X) - f^* w_2(Y)$. Then for each element $a \in K(X)$ there exists an element $f_!\, a \in K(X)$ such that*

$$\mathrm{ch} f_!\, a \cdot \sum_{i=0}^{\infty} \hat{A}_i(p_1(Y), \ldots, p_i(Y)) = f_* \left(\mathrm{ch}\, a \cdot e^{\frac{1}{2}d} \cdot \sum_{i=0}^{\infty} \hat{A}_i(p_1(X), \ldots, p_i(X))\right)$$

where $f_*: H^*(X, \mathbf{Q}) \to H^*(Y, \mathbf{Q})$ is the Gysin *homomorphism.*

Proof: Factorise f as the composition of an embedding $X \to Y \times \mathbf{S}^{2N}$ and a product projection $Y \times \mathbf{S}^{2N} \to Y$. The theorem is true for the embedding (26.5.1) and for the projection (24.5.1). Hence it is true for f.

In the special case in which X, Y are connected almost complex manifolds and $d = c_1(X) - f^* c_1(Y)$, Theorem 26.5.2 gives the following differentiable analogue of G-R-R: *for each element $a \in K(X)$ there exists an element $f_! a \in K(Y)$ such that*

$$\operatorname{ch} f_! a \cdot \operatorname{td}(Y) = f_*(\operatorname{ch} a \cdot \operatorname{td}(X)) .$$

Bibliographical note

Where no other reference is given the material in this appendix is based either on the appendix to the second German edition or on one of the following mimeographed lecture notes: *Lectures on characteristic classes*, Princeton 1957, by J. Milnor; *Lectures on $K(X)$*, Harvard 1962, by R. Bott; *Topology seminar*, Harvard 1962, lectures by M. F. Atiyah, R. Bott and I. M. Singer; *Seminar*, Bonn 1963, lectures by F. Hirzebruch, E. Brieskorn, K. Lamotke and K. H. Mayer; *Seminar on the* Atiyah-Singer *index theorem*, Institute for Advanced Study, Princeton 1964, lectures by A. Borel, R. Palais and R. Solovay; *Lectures on K-theory*, Harvard 1965, by M. F. Atiyah; *Equivariant K-theory*, Oxford 1965, by M. F. Atiyah and G. B. Segal. An excellent survey of much of the work described in this appendix is given in a series of reviews by Bott [especially Math. Rev. **22**, 171—174 (1961); **22**, 1153—1155 (1961) and **28**, 129—130 (1964)].

The Atiyah-Singer index theorem for actions of a compact Lie group G, and the Atiyah-Bott fixed point formula, have been mentioned briefly in 25.6. Until complete published versions become available, the following temporary references will be found useful: Atiyah-Bott [3] and a lecture by Bott in the Séminaire Bourbaki, **18** (1965/66). In addition the notes *Equivariant K-theory* give the explicit construction of the topological index $\gamma(D)$ in the case that G is a torus or cyclic group, with remarks on the definition for an arbitrary compact Lie group. A survey lecture by Hirzebruch (Elliptische Differentialoperatoren auf Mannigfaltigkeiten, Weierstrass Festband, Westdeutscher Verlag, Opladen 1966) states the theorems for differentiable maps $f: X \to X$ in the two cases: *f has only simple fixed points* (fixed point formula; see 25.6) and *f has finite order* (index theorem for G cyclic). The main part of the lecture contains applications when V_n is a compact complex manifold, $f: V \to V$ is a holomorphic map, K is the canonical line bundle, and $f^{(i)}: H^i(V, K^r) \to H^i(V, K^r)$ is the homomorphism induced by f. The above theorems then give an explicit formula, in terms of the characteristic classes of V and the fixed point set of f, for the complex number

$$\chi(V, K^r, f) = \sum_{i=0}^{n} (-1)^i \operatorname{trace} f^{(i)}$$

which reduces to the Riemann-Roch theorem $\chi(V, K^r) = T(V, K^r)$ when f is the identity. An application, due jointly to Atiyah, Bott and Hirzebruch, is sketched in which M is a bounded homogeneous symmetric domain, and Δ is a group satisfying properties (a), (b) of 22.2. The formula for $\chi(V, K^r, f)$ is applied with $V = M/\Gamma$ where Γ is a subgroup of Δ given by Theorem 22.2.2. The method of 22.3 is then used to compute the dimension $\Pi_r(M, \Delta)$ of the space of automorphic forms of weight r. The results agree with those originally proved by Langlands [1], and reduce to those given in 22.3 when Δ acts freely on M.

Appendix Two

A spectral sequence for complex analytic bundles

by ARMAND BOREL

The spectral sequence to be discussed here relates the $\bar{\partial}$-cohomology of the total space, the base space and the typical fibre of a complex fibre bundle with compact connected fibres. In addition to the usual fibre- and base-degrees, it carries a bigrading stemming from the type of differential forms. The precise statement is given in 2.1, the proof in Sections 3 to 6. The latter proceeds along more or less expected lines, albeit in a rather cumbersome notation, one point of interest however being the exactness of the sequence 3.7 (4), which is essentially a consequence of smoothness properties of a GREEN operator. The main applications of 2.1 given here concern the multiplicative behaviour of the χ_y-genus (8.1) and the $\bar{\partial}$-cohomology of the CALABI-ECKMANN manifolds (9.5).

Familiarity with spectral sequences of fibre bundles is assumed. As to the rest, we follow the notation and conventions of this book, with some minor deviations to be mentioned explicitly. References to sections of this paper are in ordinary type; those to other sections of this book in boldface.

This is a revised version of a paper written in 1953, quoted in the bibliography of the first edition of this book, but not published.

§ 1. Preliminaries

1.1. Manifolds are HAUSDORFF and paracompact; smooth means differentiable of class C^∞. The sheaves on a manifold M are always $\mathbf{C}_\mathfrak{b}(M)$ modules, [where $\mathbf{C}_\mathfrak{b}(M)$ is the sheaf of germs of smooth complex valued functions on M], and tensor products of sheaves are over $\mathbf{C}_\mathfrak{b}(M)$.

1.2. Let M be a complex manifold, W a complex vector bundle over M, and $\mathfrak{W}$ the sheaf of germs of smooth sections of W. Then $A^{p,q}_M(W)$ denotes the space of smooth exterior differential forms on M, of type (p, q), with coefficients in W (see **15.4**), and $\mathfrak{A}^{p,q}_M(W)$ is the sheaf of germs of such forms. If $W = \mathbf{1}$ is the trivial bundle $M \times \mathbf{C}$, then we omit (W) in the preceding notation. We have

$$\mathfrak{A}^{p,q}_M(W) \cong \mathfrak{W} \otimes \mathfrak{A}^{p,q}_M, \quad A^{p,q}_M(W) \cong \Gamma(\mathfrak{A}^{p,q}_M(W)) . \tag{1}$$

$A^i_M(W)$ denotes the sum of the $A^{p,q}_M(W)$, where $p + q = i$, $A_M(W)$ the sum of the $A^i_M(W)$, and similarly for the corresponding sheaves.

Let U be an open subset of M over which W may be (and has been) identified with the trivial bundle $U \times \mathbf{C}_d$. We recall that $A_U^{p,q}(W|_U)$ is canonically identified to the set of d-ples of ordinary exterior differential (p, q)-forms on U. If $\omega \in A_U^{p,q}(W|_U)$ corresponds to $(\omega_1, \ldots, \omega_d)$, then $\bar\partial \omega$ corresponds to $(\bar\partial \omega_1, \ldots, \bar\partial \omega_d)$. Assume moreover that U is a coordinate neighbourhood, with local coordinates $z_1, \ldots, z_n$. For a subset $I = \{i_1, \ldots, i_k\}$ of $\{1, \ldots, n\}$, we put

$$d z_I = d z_{i_1} \wedge \cdots \wedge d z_{i_k}, \quad d \bar z_I = d \bar z_{i_1} \wedge \cdots \wedge d \bar z_{i_k}.$$

Then the above form ω_i may be written uniquely

$$\omega_i = \sum_{I,J} f_{i,I,J} \cdot d z_I \wedge d \bar z_J, \tag{2}$$

where I (resp. J) runs through the subsets of p (resp. q) elements of $\{1, \ldots, n\}$, and $f_{i,I,J}$ is a smooth complex valued function on U.

1.3. The direct sum of the spaces $H^{p,q}(M)$ [resp. $H^{p,q}(M, W)$, see **15.4**] is denoted $H_{\bar\partial}(M)$ [resp. $H_{\bar\partial}(M, W)$], and $h^{p,q}$ or $h^{p,q}(M)$ [resp. $h^{p,q}(W)$ or $h^{p,q}(M, W)$] is the dimension of $H^{p,q}(M)$ [resp. $H^{p,q}(M, W)$]. The space $H_{\bar\partial}(M)$ is in a natural way an anticommutative bigraded algebra. If $W = M \times F$ is a trivial bundle, then $H_{\bar\partial}(M, W) \cong H_{\bar\partial}(M) \otimes F$, as follows directly from the definitions.

1.4. Let now M be compact. The spaces $H^{p,q}(M, W)$ are then finite dimensional (**15.4.2**). Let further G be a Lie group operating continuously on M, by means of bi-holomorphic transformations, and let $\varphi : G \to \mathrm{Aut} M$ be the map defined by this action. Then φ induces a continuous representation φ^0 of G into $H^{p,q}(M)$. *If M is kählerian, then φ^0 is constant on each connected component of G*; in fact, in this case, $H_{\bar\partial}(M)$ may be canonically identified with the usual cohomology algebra $H^*(M, \mathbf{C})$ of M (see e. g. Weil [2], Chap. IV), by an isomorphism which clearly commutes with the natural representations of G in $H_{\bar\partial}(M)$ and $H^*(M, \mathbf{C})$; our assertion is then a consequence of the homotopy axiom. In the non-kählerian case however, this need not be true, as is shown by an example of Kodaira [cf. Gugenheim and Spencer, Proc. A. M. S. **7** (1956), 144—152].

1.5. Let $\xi = (E, B, F, \pi)$ be a complex analytic bundle (**3.2**), where E is the total space, B the base space, F the standard fibre and $\pi : E \to B$ the projection map. We assume F to be compact connected. By definition (loc. cit.) the structure group G of ξ is a complex Lie group, acting on F by means of a holomorphic map $\psi : G \times F \to F$. Let ξ be defined by means of the transition functions $f_{\alpha\beta} : U_\alpha \cap U_\beta \to G$, where $(U_\alpha)_{\alpha \in \mathscr{A}}$ is a suitable covering of B. It is clear that $\bigcup_{b \in B} H^{p,q}(F_b)$, $(F_b = \pi^{-1}(b)$, $b \in B)$, is in a natural way the total space of a smooth vector bundle over B, whose transition functions $f^0_{\alpha\beta}$ are obtained by composing the

$f_{\alpha\beta}$ with the given representation φ^0 of G in $\mathbf{GL}(H^{p,q}(F))$. This bundle is denoted $\mathbf{H}^{p,q}(F)$, and $\mathbf{H}_{\bar\partial}(F)$ is the direct sum of the $\mathbf{H}^{p,q}(F)$.

If φ^0 is constant on the connected components of G, in particular if F is kählerian, then $\mathbf{H}_{\bar\partial}(F)$ is a holomorphic complex vector bundle over B (with locally constant transition functions).

In fact, the $f^0_{\alpha\beta}$ are then locally constant functions, hence may be viewed as holomorphic maps of $U_\alpha \cap U_\beta$ into $\mathbf{GL}(H_{\bar\partial}(F))$.

§ 2. The spectral sequence

2.1. Theorem. *Let $\xi = (E, B, F, \pi)$ be a complex analytic fibre bundle, where E, B, F are connected and F is compact. Let W be a complex vector bundle on B, and $\hat{W} = \pi^* W$ its inverse image on E. Assume that every connected component of the structure group G of ξ acts trivially on $H_{\bar\partial}(F)$. Then there exists a spectral sequence (E_r, d_r), $(r \geqq 0)$, with the following properties:*

(i) *E_r is 4-graded, by the fibre-degree, the base-degree and the type. Let ${}^{p,q}E_r^{s,t}$ be the subspace of elements of E_r of type (p, q), fibre-degree s, base degree t. We have ${}^{p,q}E_r^{s,t} = 0$ if $p + q \neq s + t$, or if one of p, q, s, t is <0. The differential d_r maps ${}^{p,q}E_r^{s,t}$ into ${}^{p,q+1}E_r^{s+r,t-r+1}$.*

(ii) *If $p + q = s + t$, we have*

$$ {}^{p,q}E_2^{s,t} \cong \sum_{i \geqq 0} H^{i,s-i}(B, W \otimes \mathbf{H}^{p-i,q-s+1}(F)) \,. $$

(iii) *The spectral sequence converges to $H_{\bar\partial}(E, \hat{W})$. For all $p, q \geqq 0$, we have*

$$ \operatorname{Gr} H^{p,q}(E, \hat{W}) = \sum_{s+t=p+q} {}^{p,q}E_\infty^{s,t} \,, $$

for a suitable filtration of $H^{p,q}(E, \hat{W})$.

(iv) *If $W = \mathbf{1}$, then (E_r, d_r) consists of differential anticommutative algebras, and the isomorphism of* (iii) *is compatible with the products.*

2.2. Remarks. (1) Under our assumption on G, the bundle $\mathbf{H}_{\bar\partial}(F)$ is holomorphic so that (ii) makes sense. This condition is automatically fulfilled if F is kählerian (1.4).

(2) 2.1 (ii) shows that E_2 has a 4-grading which is finer than the one mentioned in 2.1 (i), namely the 4-grading given by the type of differential forms on B and on F. The proof will show that this 4-grading is also present in E_0, E_1.

Since ${}^{p,q}E_r^{s,t} = 0$ unless $p + q = s + t$, the superscript t is in fact redundant and it would be more correct to say that the spectral sequence is trigraded by the type (p, q) and s, where s will turn out to be the degree associated to the filtration underlying the spectral sequence. The total degree is of course $p + q$. The degree t has been added however to bring closer the analogy with the usual spectral sequence of fibre bundles, but it will be omitted in §§ 4, 5, 6.

§ 3. Auxiliary sheaves and exact sequences

3.1. Until § 6 inclusive, ξ, W, $\hat{W}$, G are as in 2.1, $\mathfrak{W}$ is the sheaf of germs of smooth sections of W, and $\mathbf{C}_\mathfrak{b}$ stands for $\mathbf{C}_\mathfrak{b}(B)$. We remark however that no assumption about the action of G on $H_{\bar\partial}(F)$ is needed before 6.1.

$\mathscr{U} = (U_\alpha)_{\alpha \in \mathscr{A}}$ is a locally finite open covering of B by coordinate neighbourhoods over which W and ξ are trivial. We let

$$\varphi_\alpha : W|_{U_\alpha} \to U_\alpha \times \mathbf{C}_m \quad \text{and} \quad \psi_\alpha : \pi^{-1}(U_\alpha) \to U_\alpha \times F\,, \quad (\alpha \in \mathscr{A})$$

be allowable trivialisations and

$$\varphi_{\alpha\beta} : U_\alpha \cap U_\beta \to \mathbf{GL}(m, \mathbf{C}) \quad \text{and} \quad \psi_{\alpha,\beta} : U_\alpha \cap U_\beta \to G\,, \quad (\alpha, \beta \in \mathscr{A}),$$

be the corresponding transition functions.

For every $z \in U_\alpha \cap U_\beta$, the map $\psi_{\alpha\beta}(z)$ induces an automorphism of A_F, to be denoted sometimes by $\psi'_{\alpha\beta}(z)$.

$(z_1^\alpha, \ldots, z_n^\alpha)$ is a set of local coordinates on U_α, and $\eta_{\alpha\beta}$ is the change of local coordinates in $U_\alpha \cap U_\beta (\alpha, \beta \in \mathscr{A})$.

3.2. Let φ be the complex tangent vector bundle along the fibres of ξ (Borel-Hirzebruch [1], § 7.4). We let $\varphi^{a,b}$ be the bundle of (a, b)-forms associated to φ. Thus $\varphi^{a,b} = (\lambda^a \varphi) \wedge (\lambda^b \bar\varphi)$, where $\bar\varphi$ is the conjugate bundle to φ. Let $\mathscr{F}^{a,b}$ be the space of smooth sections of $\varphi^{a,b}$. For every $z \in B$, the restriction x_z of an element $x \in \mathscr{F}^{a,b}$ to the fibre $F_z = \pi^{-1}(z)$ is an (a, b)-form on F_z, and thus x may be viewed as a family of (a, b)-forms on the fibres, parametrized by B, and smooth in an obvious sense. x is called a *fibre* (a, b)-*form* (on B). There is a $\mathbf{C}_\mathfrak{b}$-linear map $\bar\partial_F : \mathscr{F}^{a,b} \to \mathscr{F}^{a,b+1}$ characterised by $r_z(\bar\partial_F x) = \bar\partial(r_z x)$, $(z \in B, x \in \mathscr{F}^{p,q})$ (for all this, see Kodaira-Spencer [5], I, § 2).

Let $\mathfrak{F}^{a,b}$ be the sheaf of germs on B of fibre (a, b)-forms. We have $\Gamma(\mathfrak{F}^{a,b}) = \mathscr{F}^{a,b}$, and $\bar\partial_F$ is the map of sections induced by a homomorphism of $\mathbf{C}_\mathfrak{b}$-modules of $\mathfrak{F}^{a,b}$ into $\mathfrak{F}^{a,b+1}$, also denoted by $\bar\partial_F$. Let $\mathfrak{Z}^{a,b} \subset \mathfrak{F}^{a,b}$ be its kernel. By definition, the *sequence*

$$0 \to \mathfrak{Z}^{a,b} \xrightarrow{i} \mathfrak{F}^{a,b} \xrightarrow{\bar\partial_F} \bar\partial_F(\mathfrak{F}^{a,b}) \to 0\,, \tag{1}$$

where i is the inclusion map, *is exact*.

3.3. More generally we shall consider the fibre $\hat{W}$-(a, b)-forms on B. They may be defined first as the smooth sections of $W \otimes \varphi^{a,b}$ (see Kodaira-Spencer [5], I, § 2); for this, $\hat{W}$ could of course be any complex vector bundle on E. If x is such a form, then $r_z(x)$ is a (a, b)-form on F_z, with coefficients in the trivial bundle $V_z \times F_z$, where V_z is the fibre over z of W. Clearly, we may identify these forms with the sections of the sheaf $\mathfrak{W} \otimes \mathfrak{F}^{a,b}$.

3.4. Let

$$M^{a,b,c,d} = \Gamma(\mathfrak{W} \otimes \mathfrak{F}^{a,b} \otimes \mathfrak{A}_B^{(c,d)}), \quad (a, b, c, d \in \mathbf{Z};\ a, b, c, d \geqq 0).$$

Its elements are to be thought of as "(c, d)-forms on B with coefficients in the fibre $\hat{W}$-(a, b)-forms". In the notation of 3.1, an element $h \in M^{a,b,c,d}$ is given by its restrictions h_α to the open subsets U_α, and h_α is an array of differential m-forms $h_{\alpha,i}$ which may be written

$$h_{\alpha,i} = \sum_{I,J} h_{\alpha,i,I,J}\, dz_I^\alpha \wedge d\bar{z}_J^\alpha,$$

where I and J run respectively through the subsets of c and d elements of $\{1, \ldots, n\}$, and where $h_{\alpha,i,I,J} \in \mathscr{F}^{a,b}(U_\alpha)$, is a fibre (a, b)-form on U_α.

Thus h_α is identified with a $\hat{W}$-$(a + c,\ b + d)$ differential form on $\pi^{-1}(U_\alpha)$. But of course, this identification depends essentially on the local trivialisations, and h itself cannot be viewed as a differential form. To be more precise, h_α and h_β are related by the transformations defined by $\varphi_{\alpha\beta}$, $\psi_{\alpha\beta}$, and $\eta_{\alpha\beta}$. However, if we want to describe the differential form h_α in $U_\beta \cap U_\alpha$ by means of the local coordinates (z_i^β) and the local trivialisations over U_β, then we have also to take into account the derivatives of the $\psi_{\alpha\beta}$ with respect to z. This implies that in these new coordinates, h_α will be equal to the sum of h_β and of differential forms of base-degree $> c + d$.

3.5. Although this is not needed in the sequel, we remark here, without going into details, that if we allow vector bundles to have infinite dimensional fibres, we may also view the elements of $M^{a,b,c,d}$ as (c, d)-forms on B with coefficients in a vector bundle.

In fact, $A_F^{a,b}$ is a FRECHET space in a natural way (SERRE [3]), and any automorphism of F induces a homeomorphism of $A_F^{a,b}$. Thus the transition functions $\psi'_{\alpha,\beta}: U_\alpha \cap U_\beta \to \operatorname{Aut} A_F^{a,b}$ allow one to define over B an associated bundle $\mu^{a,b}$, with standard fibre $A_F^{a,b}$. Furthermore, the transition functions are smooth in the sense that if $\varrho: U_\alpha \cap U_\beta \to A_F^{a,b}$ is smooth, then $\psi_{\alpha\beta} \circ \varrho$ is also smooth. Thus it makes sense to speak of the smooth sections of $\mu^{a,b}$. It may then be seen that the elements of $M^{a,b,c,d}$ are just the (c, d)-forms on B, with coefficients in $W \otimes \mu^{a,b}$.

3.6. The sheaf $\mathfrak{W} \otimes \mathfrak{A}_B^{c,d}$ is locally free over $\mathbf{C}_\mathfrak{b}$, therefore the sequence

$$0 \to \mathfrak{W} \otimes \mathfrak{Z}^{a,b} \otimes \mathfrak{A}_B^{c,d} \to \mathfrak{W} \otimes \mathfrak{F}^{a,b} \otimes \mathfrak{A}_B^{c,d} \to \mathfrak{W} \otimes \bar{\partial}_F(\mathfrak{F}^{a,b}) \otimes \mathfrak{A}_B^{c,d} \to 0, \quad (2)$$

obtained by tensoring 3.2 (1) by $\mathfrak{W} \otimes \mathfrak{A}_B^{c,d}$ is also exact. Moreover, since $\mathfrak{A}_B^{c,d}$ is fine (3.5), the sequence

$$\begin{aligned} 0 \to \Gamma(\mathfrak{W} \otimes \mathfrak{Z}^{a,b} \otimes \mathfrak{A}_B^{(c,d)}) &\to \Gamma(\mathfrak{W} \otimes \mathfrak{F}^{a,b} \otimes \mathfrak{A}_B^{(c,d)}) \to \\ &\to \Gamma(\mathfrak{W} \otimes \bar{\partial}_F(\mathfrak{F}^{a,b}) \otimes \mathfrak{A}_B^{(c,d)}) \to 0, \end{aligned} \quad (3)$$

derived from (2), is exact (**2.10.1, 2.11.1**).

3.7. Let σ be the map which sends a $\bar{\partial}$-closed form on F into its $\bar{\partial}$-cohomology class. This map defines a $\mathbf{C}_b$-homomorphism, also to be denoted σ, of $\mathfrak{Z}^{a,b}$ into the sheaf $\mathfrak{H}^{a,b}(F)$ of germs of smooth sections of the bundle $\mathbf{H}^{a,b}(F)$ defined in 1.5. We claim that *the sequence*

$$0 \to \bar{\partial}_F(\mathfrak{F}^{a,b-1}) \xrightarrow{i} \mathfrak{Z}^{a,b} \xrightarrow{\sigma} \mathfrak{H}^{a,b}(F) \to 0\,, \quad (a \geqq 0, b \geqq 1)\,, \tag{4}$$

is exact.

That $\sigma \circ i = 0$ is clear. Furthermore, since $H^{a,b}(F)$ is finite dimensional, it is readily seen that σ is surjective. There remains to prove that $\operatorname{im} i \supset \ker \sigma$. This amounts to the following assertion:

Let $z \in B$ and U an open neighbourhood of z in B, ω a fibre (a, b)-form over U, *i. e.* a map assigning to $x \in U$ an (a, b)-form $\omega(x)$ on F depending smoothly on x. Assume that for each x, there exists an $(a, b-1)$-form ν_x on F such that $\omega(x) = \bar{\partial}\nu_x$. Then there exists a neighbourhood V of z in U and a fibre $(a, b-1)$-form τ on V such that $\omega(x) = \bar{\partial}\tau(x)$ for all $x \in V$.

In other words, we may choose ν_x so as to depend smoothly on x. This assertion is contained in Theorems 7, 8 of Kodaira-Spencer [7].

3.8. In the same way as the exactness of (3) was deduced from that of (1), it follows from 3.7 that the sequence

$$\begin{aligned} 0 &\to \Gamma(\mathfrak{W} \otimes \bar{\partial}_F(\mathfrak{F}^{a,b-1}) \otimes \mathfrak{A}_B^{c,d}) \to \\ &\to \Gamma(\mathfrak{W} \otimes \mathfrak{Z}^{a,b} \otimes \mathfrak{A}_B^{c,d}) \xrightarrow{\sigma} \Gamma(\mathfrak{W} \otimes \mathfrak{H}^{a,b}(F) \otimes \mathfrak{A}_B^{c,d}) \to 0 \end{aligned} \tag{5}$$

is exact. On the other hand, there is a natural isomorphism

$$\mathfrak{W} \otimes \mathfrak{H}^{a,b}(F) = \mathfrak{S}(W \otimes \mathbf{H}^{a,b}(F))\,, \tag{6}$$

where $\mathfrak{S}(W \otimes \mathbf{H}^{a,b}(F))$ is the sheaf of germs of smooth sections of the tensor product bundle $W \otimes \mathbf{H}^{a,b}(F)$. Therefore, we also have

$$\Gamma(\mathfrak{W} \otimes \mathfrak{H}^{a,b}(F) \otimes \mathfrak{A}_B^{c,d}) \cong A_B^{c,d}(W \otimes \mathbf{H}^{a,b}(F)) \cong \Gamma\big(\mathfrak{S}(W \otimes \mathbf{H}^{a,b}(F)) \otimes \mathfrak{A}_B^{c,d}\big). \tag{7}$$

§ 4. The filtration. Proof of 2.1 (i), (iii), (iv)

4.1. Let us say that an open subset $U \subset E$ is *small* if ξ and W are trivial over $\pi(U)$ and if an allowable trivialisation of ξ over $\pi(U)$ carries U into the product of coordinate neighbourhoods of B and F. For every small open set U and positive integer k, let $L_k(U)$ be the set of elements of $A_U(\hat{W}|_U)$ which, when expressed in terms of local coordinates (z_i) on B and (y_j) on F, are sums of monomials $dz_I \wedge d\bar{z}_J \wedge dy_{I'} \wedge d\bar{y}_{J'}$, in which $|I| + |J| \geqq k$, where $|A|$ denotes the number of elements in a finite set A. It is clear that $L_k(U)$ is invariant under change of coordinates (but the set of elements for which $|I| + |J| = k$ is not, and consequently the filtration introduced below is not associated to a grading). Let

$$L_k = \{\omega \in A_E(\hat{W})\,;\ \omega|_U \in L_k(U) \text{ for every small open subset } U \text{ of } E\}. \quad (1)$$

It is of course enough to check this condition when U runs through the elements of one open covering of E by small sets. We have

$$L_0 = A_E(\hat{W}),\ L_k = 0\,(k > \dim_{\mathbf{R}} B)\,;\ L_k \supset L_{k+1},\ \bar{\partial}(L_k) \subset L_k\ \ (k \geqq 0)\,, \quad (2)$$

which shows that the L_k define a bounded decreasing filtration of the differential $\mathbf{C}$-module $(A_E(\hat{W}), \bar{\partial})$ by submodules stable under $\bar{\partial}$. The corresponding spectral sequence is by definition the spectral sequence (E_r, d_r) of 2.1. We have clearly

$$L_k = \sum_{p,q} {}^{p,q}L_k\,, \quad ({}^{p,q}L_k = L_k \cap A_E^{p,q}(\hat{W}))\,,$$

which means that the filtration is compatible with the bigrading provided by the type, hence also with the total degree. Moreover, $\bar{\partial}$ is homogeneous of degree 1 in q, of degree 0 in p, hence this bigrading is also present in the spectral sequence. We denote by ${}^{p,q}E_r^{s,t}$ or ${}^{p,q}E_r^s$ [see 2.2 (2)] the space of elements of E_r of type (p, q), total degree $s + t$, and degree s in the grading defined by the filtration. As is usual, s and t will be called respectively base-degree and fibre-degree. Of course, ${}^{p,q}E_r^{s,t} = 0$ if $p + q \neq s + t$.

The assertions 2.1 (i), (iii) then follow from standard general facts about convergent spectral sequences of filtered-graded differential modules.

If now $\hat{W} = \mathbf{1}$ is the trivial bundle with fibre $\mathbf{C}$, then $A_E(\hat{W})$ is an anticommutative differential algebra. Again from general principles, this product shows up in the spectral sequence, and we have 2.1 (iv). There remains to prove 2.1 (ii).

4.2. We give here a slight reformulation of the definition of the filtration which will be useful below.

Let $V_\alpha = \pi^{-1}(U_\alpha)$, and identify V_α to $U_\alpha \times F$ by means of ψ_α. We denote by $M_\alpha^{a,b,c,d}$ the space of W-(c, d)-forms on B with coefficients in the (a, b)-forms of the fibre (DE RHAM [1], Chap. II, § 7). Using ψ_α, we see that

$$M_\alpha^{a,b,c,d} \cong \Gamma_{U_\alpha}(\mathfrak{W} \otimes \mathfrak{F}^{a,b} \otimes \mathfrak{A}_B^{(c,d)})\,, \quad (3)$$

$$A_{V_\alpha}(\hat{W}|_{U_\alpha}) = \sum_{a,b,c,d} M_\alpha^{a,b,c,d}\,. \quad (4)$$

Then

$$L_s = \{\omega \in A_E(\hat{W})\,;\ \omega|_{V_\alpha} \in L_{s,\alpha}(\alpha \in \mathscr{A})\}\,. \quad (5)$$

where

$$L_{s,\alpha} = \sum_{c+d \geqq s} M_\alpha^{a,b,c,d}\,. \quad (6)$$

We remark that the isomorphism (3) and the direct sum decomposition (4) depend on the trivialisation ψ_α but, as before, the condition $\omega|_{V_\alpha} \in L_{s,\alpha}$ does not.

§ 5. The terms E_0, E_1

5.1. Lemma. *There exists a canonical isomorphism*

$$^{p,q}k_0^s : {}^{p,q}E_0^s \xrightarrow{\sim} \sum_i \Gamma(\mathfrak{W} \otimes \mathfrak{F}^{p-i,q-s+i} \otimes \mathfrak{A}_B^{i,s-i}) \quad (p, q, s \geqq 0) .$$

The sum k_0 of the maps $^{p,q}k_0^s$ carries d_0 onto $\bar{\partial}_F$.

We keep the previous notation. Let $\omega \in {}^{p,q}L_s$ and ω_α be its restriction to $\pi^{-1}(U_\alpha)$ $(\alpha \in \mathscr{A})$. We may write (4.2):

$$\omega_\alpha = \sum_i \omega_\alpha^{p-i,q-s+i,i,s-i} \bmod L_{s+1,\alpha} , \tag{1}$$

where

$$\omega_\alpha^{a,b,c,d} \in M_\alpha^{a,b,c,d} = \Gamma_{U_\alpha}(\mathfrak{W} \otimes \mathfrak{F}^{a,b} \otimes \mathfrak{A}_B^{c,d}) . \tag{2}$$

We claim that, for each i, the forms $\omega_\alpha^{p-i,q-s+i,i,s-i}$ $(\alpha \in \mathscr{A})$ match so as to define a section $\omega^{p-i,q-s+i,i,s-i}$ of $\mathfrak{W} \otimes \mathfrak{F}^{p-i,q-s+i} \otimes \mathfrak{A}_B^{i,s-i}$. In fact, let $\alpha, \beta \in \mathscr{A}$ be such that $U_\alpha \cap U_\beta \neq \emptyset$. The elements ω_α and ω_β represent the same differential form on $U_\alpha \cap U_\beta$, hence are related by a transformation $f_{\alpha\beta}$ associated to the coordinate transformations $\psi_{\alpha,\beta}$, $\varphi_{\alpha,\beta}$, $\eta_{\alpha,\beta}$. Now $f_{\alpha\beta}$ also involves the derivatives of the $\psi_{\alpha,\beta}$ with respect to the local coordinates on B. However, as was already pointed out (3.4), each term in which such a derivative occurs has a strictly bigger total base-degree, hence belongs to $L_{s+1,\alpha}$. Thus to go from $\omega_\alpha^{p-i,q-s+i,i,s-i}$ to $\omega_\beta^{p-i,q-s+i,i,s-i}$, one may neglect these derivatives and just apply the transformation defined by $\psi_{\alpha,\beta}$, $\varphi_{\alpha,\beta}$, $\eta_{\alpha,\beta}$; but this is precisely how sections of the sheaf $\mathfrak{W} \otimes \mathfrak{F}^{p-i,q-s+i} \otimes \mathfrak{A}_B^{i,s-i}$ over U_α and U_β have to match in order to define a section over $U_\alpha \cup U_\beta$.

We now associate to ω the sum of the $\omega^{p-i,q-s+i,i,s-i}$. This defines a map

$$^{p,q}k^s : {}^{p,q}L_s \to \sum_i \Gamma(\mathfrak{W} \otimes \mathfrak{F}^{p-i,q-s+i} \otimes \mathfrak{A}_B^{i,s-i}) ,$$

which is obviously linear, with kernel $^{p,q}L_{s+1}$, whence an injective linear map

$$^{p,q}k_0^s : {}^{p,q}E_0^s \cong {}^{p,q}L_s/{}^{p,q}L_{s+1} \to \sum_i \Gamma(\mathfrak{W} \otimes \mathfrak{F}^{p-i,q-s+i} \otimes \mathfrak{A}_B^{i,s-i}) .$$

To compute $d_0(\bar{\omega})$, $(\bar{\omega} \in {}^{p,q}E_0^s)$, we have to apply $\bar{\partial}$ to a representative ω of $\bar{\omega}$ in L_s, and reduce $\bmod L_{s+1}$. In local coordinates, this means that we may disregard differentiation with respect to local coordinates on B, and take into account only the coordinates on the fibres. But this is how $\bar{\partial}_F$ is defined, whence

$$^{p,q}k_0^s(d_0\,\bar{\omega}) = \bar{\partial}_F({}^{p,q}k_0^s(\bar{\omega})) .$$

There remains to show that ${}^{p,q}k_0^s$ is surjective. Let $u \in M^{a,b,c,d}$. Put $p = a + c$, $q = b + d$, $s = c + d$. We have to find $\omega \in {}^{p,q}L_s$ such that ${}^{p,q}k^s(\omega) = u$.

There exists a countable locally finite covering $\mathscr{V} = (V_j)$ $(j = 1, 2, \ldots)$ of E by small open subsets (see 4.1) such that for each j there exists $\alpha = \alpha(j) \in \mathscr{A}$ for which $\pi(V_j) \subset U_\alpha$. Since E is paracompact, we may further find a sequence of open coverings $\mathscr{V}^{(l)} = (V_j^{(l)})$ $(l = 1, 2, \ldots)$ such that

$$V_j^{(1)} = V_j, \ \overline{V}_j^{(l)} \subset V_j^{(l-1)} \quad (j, l \geqq 1) .$$

Let us put

$$V_j^{(\infty)} = \bigcap_{l \geqq 1} V_j^{(l)} \quad (j \geqq 1) .$$

Since $\mathscr{V}$ is locally finite, it is clear that the $V_j^{(\infty)}$ also form a covering of E (not necessarily open of course). Therefore, if (n_j) is a sequence of strictly positive integers, the union of the $V_j^{(n_j)}$ is also an open covering. The form ω will be defined by means of its restrictions to the elements of such a covering.

In each V_j we choose local coordinates once and for all. The restriction u_j of u to V_j may then be identified with a differential form, with coefficients in the typical fibre of W, also denoted u_j. By definition $\omega^{(1)} = u_1$ on V_1. If $V_1 \cap V_2^{(2)} = \emptyset$, we put $\omega^{(2)} = u_1$ on V_1, $\omega^{(2)} = u_2$ on $V_2^{(2)}$. Suppose now $V_1 \cap V_2^{(2)} \neq \emptyset$. In that intersection, we have $\omega^{(1)} = u_2 + \sigma$, where σ is a form whose base-degree (i. e. degree in the differentials of local coordinates on B) is $> c + d$. We can find a form τ on $V_2^{(1)}$ which coincides with σ on $\overline{V}_1^{(2)} \cap \overline{V}_2^{(2)}$ (this is a trivial extension problem, since σ is already defined in an open neighbourhood of $V_1^{(2)} \cap V_2^{(2)}$). We then let $\omega^{(2)}$ be the differential form on $V_1^{(2)} \cup V_2^{(2)}$ which is equal to u_1 on $V_1^{(2)}$, to $u_2 + \tau$ on $V_2^{(2)}$.

Let now $l \geqq 2$. Assume that there is a sequence of l strictly positive integers $n_{j,l}$ $(j = 1, \ldots, l)$ and a differential form $\omega^{(l)}$ defined on

$$V_{(l)} = \bigcup_{1 \leqq j \leqq l} V_j^{(n_{j,l})} ,$$

such that

$$(\omega^{(l)} - u_j)\big|_{V_j^{(n_{j,l})}} \in L_{s+1}(V_j^{(n_{j,l})}) \quad (1 \leqq j \leqq l) . \tag{3}$$

Let now I be the set of integers j between 1 and l for which

$$V_{l+1} \cap V_j^{(n_{j,l})} \neq \emptyset .$$

In the intersection

$$V_{l+1} \cap V_{(l)} = V_{l+1} \cap \Big(\bigcup_{j \in I} V_j^{(n_{j,l})}\Big),$$

the difference $\sigma = \omega^{(l)} - u_{l+1}$ belongs to L_{s+1}. As before, we may find a form τ on V_{l+1} which coincides with σ on

$$\overline{V}_{l+1}^{(2)} \cap \Big(\bigcup_{j \in I} \overline{V}_j^{(n_{j,l}+1)}\Big).$$

Let us define a sequence $(n_{j,l+1})$ of $l+1$ integers by

$$n_{j,l+1} = n_{j,l} + 1,\ (j \in I);\ n_{j,l+1} = n_{j,l},\ (1 \leqq j \leqq l, j \notin I);\ n_{l+1,l+1} = 2\,.$$

We let then $\omega^{(l+1)}$ be the form on

$$V_{(l+1)} = \bigcup_{1 \leqq j \leqq l+1} V_j^{(n_{j,l+1})}\,,$$

which is equal to $\omega^{(l)}$ on $V_j^{(n_{j,l+1})}$ for $j \leqq l$ and to $u_{l+1} + \tau$ on $V_{l+1}^{(2)}$. Then it satisfies the condition (3) with l replaced by $l+1$.

In order to go from the domain of definition of $\omega^{(l)}$ to that of $\omega^{(l+1)}$ we may have to shrink some of the $V_j^{(n_{j,l})}$, but not if $V_j \cap V_{l+1} = \emptyset$. Our covering being locally finite, given $m \geqq 1$, there exists $l(m)$ such that $V_m \cap V_l = \emptyset$ for all $l \geqq l(m)$. As a consequence, for fixed j, the sequence $n_{j,l}$ becomes stationary and there exists an integer n_j such that $V_j^{(n_j)}$ belongs to the domain of definition of $\omega^{(l)}$ for all $l \geqq 1$. By construction, we have then $\omega^{(l)} = \omega^{(l')}$ on $V_j^{(n_j)}$ for $l, l' \geqq n_j$. There exists therefore a differential form ω on E such that $\omega = \omega^{(n_j)}$ on $V_n^{(n_j)}$ for all j. It follows then from (3) that ${}^{p,q}k^s(\omega) = u$.

5.2. Lemma. *The map ${}^{p,q}k_0^s$ of 5.1 induces an isomorphism ${}^{p,q}k_1^s$ of ${}^{p,q}E_1^s$ onto $\sum_i A_B^{i,s-i}(W \otimes \mathbf{H}^{p-i,q-s+i}(F))$.*

This follows from 5.1, from the exactness of the sequences (3), (5) of § 3, and from the isomorphisms 3.8 (7).

§ 6. The term E_2. Proof of 2.1 (ii)

We let k_0 (resp. k_1) be the direct sum of the maps ${}^{p,q}k_0^s$ (resp. ${}^{p,q}k_1^s$). In view of our assumption on the structure group of ξ, the image space of k_1 is the space of forms on B with coefficients in a holomorphic vector bundle (1.5), therefore it is a differential module under $\bar\partial$. The assertion 2.1 (ii) will then be a consequence of the

6.1. Lemma. *The map k_1 carries d_1 onto $\bar\partial$. For all p, q, s, it induces an isomorphism*

$${}^{p,q}k_2^s : {}^{p,q}E_2^s \xrightarrow{\sim} \sum_i H^{i,s-i}(B, W \otimes \mathbf{H}^{p-i,q-s+i}(F))\,.$$

The second assertion follows directly from the first one, which we prove now.

Let $\varkappa_1^0$ be the map of the space $Z(E_0)$ of d_0-cocycles of E_0 onto E_1. In view of 5.1, 5.2, we have the following commutative diagram

$$\begin{array}{ccc} \sum\limits_{c+d \geqq s} \Gamma(\mathfrak{W} \otimes \mathfrak{Z} \otimes \mathfrak{A}_B^{c,d}) & \xrightarrow{\sigma} & \sum\limits_{c+d \geqq s} \Gamma(\mathfrak{W} \otimes \overline{\mathfrak{H}}(F) \otimes \mathfrak{A}_B^{c,d}) \\ \uparrow{\scriptstyle k_0} & & \uparrow{\scriptstyle k_1} \\ Z(E_0^s) & \xrightarrow{\varkappa_1^0} & E_1^s \end{array} \tag{1}$$

where

$$\mathfrak{Z} = \sum \mathfrak{Z}^{a,b}, \quad E_i^s = \sum_{p,q} {}^{p,q}E_i^s \quad (i = 0, 1, \ldots), \tag{2}$$

and σ is as in 3.8 (5). We denote μ_u the projection of L_u onto $E_0^u = L_u/L_{u+1}$ $(u = 0, 1, \ldots)$.

Let a, b, c, d be positive integers, and set $p = a + c$, $q = b + d$, $s = c + d$. Let $u \in \Gamma(\mathfrak{W} \otimes \mathfrak{H}^{a,b}(F) \otimes \mathfrak{A}_B^{c,d})$ and u' an element of $\Gamma(\mathfrak{W} \otimes \mathfrak{Z}^{a,b} \otimes \mathfrak{A}_B^{c,d})$ such that $\sigma(u') = u$, which exists by 3.8. Let $v = k_1^{-1}(u)$ and $v' = k_0^{-1}(u')$. We have to prove:

$$k_1(d_1 v) = \bar{\partial} u . \tag{3}$$

By definition, $v' \in Z(E_0^s)$. There exists therefore $v'' \in L_s$ such that $\bar{\partial}(v'') \in L_{s+1}$ and $\mu_s(v'') = v'$. By the above, we have then

$$u = k_1 \cdot \varkappa_1^0 \cdot \mu_s(v'') = \sigma \cdot k_0 \cdot \mu_s(v'') . \tag{4}$$

On the other hand, the definition of d_1 gives $d_1 v = \varkappa_1^0 \cdot \mu_{s+1}(\bar{\partial} v'')$ hence, also, by (1),

$$k_1(d_1 v) = \sigma \cdot k_0 \cdot \mu_{s+1}(\bar{\partial} v'') . \tag{5}$$

Therefore, (3) is equivalent to

$$\sigma \cdot k_0 \cdot \mu_{s+1}(\bar{\partial} v'') = \bar{\partial} u . \tag{6}$$

It is enough to prove this for the restriction of v'' to $\mu^{-1}(U_\alpha)$, for all $\alpha \in \mathscr{A}$. We may write (4.2):

$$v'' = v^{a,b,c,d} + v^{a-1,b,c+1,d} + v^{a,b-1,c,d+1} \bmod L_{s+2,\alpha} ,$$

where $v^{e,f,j,k} \in \Gamma_{U_\alpha}(\mathfrak{W} \otimes \mathfrak{F}^{e,f} \otimes \mathfrak{A}_B^{j,k})$; by construction, $v^{a,b,c,d}$ may be identified with u'. We have then

$$\bar{\partial} v'' = \bar{\partial} u' + \bar{\partial}(v^{a-1,b,c+1,d} + v^{a,b-1,c,d+1}) \bmod L_{s+2,\alpha} .$$

Since we compute $\bmod L_{s+2,\alpha}$, we may neglect all terms of base degree $> c + d + 1$; this means that we also have:

$$\bar{\partial} v'' = \bar{\partial} u' + \bar{\partial}_F(v^{a-1,b,c+1,d} + v^{a,b-1,c,d+1}) \bmod L_{s+2,\alpha} ,$$

$$k_0 \mu_{s+1} \bar{\partial}(v'') = \bar{\partial} u' + \bar{\partial}_F(v^{a-1,b,c+1,d} + v^{a,b-1,c,d+1}) .$$

The second term on the right hand side, being a $\bar{\partial}_F$-coboundary, is annihilated by σ, hence

$$\sigma k_0 \mu_{s+1} \bar{\partial}(v'') = \sigma \bar{\partial} u' .$$

But it is clear that

$$\sigma \cdot \bar{\partial}(u') = \bar{\partial}(\sigma(u')) = \bar{\partial} u ,$$

whence the equality (6).

Remark. A similar proof yields a construction by means of differential forms of the spectral sequence in real cohomology of a differentiable fibre bundle. The differential algebra is the space of real valued differential forms on E, filtered by the degree in base coordinates, as in 4.1. The proof is basically the same, simpler in notation since we dispense with W and the type. If F is compact, the exactness of the sequence corresponding to 3.7 (4) follows again from the smoothness properties of the GREEN operator (DE RHAM [1], p. 157). In the general case, it is a consequence of a result of VAN EST [Proc. Konikl. Neder. Ak. van Wet. Series A, **61** (1958), 399–413, Cor. 1 to Thm. 1].

§ 7. Elementary properties and applications of the spectral sequence

We keep the notation and assumptions of 2.1.

7.1. *If the bundle* $\mathbf{H}_{\bar\partial}(F)$ *is trivial, in particular if the structure group of* ξ *is connected, then*

$$ {}^{p,q}E_2^{s,t} \cong \sum_i H^{i,s-i}(B, W) \otimes H^{p-i,q-s+i}(F) . $$

This follows from 2.1 (ii) and 1.3.

7.2. *The space* ${}^{p,q}E_r^{p+q,0}$ *is a quotient of* ${}^{p,q}E_{r-1}^{p+q,0}$ $(r \geqq 3)$. *The composition of the natural maps*

$$ H^{p,q}(B, W) \cong {}^{p,q}E_2^{p+q,0} \to {}^{p,q}E_\infty^{p+q,0} \subset H^{p,q}(E, \hat W) , $$

is π^*. *It is injective if* $q = 0$.

The first assertion follows in the usual way from the construction of the spectral sequence and from standard facts about "edge homomorphisms". Since no element of type $(p, 0)$ can be a d_r-coboundary $(r \geqq 0)$, the second one is then obvious.

7.3. By our assumption on G, the bundle $\mathbf{H}_{\bar\partial}(F)$ has the discrete structure group G/G^0, where G^0 is the identity component of G. There is then, in the usual manner, a homomorphism of the fundamental group $\pi_1(B)$ of B into $\operatorname{Aut} H_{\bar\partial}(F)$, and $\mathbf{H}_{\bar\partial}(F)$ may be viewed as a local system of coefficients. From this it is easily seen that if B is compact, then $H^{0,0}(B, \mathbf{H}^{p,q}(F))$ is isomorphic to the space $H^{p,q}(F)^\pi$ of fixed points of $\pi_1(B)$ under the above action. Thus

$$ {}^{p,q}E_2^{0,p+q} \cong H^{p,q}(F)^\pi . $$

7.4. *The space* ${}^{p,q}E_r^{0,p+q}$ *may be identified with the space of* d_{r-1}-*cocycles of* ${}^{p,q}E_{r-1}^{0,p+q}$ $(r \geqq 3)$. *If* $W = \mathbf{1}$ *and* B *is compact, the composition of the natural maps*

$$ H^{p,q}(E) \to {}^{p,q}E_\infty^{0,p+q} \subset {}^{p,q}E_2^{0,p+q} = H^{p,q}(F)^\pi \subset H^{p,q}(F) , $$

is induced by the homomorphism associated to the inclusion map of a fibre.

This follows again from elementary facts about spectral sequences.

7.5. *If the structure group of ξ is connected, then*

$$h^{p,q}(E, \hat{W}) \leqq \sum_{\substack{a+c=p\\ b+d=q}} h^{c,d}(B, W) \cdot h^{a,b}(F) .$$

This is a consequence of **7.1** and of the relations

$$h^{p,q}(E, \hat{W}) = \dim {}^{p,q}E_\infty \leqq \dim {}^{p,q}E_2 ,$$

where we have put

$${}^{p,q}E_r = \sum_{s,t \geqq 0} {}^{p,q}E_r^{s,t} .$$

7.6. Finally we note that *if G is connected, and if $i^* : H_{\bar\partial}(E) \to H_{\bar\partial}(F)$ is surjective, then $H_{\bar\partial}(E)$ is additively isomorphic to $H_{\bar\partial}(B) \otimes H_{\bar\partial}(F)$.*

In fact, E_2 may then be identified as an algebra with the tensor product of the algebras $H_{\bar\partial}(F) \otimes 1$ and $1 \otimes H_{\bar\partial}(B)$, which consist of permanent cocycles.

§ 8. The multiplicative property of the χ_y-genus

8.1. Theorem. *Let $\xi = (E, B, F, \pi)$ be a complex analytic fibre bundle with connected structure group, where E, B, F are compact, connected, and F is kählerian. Let W be a complex analytic vector bundle on B. Then $\chi_y(E, \pi^* W) = \chi_y(B, W) \cdot \chi_y(F)$.*

For the notation χ_y and χ^p, see **15.5.** Since G is connected and F is kählerian, G acts trivially on the $\bar\partial$-cohomology of the standard fibre (1.4), therefore we may apply 2.1; we have moreover (7.1):

$$E_2 \cong H_{\bar\partial}(B, W) \otimes H_{\bar\partial}(F) . \tag{1}$$

Let us put, in the notation of 7.4,

$$\chi^p(E_r) = \sum_q (-1)^q \dim {}^{p,q}E_r ,$$

$$\chi_y(E_r) = \sum_p \chi^p(E_r) \cdot y^p .$$

It follows from 2.1 (iii) that

$$\chi_y(E, \pi^* W) = \chi_y(E_\infty) . \tag{2}$$

A simple calculation, using (1), yields

$$\chi_y(E_2) = \chi_y(B, W) \cdot \chi_y(F) . \tag{3}$$

Let ${}^{(p)}E_r = \sum_q {}^{p,q}E_r$ $(r \geqq 2)$. This is a graded space, whose Euler characteristic $\chi({}^{(p)}E_r)$ is equal to $\chi^p(E_r)$; it is stable under d_r, and its derived group is ${}^{(p)}E_{r+1}$. By a well-known and elementary fact, we have then $\chi({}^{(p)}E_r) = \chi({}^{(p)}E_{r+1})$, hence $\chi^p(E_r) = \chi^p(E_{r+1})$, $r \geqq 2$, which, together with (2) and (3), ends the proof.

§ 9. The $\bar{\partial}$-cohomology of the CALABI-ECKMANN manifolds

9.1. We shall denote by $A^0(X)$ the identity component of the group $A(X)$ of complex analytic homeomorphisms of a compact connected complex manifold X. Although this is not really needed below, we recall that, by a well-known result of BOCHNER-MONTGOMERY, $A(X)$ is a complex LIE group. If X is the total space of a complex analytic fibering (X, Y, F, π), then every element of $A^0(X)$ commutes with π, whence a natural homomorphism $\pi^0: A^0(X) \to A^0(Y)$ (BLANCHARD [3], Prop. I. 1, p. 160). In particular, if M and N are connected complex analytic compact manifolds, then $A^0(M \times N) = A^0(M) \times A^0(N)$ (BLANCHARD [3], p. 161).

9.2. We let $\mathbf{M}_{u,v}$, $(u, v \in \mathbf{Z};\ u, v \geqq 0)$ be the product $\mathbf{S}^{2u+1} \times \mathbf{S}^{2v+1}$ endowed with one of the complex structures of CALABI-ECKMANN [1]. It is the total space of a principal complex analytic fibre bundle $\xi_{u,v}$ over $\mathbf{B}_{u,v} = \mathbf{P}_u(\mathbf{C}) \times \mathbf{P}_v(\mathbf{C})$, with standard fibre and structure group a complex torus $\mathbf{T}$ of complex dimension one. We have

$$A^0(\mathbf{M}_{u,v}) = (\mathbf{GL}(u+1, \mathbf{C}) \times \mathbf{GL}(v+1, \mathbf{C}))/\Gamma,$$

where Γ is an infinite cyclic discrete central subgroup (BLANCHARD [1]), and the map

$$\nu_{u,v}: A^0(\mathbf{M}_{u,v}) \to A^0(\mathbf{B}_{u,v}) = \mathbf{PGL}(u+1, \mathbf{C}) \times \mathbf{PGL}(v+1, \mathbf{C}),$$

associated to the projection $\pi_{u,v}$ of $\mathbf{M}_{u,v}$ onto $\mathbf{B}_{u,v}$ (9.1) is the obvious homomorphism.

For $u = 0$ or $v = 0$, $\mathbf{M}_{u,v}$ is a HOPF manifold. Let $\sigma_{u,v}$ be the projection $\pi_{u,v}$ followed by the projection of $\mathbf{B}_{u,v}$ on its first factor. Then $\sigma_{u,v}$ is the projection of a complex analytic fibering $\eta_{u,v}$ with typical fibre $\mathbf{M}_{0,v}$. To see this, one may for instance use the fact (BLANCHARD [1]) that $\mathbf{M}_{u,v}$ is the base space of a complex analytic principal bundle with total space $\mathbf{M}_{u,0} \times \mathbf{M}_{0,v}$, structure group a complex 1-dimensional torus, and projection map ν, such that

$$\pi_{u,v} \circ \nu = \pi_{u,0} \times \pi_{0,v}: \mathbf{M}_{u,0} \times \mathbf{M}_{0,v} \to \mathbf{B}_{u,v}.$$

9.3. Lemma. *The group $A^0(\mathbf{M}_{u,v})$ acts trivially on $H_{\bar{\partial}}(\mathbf{M}_{u,v})$.*

The fibre bundle has a connected structure group and a kählerian fibre, namely $\mathbf{T}$, and 2.1 applies. By 9.1, the group $A^0(\mathbf{M}_{u,v})$ is an automorphism group of the fibred structure, hence it operates on the spectral sequence. This action is trivial on $E_2 = H_{\bar{\partial}}(\mathbf{B}_{u,v}) \otimes H_{\bar{\partial}}(\mathbf{T})$, since both $\mathbf{B}_{u,v}$ and $\mathbf{T}$ are kählerian (1.4), hence also on E_∞. But $E_\infty = \mathrm{Gr}(H_{\bar{\partial}}(\mathbf{M}_{u,v}))$. By full reducibility, any compact subgroup of $A^0(\mathbf{M}_{u,v})$ acts trivially on $H_{\bar{\partial}}(\mathbf{M}_{u,v})$. The kernel of the action of $A^0(\mathbf{M}_{u,v})$ on $H_{\bar{\partial}}(\mathbf{M}_{u,v})$ is then a normal subgroup which contains all compact subgroups, hence is equal to the whole group.

9.4. Lemma. *We have* $H^{1,0}(\mathbf{M}_{0,v}) = 0$, $(v \geqq 1)$.

This lemma is known. We recall a proof for the sake of completeness. $\mathbf{M}_{0,v}$ may be defined as the quotient of $\mathbf{C}_{v+1} - \{0\}$ by the discrete group generated by a homothetic transformation $\gamma: z \to c \cdot z$ $(c \neq 1)$. Let ω be a holomorphic differential on $\mathbf{M}_{0,v}$. Its inverse image ω^* in $\mathbf{C}_{v+1} - \{0\}$ may be written as $\omega^* = g_1 \cdot dz_1 + \cdots + g_{v+1} \cdot dz_{v+1}$ where the z_i's are coordinates and the g_i's are holomorphic in $\mathbf{C}_{v+1} - \{0\}$. The form ω^* is invariant under γ; this implies

$$g_i(c^n \cdot z) = c^{-n} \cdot g_i(z) \quad (n \in \mathbf{Z}, i = 1, \ldots, v+1),$$

and shows that if $g_i \not\equiv 0$, then g_i is not bounded near the origin, in contradiction with HARTOG's theorem.

The $\bar{\partial}$-cohomology of $\mathbf{M}_{u,v}$ will be generated by pure elements, and subscripts will indicate the type.

9.5. Theorem. *Let* $u \leqq v$. *Then*

$$H_{\bar{\partial}}(\mathbf{M}_{u,v}) \cong \mathbf{C}[x_{1,1}]/(x_{1,1}^{u+1}) \otimes \Lambda(x_{v+1,v}, x_{0,1}).$$

We consider first the case where $u = 0$. In the spectral sequence of the fibering $\xi_{0,v}$ we have

$$E_2 \cong \mathbf{C}[x_{1,1}]/(x_{1,1}^{v+1}) \otimes \Lambda(x_{1,0}, x_{0,1}),$$

where the first factor on the right hand side represents the cohomology of the base $\mathbf{P}_v(\mathbf{C})$, and the second one the cohomology of the fibre $\mathbf{T}$. The element $x_{0,1}$ generates ${}^{0,1}E_2^{0,1}$ and is mapped by d_2 into ${}^{0,2}E_2^{2,0}$, which is zero, hence $d_2(x_{0,1}) = 0$. If $d_2(x_{1,0}) = 0$, then $x_{1,0}$ would be a permanent cocycle and would show up in E_∞ (see 7.4), which would contradict 9.4. We may therefore assume that $d_2(x_{1,0}) = x_{1,1}$. A routine computation then yields:

$$E_3 \cong \Lambda(y_{v+1,v}, x_{0,1}),$$

where

$$y_{v+1,v} = \varkappa_3^2(x_{1,1}^{v+1} \otimes x_{1,0}) \in {}^{v+1,v}E_3^{2v,1}.$$

$y_{v+1,v}$ and $x_{0,1}$ have fibre degree one, hence are d_r-cocycles for all $r \geqq 3$, whence $E_3 \cong E_\infty$. Since E_∞ is a free anticommutative graded algebra, we have $E_\infty \cong H_{\bar{\partial}}(\mathbf{M}_{u,v})$ also multiplicatively.

If now $0 < u \leqq v$, consider the fibering $\eta_{u,v}$ of $\mathbf{M}_{u,v}$ over $\mathbf{P}_u(\mathbf{C})$, with fibre $\mathbf{M}_{0,v}$ (9.2). Its structure group is connected, since the base is simply connected, and it acts trivially on the $\bar{\partial}$-cohomology of the standard fibre (9.3). We may therefore apply 2.1, and we have

$$E_2 = \mathbf{C}[x_{1,1}]/(x_{1,1}^{u+1}) \otimes \Lambda(x_{v+1,v}, x_{0,1}).$$

As before, it is seen that $x_{0,1}$ is a d_2-cocycle, hence a permanent cocycle. Since $u \leqq v$, there is no element of type $(v+1, v+1)$ with a strictly positive base degree in the spectral sequence, hence $x_{v+1,v}$ is a permanent cocycle too. The base terms being always permanent cocycles, it follows that $d_r = 0$, $(r \geqq 2)$, and that $E_2 \cong E_\infty$. We have therefore $E_\infty \cong H_{\bar{\partial}}(\mathbf{M}_{u,v})$ at least additively. But representatives of $x_{v+1,v}$, $x_{0,1}$ in $H_{\bar{\partial}}(\mathbf{M}_{u,v})$ are always of square zero, and there is a representative $y_{1,1}$ of $x_{1,1}$, namely $\pi^*_{u,v}(x_{1,1})$, such that $y_{1,1}^{u+1} = 0$. From this it follows immediately that E_∞ and $H_{\bar{\partial}}(\mathbf{M}_{u,v})$ are also isomorphic as algebras, which proves the theorem.

Remark. The $\bar{\partial}$-cohomology of the HOPF manifold $M_{0,v}$ is computed in KODAIRA-SPENCER [5], § 15 for $v = 1$, in ISE [1] for any v. Theorem 4 of ISE [1] also describes the $\bar{\partial}$-cohomology of a HOPF manifold with coefficients in a line bundle.

Bibliography

ADAMS, J. F.: [1] On CHERN characters and the structure of the unitary group. Proc. Cambridge Phil. Soc. **57**, 189–199 (1961). [2] On formulae of THOM and WU. Proc. London Math. Soc. **11**, 741–752 (1961). [3] Vector fields on spheres. Ann. Math. **75**, 603–632 (1962).

AGRANOVIC, M. S.: [1] On the index of elliptic operators (Russian). Dokl. Akad. Nauk. S. S. S. R. **142**, 983–985 (1962). Sov. Math. Dokl. **3**, 194–197 (1962).

AKIZUKI, Y.: [1] Theorems of BERTINI on linear systems. J. Math. Soc. Japan 3, 170–180 (1951).

AKIZUKI, Y., and S. NAKANO: [1] Note on KODAIRA-SPENCER's proof of LEFSCHETZ theorems. Proc. Japan. Acad. **30**, 266–272 (1954).

ATIYAH, M. F.: [1] Complex fibre bundles and ruled surfaces. Proc. London Math. Soc. **5**, 407–434 (1955). [2] Vector bundles over an elliptic curve. Proc. London Math. Soc. **7**, 414–452 (1957). [3] Complex analytic connections in fibre bundles. Trans. Am. Math. Soc. **85**, 181–207 (1957). [4] Bordism and cobordism. Proc. Cambridge Phil. Soc. **57**, 200–208 (1961). [5] THOM complexes. Proc. London Math. Soc. **11**, 291–310 (1961). [6] Vector bundles and the KÜNNETH formula. Topology **1**, 245—248 (1962). [7] On the K-theory of compact LIE groups. Topology **4**, 95—99 (1965).

ATIYAH, M. F., and R. BOTT: [1] On the periodicity theorem for complex vector bundles. Acta Math. **112**, 229–247 (1964). [2] The index problem for manifolds with boundary. Bombay Colloquium on Differential Analysis, 175–186 (1964). [3] A LEFSCHETZ fixed point formula for elliptic differential operators. Bull. Am. Math. Soc. (to appear).

ATIYAH, M. F., R. BOTT, and A. SHAPIRO: [1] CLIFFORD modules. Topology 3, Suppl. **1**, 3–38 (1964).

ATIYAH, M. F., and F. HIRZEBRUCH: [1] RIEMANN-ROCH theorems for differentiable manifolds. Bull. Am. Math. Soc. **65**, 276–281 (1959). [2] Quelques théorèmes de non-plongement pour les variétés différentiables. Bull. soc. math. France **87**, 383–396 (1959). [3] Vector bundles and homogeneous spaces. Proceedings of Symposia in Pure Mathematics. Vol. 3, p. 7–38. Am. Math. Soc. 1961. [4] Cohomologie-Operationen und charakteristische Klassen. Math. Z. **77**, 149 bis 187 (1961). [5] BOTT periodicity and the parallelisability of the spheres. Proc. Cambridge Phil. Soc. **57**, 223–226 (1961). [6] Charakteristische Klassen und Anwendungen. Enseignement Mathématique **7**, 188–213 (1961). [7] Analytic cycles on complex manifolds. Topology **1**, 25–45 (1962). [8] The RIEMANN-ROCH theorem for analytic embeddings. Topology **1**, 151–166 (1962).

ATIYAH, M. F., and I. M. SINGER: [1] The index of elliptic operators on compact manifolds. Bull. Am. Math. Soc. **69**, 422–433 (1963).

BLANCHARD, A.: [1] Automorphismes des variétés fibrées analytiques complexes. C. R. Acad. Sci. (Paris) **233**, 1337–1339 (1951). [2] Espaces fibrés kählériens compacts. C. R. Acad. Sci. (Paris) **238**, 2281–2283 (1954). [3] Sur les variétés analytiques complexes. Ann. sci. école norm. super. **73**, 157–202 (1956).

BOREL, A.: [1] Les fonctions automorphes de plusieurs variables complexes. Bull. soc. math. France **80**, 167–182 (1952). [2] Sur la cohomologie des espaces fibrés principaux et des espaces homogènes de groupes de LIE compacts. Ann. Math. **57**, 115–207 (1953). [3] Topology of LIE groups and characteristic classes. Bull. Am. Math. Soc. **61**, 397—432 (1955). [4] Compact CLIFFORD-KLEIN forms of symmetric spaces. Topology **2**, 111—122 (1963).

BOREL, A., and F. HIRZEBRUCH: [1] Characteristic classes and homogeneous spaces I, II, III. Am. J. Math. **80**, 458—538 (1958); **81**, 315—382 (1959); **82**, 491—504 (1960).

BOREL, A., and J.-P. SERRE: [1] Groupes de LIE et puissances reduites de STEENROD. Am. J. Math. **75**, 409—448 (1953). [2] Le théorème de RIEMANN-ROCH (d'après GROTHENDIECK). Bull. soc. math. France **86**, 97—136 (1958).

BOTT, R.: [1] Homogeneous vector bundles. Ann. Math. **66**, 203—248 (1957). [2] The stable homotopy of the classical groups. Ann. Math. **70**, 313—337 (1959). [3] The space of loops on a LIE group. Mich. Math. J. **5**, 35—61 (1958). [4] On a theorem of LEFSCHETZ. Mich. Math. J. **6**, 211—216 (1959). [5] Quelques remarques sur les théorèmes de periodicité. Bull. soc. math. France **89**, 293—310 (1959).

BRIESKORN, E.: [1] Ein Satz über die komplexen Quadriken. Math. Ann. **155**, 184—193 (1964). [2] Über holomorphe P_n-Bündel über P_1. Math. Ann. **157**, 343—357 (1965).

CALABI, E., and B. ECKMANN: [1] A class of compact complex manifolds. Ann. Math. **58**, 494—500 (1953).

CARTAN, H.: [1] Espaces fibrés et homotopie. Séminaire E. N. S., 1949—1950. 2nd edition, Paris 1956. [2] Cohomologie des groupes, suite spectrale, faisceaux. Séminaire E. N. S., 1950—1951. 2nd edition, Paris 1955. [3] Théorie des fonctions de plusieurs variables. Séminaire E. N. S., 1951—1952. Paris 1952. [4] Théorie des fonctions automorphes et des espaces analytiques. Séminaire E. N. S., 1953—1954. Paris 1954. [5] Variétés analytiques complexes et cohomologie. Centre Belge Rech. Math., Colloque sur les fonctions de plusieurs variables p. 41—55. Liège: Georges Thone, and Paris: Masson & Cie. 1953. [6] Fonctions automorphes. Séminaire E. N. S., 1957—1958. Paris 1958.

CARTAN, H., and J.-P. SERRE: [1] Un théorème de finitude concernant les variétés analytiques compactes. C. R. Acad. Sci. (Paris) **237**, 128—130 (1953).

CARTAN, H., and L. SCHWARTZ: [1] Le théorème d' ATIYAH-SINGER. Séminaire E. N. S., 1963—1964. Paris 1964.

CHERN, S. S.: [1] Characteristic classes of Hermitian manifolds. Ann. Math. **47**, 85—121 (1946). [2] On the characteristic classes of complex sphere bundles and algebraic varieties. Am. J. Math. **75**, 565—597 (1953).

CHERN, S. S., F. HIRZEBRUCH, and J.-P. SERRE: [1] On the index of a fibered manifold. Proc. Am. Math. Soc. **8**, 587—596 (1957).

CHOW, W. L.: [1] On compact complex analytic varieties. Am. J. Math. **71**, 893—914 (1949).

CONNER, P. E., and E. E. FLOYD: [1] Differentiable periodic maps. Ergebnisse der Mathematik und ihrer Grenzgebiete 33. Berlin-Göttingen-Heidelberg: Springer 1964.

DIEUDONNÉ, J.: [1] Une généralisation des espaces compacts. J. Math. Pures Appl. **23**, 65—76 (1944).

DOLBEAULT, P.: [1] Sur la cohomologie des variétés analytiques complexes. C. R. Acad. Sci. (Paris) **236**, 175—177 (1953). [2] Formes différentielles et cohomologie sur une variété analytique complexe I, II. Ann. Math. **64**, 83—130 (1956); **65**, 282—330 (1957).

DOLD, A.: [1] Erzeugende der THOMschen Algebra $\mathfrak{N}$. Math. Z. **65**, 25—35 (1956). [2] Vollständigkeit der WUschen Relationen zwischen den STIEFEL-WHITNEY-schen Zahlen differenzierbarer Mannigfaltigkeiten. Math. Z. **65**, 200—206 (1956). [3] Partitions of unity in the theory of fibrations. Ann. Math. **78**, 223—255 (1963).

DOMBROWSKI, P.: [1] On the geometry of the tangent bundle. J. reine angew. Math. **210**, 73—88 (1962).

DYNIN, A. S.: [1] Singular operators of arbitrary order on a manifold (Russian). Dokl. Akad. Nauk S. S. S. R. **141**, 21—23 (1961). Sov. Math. Dokl. **2**, 1375—1377 (1961).

ECKMANN, B.: [1] Quelques propriétés globales des variétés kählériennes. C. R. Acad. Sci. (Paris) **229**, **577—579** (1949).

ECKMANN, B., and H. GUGGENHEIMER: [1] Formes différentielles et métrique hermitienne sans torsion I, II. C. R. Acad. Sci. (Paris) **229**, 464—466 and 489—491 (1949). [2] Sur les variétés closes à métrique hermitienne sans torsion. C. R. Acad. Sci. (Paris) **229**, **503—505** (1949).

EILENBERG, S., and N. STEENROD: [1] Foundations of algebraic topology. Princeton Mathematical Series **15**, Princeton University Press 1952.

FRENKEL, J.: [1] Cohomologie non abélienne et espaces fibrés. Bull. soc. math. France **85**, 135—220 (1957).

GAMKRELIDZE, R. V.: [1] Computation of the CHERN cycles of algebraic manifolds (Russian). Dokl. Akad. Nauk S. S. S. R. **90**, 719—722 (1953). [2] CHERN cycles of complex algebraic manifolds (Russian). Izv. Akad. Nauk S. S. S. R. **20**, 685—706 (1956).

GARABEDIAN, P. R., and D. C. SPENCER: [1] A complex tensor calculus for KÄHLER manifolds. Acta Math. **89**, 279—331 (1953).

GELFAND, I. M.: [1] On elliptic equations (Russian). Uspehi Math. Nauk **15**, 3, 121—132 (1960). Russian Math. Surveys **15**, 3, 113—123 (1960).

GODEMENT, R.: [1] Topologie algébrique et théorie des faisceaux. Act. Sci. et Ind. 1252. Paris: Hermann 1958.

GRAUERT, H.: [1] Une notion de dimension cohomologique dans la théorie des espaces complexes. Bull. soc. math. France **87**, 341—350 (1959). [2] Ein Theorem der analytischen Garbentheorie und die Modulräume komplexer Strukturen. Institut des Hautes Études Scientifiques. Publ. Math. **5**, 1960. [3] Über Modifikationen und exzeptionelle analytische Mengen. Math. Ann. **146**, 331—368 (1962).

GRAUERT, H., and R. REMMERT: [1] Bilder und Urbilder analytischer Garben. Ann. Math. **68**, 393—443 (1958). [2] Komplexe Räume. Math. Ann. **136**, 245—318 (1958).

GRIFFITHS, P. A.: [1] On the differential geometry of homogeneous vector bundles. Trans. Am. Math. Soc. **109**, 1—34 (1963). [2] Deformations of holomorphic mappings. Illinois J. Math. **8**, 139—151 (1964). [3] Hermitian differential geometry and the theory of positive and ample holomorphic vector bundles. J. Math. Mech. **14**, 117—140 (1965).

GROTHENDIECK, A.: [1] A general theory of fibre spaces with structure sheaf (2nd edition 1958). Lawrence, Kansas: University of Kansas 1955. [2] Sur quelques points d'algèbre homologique. Tohoku Math. J. **9**, 119—221 (1957). [3] Sur la classification des fibrés holomorphes sur la sphère de RIEMANN. Am. J. Math. **79**, 121—138 (1957). [4] La théorie des classes de CHERN. Bull. soc. math. France **86**, 137—154 (1958).

GUGGENHEIMER, H.: [1] Über komplex-analytische Mannigfaltigkeiten mit KÄHLERscher Metrik. Comment. Math. Helv. **25**, 257—297 (1951). [2] Modifications of KÄHLER manifolds. Ann. Mat. Pura Appl. **41**, 87—93 (1956).

HARISH-CHANDRA: [1] Representations of semi-simple LIE groups, VI. Integrable and square integrable representations. Am. J. Math. **78**, 564—628 (1956).

HELGASON, S.: [1] Differential geometry and symmetric spaces. Pure and Applied Mathematics, Vol. XII. New York: Academic Press 1962.

HIRZEBRUCH, F.: [1] On STEENROD's reduced powers, the index of inertia, and the TODD genus. Proc. Nat. Acad. Sci. U.S.A. **39**, 951–956 (1953). [2] Arithmetic genera and the theorem of RIEMANN-ROCH for algebraic varieties. Proc. Nat. Acad. Sci. U.S.A. **40**, 110–114 (1954). [3] Der Satz von RIEMANN-ROCH in Faisceau-theoretischer Formulierung. Proc. Intern. Congress Math. 1954, Vol. III, p. 457–473. Amsterdam: North Holland Publishing Co. 1956. [4] Characteristic numbers of homogeneous domains. Seminars on analytic functions, Vol. 2, p. 92–104. Institute for Advanced Study 1957. [5] Automorphe Formen und der Satz von RIEMANN-ROCH. Symp. Intern. Top. Alg. 1956, p. 129–144. Universidad de Mexico 1958. [6] Komplexe Mannigfaltigkeiten. Proc. Intern. Congress Math. 1958, p. 119–136. Cambridge University Press 1960.

HIRZEBRUCH, F., and K. KODAIRA: [1] On the complex projective spaces. J. Math. Pures Appl. **36**, 201–216 (1957).

HODGE, W. V. D.: [1] The theory and applications of harmonic integrals (2nd edition 1952). Cambridge University Press 1941. [2] A special type of KÄHLER manifold. Proc. London Math. Soc. **1**, 104–117 (1951). [3] The characteristic classes of algebraic varieties. Proc. London Math. Soc. **1**, 138–151 (1951). [4] The topological invariants of algebraic varieties. Proc. Intern. Congress Math. 1950, Vol. I, p. 182–191. Am. Math. Soc. 1952.

HODGE, W. V. D., and M. F. ATIYAH: [1] Integrals of the second kind on an algebraic variety. Ann. Math. **62**, 56–91 (1955).

HOLMANN, H.: [1] Vorlesung über Faserbündel. Münster: Aschendorff 1962. [2] SEIFERTsche Faserräume. Math. Ann. **157**, 138–166 (1964).

HORROCKS, G.: [1] Vector bundles on the punctured spectrum of a local ring. Proc. London Math. Soc. **14**, 689—713 (1964). [2] On extending vector bundles over projective space. Quart. J. Math. Oxford (to appear).

ISE, M.: [1] On the geometry of HOPF manifolds. Osaka Math. J. **12**, 387–402 (1960). [2] Generalised automorphic forms and certain holomorphic vector bundles. Am. J. Math. **86**, 70–108 (1964).

KÄHLER, E.: [1] Forme differenziali e funzioni algebriche. Mem. Accad. Ital. Mat. **3**, 1–19 (1932). [2] Über eine bemerkenswerte HERMITEsche Metrik. Abhandl. math. Seminar hamburg. Univ. **9**, 173–186 (1932).

KAHN, P. J.: [1] Characteristic numbers and oriented homotopy type. Topology **3**, 81—95 (1965). [2] Characteristic numbers and homotopy type. Mich. Math. J. **12**, 49—60 (1965).

KAUP, L.: [1] Eine KÜNNETHformel für kohärente analytische Garben. Erlangen: Thesis 1965.

KERVAIRE, M.: [1] Non-parallelizability of the n-sphere for $n > 7$. Proc. Nat. Acad. Sci. U.S.A. **44**, 280–283 (1958).

KODAIRA, K.: [1] The theorem of RIEMANN-ROCH on compact analytic surfaces. Am. J. Math. **73**, 813–875 (1951). [2] The theorem of RIEMANN-ROCH for adjoint systems on 3-dimensional algebraic varieties. Ann. Math. **56**, 288–342 (1952). [3] On cohomology groups of compact analytic varieties with coefficients in some analytic faisceaux. Proc. Nat. Acad. Sci. U.S.A. **39**, 865–868 (1953). [4] On a differential-geometric method in the theory of analytic stacks. Proc. Nat. Acad. Sci. U.S.A. **39**, 1268–1273 (1953). [5] Some results in the transcendental theory of algebraic varieties. Ann. Math. **59**, 86–134 (1954). [6] On KÄHLER varieties of restricted type (an intrinsic characterisation of algebraic varieties). Ann. Math. **60**, 28–48 (1954). [7] Characteristic linear systems of complete continuous systems. Am. J. Math. **78**, 716–744 (1956). [8] On compact complex analytic surfaces I, II, III. Ann. Math. **71**, 111–152 (1960); **77**, 563–626 (1963); **78**, 1–40 (1963). [9] On stability of compact submanifolds of complex manifolds. Am. J. Math. **85**, 79–94 (1963). [10] On the structure of compact complex analytic surfaces I. Am. J. Math. **86**, 751–798 (1964).

KODAIRA, K., and D. C. SPENCER: [1] On arithmetic genera of algebraic varieties. Proc. Nat. Acad. Sci. U.S.A. **39**, 641–649 (1953). [2] Groups of complex line bundles over compact KÄHLER varieties. Divisor class groups on algebraic varieties. Proc. Nat. Acad. Sci. U.S.A. **39**, 868–877 (1953). [3] On a theorem of LEFSCHETZ and the lemma of ENRIQUES-SEVERI-ZARISKI. Proc. Nat. Acad. Sci. U.S.A. **39**, 1273–1278 (1953). [4] On the variation of almost complex structure. Algebraic Geometry and Topology, p. 139–150. Princeton University Press 1957. [5] On deformations of complex analytic structures I, II. Ann. Math. **67**, 328–466 (1958). [6] A theorem of completeness for complex analytic fibre spaces. Acta Math. **100**, 281–294 (1958). [7] On deformations of complex analytic structures III. Stability theorems for complex structures. Ann. Math. **71**, 43–76 (1960).

LANG, S.: [1] Introduction to differentiable manifolds. New York: Interscience 1962.

LANGLANDS, R. P.: [1] The dimension of spaces of holomorphic forms. Am. J. Math. **85**, 99–125 (1963).

LERAY, J.: [1] L'anneau spectral et l'anneau filtré d'homologie d'un espace localement compact et d'une application continue. J. Math. Pures Appl. **29**, 1–139 (1950). [2] L'homologie d'un espace fibré dont la fibre est connexe. J. Math. Pures Appl. **29**, 169–213 (1950).

MAYER, K. H.: [1] Elliptische Differentialoperatoren und Ganzzahligkeitssätze für charakteristische Zahlen. Topology 4 (1965).

MILNOR, J.: [1] Construction of universal bundles I, II. Ann. Math. **63**, 272–284 and 430–436 (1963). [2] Some consequences of a theorem of BOTT. Ann. Math. **68**, 444–449 (1958). [3] On the cobordism ring Ω^* and a complex analogue. Am. J. Math. **82**, 505–521 (1960). [4] A survey of cobordism theory. Enseignement Mathématique **8**, 16–23 (1962). [5] Spin structures on manifolds. Enseignement Mathématique **9**, 198–203 (1963). [6] Microbundles I. Topology **3**, Suppl. 1, 53–80 (1964). [7] On the STIEFEL-WHITNEY numbers of complex manifolds and of spin manifolds. Topology **3**, 223–230 (1965).

NAKANO, S.: [1] On complex analytic vector bundles. J. Math. Soc. Japan **7**, 1–12 (1955). [2] Tangential vector bundle and TODD canonical systems of an algebraic variety. Mem. Coll. Sci. Univ. Kyoto (A) **29**, 145–149 (1955). [3] An example of deformations of complex analytic bundles. Mem. Coll. Sci. Univ. Kyoto (A) **31**, 181–190 (1958).

NARASIMHAN, M. S., and C. S. SESHADRI: [1] Holomorphic vector bundles on a compact RIEMANN surface. Math. Ann. **155**, 69–80 (1964). [2] Stable bundles and unitary bundles on a compact RIEMANN surface. Proc. Nat. Acad. Sci. U.S.. **52**, 207–210 (1964).

NOVIKOV, S. P.: [1] Topological invariance of rational PONTRJAGIN classes (Russian). Dokl. Akad. Nauk S. S. S. R. **163**, 298—300 (1965).

PALAIS, R. S.: [1] Seminar on the ATIYAH-SINGER index theorem, with contributions by A. BOREL, E. E. FLOYD, R. T. SEELEY, W. SHIH, and R. SOLOVAY. Annals of Mathematics Studies **57**. Princeton University Press 1965.

PONTRJAGIN, L. S.: [1] Topological groups. Princeton Mathematical Series 2. Princeton University Press 1939. [2] Characteristic cycles on differentiable manifolds (Russian). Mat. Sbornik **21**, 233–284 (1947). Am. Math. Soc. Translation 32 (1950).

PORTEOUS, I. R.: [1] Blowing up CHERN classes. Proc. Cambridge Phil. Soc. **56**, 118–124 (1960).

RHAM, G. DE: [1] Variétés différentiables. Act. Sci. et Ind. 1222. Paris: Hermann 1954.

RHAM, G. DE, and K. KODAIRA: [1] Harmonic integrals. Princeton: Institute for Advanced Study 1950.

ROBERTS, R. S.: [1] Bundles of Grassmannians and integrality theorems (to appear).

ROHLIN, V. A.: [1] New results in the theory of 4-dimensional manifolds (Russian). Dokl. Akad. Nauk S. S. S. R. **84**, 221–224 (1952). [2] On PONTRJAGIN characteristic classes (Russian). Dokl. Akad. Nauk S. S. S. R. **113**, 276–279 (1957).

ROHLIN, V. A., and A. S. ŠVARC: [1] Combinatorial invariance of PONTRJAGIN classes (Russian). Dokl. Akad. Nauk S. S. S. R. **114**, 480–493 (1957).

ROTHENBERG, M., and N. STEENROD: [1] The cohomology of the classifying space of an H-space (to appear).

SAMPSON, J. H., and G. WASHNITZER: [1] A VIETORIS mapping theorem for algebraic projective fibre bundles. Ann. Math. **68**, 348–371 (1958). [2] Cohomology of monoidal transformations. Ann. Math. **69**, 605–629 (1959). [3] A KÜNNETH formula for coherent algebraic sheaves. Illinois J. Math. **3**, 389–402 (1959).

SANDERSON, B. J., and R. L. E. SCHWARZENBERGER: [1] Non-immersion theorems for differentiable manifolds. Proc. Cambridge Phil. Soc. **59**, 319–322 (1963).

SCHWARZENBERGER, R. L. E.: [1] Vector bundles on the projective plane. Proc. London Math. Soc. **11**, 623–640 (1961).

SEELEY, R.: [1] Integro-differential operators on vector bundles. Trans. Am. Math. Soc. **117**, 167–205 (1965).

SEGRE, B.: [1] Dilatazioni e varietà canoniche sulla varietà algebriche. Ann. di Mat. **37**, 139–155 (1954).

SERRE, J.-P.: [1] Quelques problèmes globaux relatifs aux variétés de STEIN. Centre Belge Rech. Math., Colloque sur les fonctions de plusieurs variables p. 57–68. Liège: Georges Thone, and Paris: Masson & Cie. 1953. [2] Faisceaux algébriques cohérents. Ann. Math. **61**, 197–278 (1955). [3] Un théorème de dualité. Comment. Math. Helv. **29**, 9–26 (1955). [4] Géométrie algébrique et géométrie analytique. Ann. Inst. Fourier **6**, 1–42 (1956).

SEVERI, F.: [1] La géométrie algébrique Italienne. Centre Belge Rech. Math., Colloque de géométrie algébrique p. 9–55. Liège: Georges Thone, and Paris: Masson & Cie. 1950. [2] Il teorema di RIEMANN-ROCH per curve, superficie e varietà. Ergebnisse der Mathematik und ihrer Grenzgebiete **17**. Berlin-Göttingen-Heidelberg: Springer 1958.

SPENCER, D. C.: [1] Cohomology and the RIEMANN-ROCH theorem. Proc. Nat. Acad. Sci. U.S.A. **39**, 660–669 (1953). [2] DIRICHLET's principle on manifolds. Studies in mathematics and mechanics presented to RICHARD VON MISES p. 127–134. New York: Academic Press 1954.

STEENROD, N.: [1] The topology of fibre bundles. Princeton Mathematical Series 14. Princeton University Press 1951.

STONG, R. E.: [1] Relations among characteristic numbers. Topology **4** (1965).

THOM, R.: [1] Espaces fibrés en sphères et carrés de STEENROD. Ann. sci. école norm. super. **69**, 109–182 (1952). [2] Quelques propriétés globales des variétés différentiables. Comment. Math. Helv. **28**, 17–86 (1954). [3] Les classes caractéristiques de PONTRJAGIN des variétés triangulées. Symp. Intern. Top. Alg. 1956, p. 54–67. Universidad de Mexico 1958.

TODD, J. A.: [1] The arithmetical invariants of algebraic loci. Proc. London Math. Soc. **43**, 190–225 (1937). [2] Birational transformations with isolated fundamental points. Proc. Edinburgh Math. Soc. **5**, 117–124 (1938). [3] The geometric invariants of algebraic loci. Proc. London Math. Soc. **45**, 410–434 (1939). [4] Invariant and covariant systems on an algebraic variety. Proc. London Math. Soc. **46**, 199–230 (1940). [5] Birational transformations with a fundamental surface. Proc. London Math. Soc. **47**, 81—100 (1941). [6] Canonical systems on algebraic varieties. Bol. Soc. Mat. Mexicana **2**, 26—44 (1957).

TURIN, A. N.: [1] The classification of two dimensional vector bundles over algebraic curves of arbitrary genus (Russian). Izv. Akad. Nauk S. S. S. R. **28**, 21—52 (1964). [2] The classification of vector bundles over algebraic curves of arbitrary genus (Russian). Izv. Akad. Nauk S. S. S. R. **29**, 657—688 (1965).

VAN DE VEN, A. J. H. M.: [1] Characteristic classes and monoidal transformations. Indagationes math. **18**, 571—578 (1956). [2] An interpretation of the formulae of KUNDERT concerning higher obstructions. Indagationes math. **19**, 196—200 (1957).

VOLPERT, A. I.: [1] The index of systems of two dimensional integral equations (Russian). Dokl. Akad. Nauk S. S. S. R. **142**, 776—777 (1962). Sov. Math. Dokl. **3**, 154—155 (1962). [2] Elliptic systems on the sphere and two dimensional singular integral equations (Russian). Mat. Sbornik **59**, 195—214 (1962).

WAERDEN, B. L. VAN DER: [1] Birationale Transformationen von linearen Scharen auf algebraischen Mannigfaltigkeiten. Math. Z. **51**, 502—523 (1948). [2] Birational invariants of algebraic manifolds. Acta Salmantic., Sec. Mat. **2**, 1—56 (1947).

WALL, C. T. C.: [1] Determination of the cobordism ring. Ann. Math. **72**, 292—311 (1960). [2] Cobordism exact sequences for differential and combinatorial manifolds. Ann. Math. **77**, 1—15 (1963).

WASHNITZER, G.: [1] The characteristic classes of an algebraic fiber bundle. Proc. Nat. Acad. Sci. U.S.A. **42**, 433—436 (1956). [2] Geometric syzygies. Am. J. Math. **81**, 171—248 (1959).

WEIL, A.: [1] Généralisation des fonctions abéliennes. J. Math. Pures Appl. **17**, 47—87 (1938). [2] Variétés kählériennes. Act. Sci. et Ind. 1267. Paris: Hermann 1958.

WEYL, H.: [1] Die Idee der RIEMANNschen Fläche. Berlin: Teubner 1913. (3rd edition, Stuttgart: Teubner 1955.)

WOLF, J. A.: [1] On the classification of hermitian symmetric spaces. J. Math. Mech. **13**, 489—495 (1964).

WOOD, R.: [1] BANACH algebras and BOTT periodicity. Topology **4** (1965).

ZAPPA, G.: [1] Sopra una probabile diseguaglianza tra i caratteri invariantivi di una superficie algebrica. Rend. Mat. e Appl. **14**, 455—464 (1955).

ZARISKI, O.: [1] Algebraic surfaces. Ergebnisse der Mathematik und ihrer Grenzgebiete **5**, Berlin: Springer 1935 (reprinted: New York: Chelsea 1948). [2] Pencils on an algebraic variety and a new proof of a theorem of BERTINI. Trans. Am. Math. Soc. **50**, 48—70 (1941). [3] The theorem of BERTINI on the variable singular points of a linear system of varieties. Trans. Am. Math. Soc. **56**, 130—140 (1944). [4] Complete linear systems on normal varieties and a generalization of a lemma of ENRIQUES-SEVERI. Ann. Math. **55**, 552—592 (1952).

Index

For remarks on standard notations and terminology see the Introduction (§ 0.9)

Druck: Brühlsche Universitätsdruckerei Gießen